D1195022

CONSTRUCTION DELAYS

CONSTRUCTION DELAYS

Third Edition

MARK F. NAGATA, PSP, CDT
WILLIAM A. MANGINELLI
J. SCOTT LOWE, P.E., CDT
THEODORE J. TRAUNER, P.E.
Trauner Consulting Services, Inc., Philadelphia, Pennsylvania

Butterworth-Heinemann is an imprint of Elsevier
The Boulevard, Langford Lane, Kidlington, Oxford OX5 1GB, United Kingdom
50 Hampshire Street, 5th Floor, Cambridge, MA 02139, United States

British Library Cataloguing-in-Publication Data
A catalogue record for this book is available from the British Library

Library of Congress Cataloging-in-Publication Data
A catalog record for this book is available from the Library of Congress

ISBN: 978-0-12-811244-1

For Information on all Butterworth-Heinemann publications
visit our website at https://www.elsevier.com/books-and-journals

Publisher: Matthew Deans
Acquisition Editor: Ken McCombs
Editorial Project Manager: Peter Jardim
Production Project Manager: Sruthi Satheesh
Cover Designer: Victoria Pearson

Typeset by MPS Limited, Chennai, India

DEDICATION

This book is dedicated to Theodore Joseph Trauner; founder of Trauner Consulting Services, our mentor, and a good friend. He wrote the original book and contributed in so many ways to each edition, including this one. Though this book now represents the collective wisdom of the authors and staff of TRAUNER, at its core this book is still the product of Ted's determination to understand construction delays and to share what he has learned with his colleagues and with the industry. It remains the embodiment of his belief that "true" experts make lasting contributions to their industry by advancing the state of knowledge and practice.

CONTENTS

FOREWORD

This book was first written and published in 1990. The Second Edition was published in 2009. Over the past 8 years, the industry has continued to mature. Not too long ago, one knowledgeable observer described the world of delay analysis as "the Wild West." While it did not look that way from our perspective, it was probably true that too many analysts had their own way of analyzing delays.

But the last few years have seen a significant movement, if not consensus, towards some shared ideals. This Third Edition is devoted to helping readers appreciate how far we have come and, to some extent, what is left to be done. It is certainly not the Wild West anymore, though there may still be the occasional act of lawlessness.

For the most part, the industry has coalesced around the basic principles espoused in the First Edition of this book:

- There is no question that the only way to delay the scheduled project completion date is to delay the critical path of work.
- There is also no debate that every project, whether a CPM schedule was used or not, has a critical path.
- There is wide agreement that the analysis of delays must consider the contemporaneous project schedules and schedule updates. There may be some debate as to exactly what "consider" means, but no analysis can simply ignore the schedules prepared during the course of the project without good reason.

As they did after our First Edition, our readers continue to ask questions and make suggestions concerning the content. We trust that many of these questions and comments will be answered in this new edition.

We have also provided many more and more complex examples of the proper approach to analyzing delays. You will find entirely new chapters devoted to analyzing delays step by step.

Construction contracts have become much more sophisticated with regard to scheduling, determining time extensions, and providing compensation for delay. This Third Edition was greatly expanded to include examples of these more sophisticated provisions.

When a construction project is delayed beyond the contract completion date or beyond the contractor's scheduled completion date, significant additional costs can be experienced by the contractor, the owner, or both. Because delays can be costly, more and more projects end up in arbitration, litigation, or some form of dispute concerning time-related questions. A judge, jurors, or arbitrators are then faced with the task of sorting out "who shot whom" from a complex collection of facts and dates. Oftentimes, experts are required both to perform an analysis of the delays that occurred, and to provide testimony to explain the analysis. One of the most difficult tasks of the expert is to educate the parties involved so that an understanding can be reached concerning the delays that occurred and who is responsible for them.

This book provides the background information necessary to understand delays. This understanding is not geared solely to the context of disputes, but rather provides a framework to help prevent disputes from occurring, and to resolve questions of time as they arise during the project.

Chapter 1, "Project Scheduling," provides an overview and definitions of basic scheduling concepts and terms that will be referred to throughout the book. It is not intended as a CPM scheduling primer. Rather, it addresses important basic concepts required for using project schedules to identify and measure delay.

Chapter 2, "Float and the Critical Path," is a new chapter devoted to defining and discussing float and the critical path. These two concepts are crucial to understanding how to analyze construction project delays. Included within this chapter is also a discussion of float ownership.

Chapter 3, "Reviewing the Project Schedule," is another chapter new to the Third Edition. Given the importance of the contemporaneous project schedules to delay analysis, the purpose of this chapter is to provide guidance regarding the preparation and review of project schedules.

Chapter 4, "Types of Construction Delays," explains the basic categories of excusable and nonexcusable delays, and the subcategories of compensable and noncompensable delays. It addresses the concept of concurrency and also noncritical delays. This primer in delays prepares the reader for the specific issues covered in succeeding chapters.

Chapter 5, "Measuring Delays—The Basics," explains how to approach the analysis, including the starting points of as-planned schedules and as-built diagrams, and how one must compare the two in order to quantify the delays that have occurred. The question of liability is addressed separately, because this determination is made most effectively after the specific delays have been identified.

Chapters 6–8 travel through the actual process of analyzing delays with bar chart schedules, CPM schedules, and when no schedule is available.

Recognizing that there are numerous approaches used to analyze delays, Chapter 9, Other Retrospective Delay Analysis Techniques—Their Strengths and Weaknesses, explores some of the more common approaches used and describes their strengths and weaknesses.

Added costs incurred by the owner and contractor are addressed in Chapters 10–15. Since inefficiency and acceleration costs are often time-related issues associated with delay, they have been addressed separately in the hopes that some of the myth and magic that surrounds them may be cleared away. Similarly, the topic of costs associated with noncritical delays has been given special attention, because many projects experience these with little or no recognition of the problem.

Chapter 16, "Determining Responsibility for Delay," explains the process used to assess the party who caused the delay. The responsibility for delays is addressed separately from the delay analysis because we believe that this is the proper approach to use—first determine the activities that are delayed and the magnitude of the delay—and then address responsibility or liability.

Chapter 17, "Delay–Risk Management," could also be called "Prevention of Time-Related Problems," since it focuses on the delay-related risks of the various parties in a construction project. By maintaining this focus, each of the parties has a tendency to better control time and resolve delay problems as they occur.

Chapter 18, "Delays and the Contract," is a new chapter that provides examples and discussion regarding contract provisions related to scheduling, time extensions, and the pricing of delays.

This book has been written with the hope that a better understanding of delays, time extensions, and delay costs will help to prevent problems rather than foster and fuel the already litigious atmosphere that exists in construction.

Bear in mind that the methodology described herein can be applied to any type of project that (1) has a time constraint and (2) is amenable to scheduling and the monitoring and control of time. This could include supply contracts, manufacturing projects, and research and development projects, as well as traditional construction projects. The approach will be the same for all situations, given a logical and reasoned application within the context of the existing facts.

ACKNOWLEDGMENTS

This book would not have been possible without the efforts of the entire staff of Trauner Consulting Services, all of whom have contributed to the experience and knowledge that drives the pages of each chapter. The authors wish to specifically acknowledge Janet Montgomery for her exacting efforts in ensuring that this Third Edition came together in an organized and timely fashion.

INTRODUCTION TO THIRD EDITION

Construction is risky business. And, the financial risks are great. Construction projects are bid under fierce competition with small margins. Even simple projects require the coordination of many trades under demanding conditions and challenging timeframes. This is rendered all the more challenging because most projects do not go exactly according to plan. The competition, the big risks, the small margins, and the project complexity are a recipe for conflict.

One of the biggest risks is that the project will not complete on time. The cost of a lost day on a construction project can be staggering. Unfortunately, even the most sophisticated practitioners can find it difficult to identify and quantify delays. This is true despite the fact that we have ever more sophisticated scheduling tools available to plan and manage project time.

While the parties' management teams are able to analyze and assess most factors related to a change, delays remain difficult for the parties to the construction project to understand and accurately measure. Consequently, even though the parties can often reach compromises related to most aspects of a change, the delay component sometimes keeps them from reaching resolution. As a result, most construction claims include a component related to delays. Because the expertise required to reliably and convincingly assess delays and delay damages often goes beyond that of the participants, experts are hired to analyze delays to help the parties resolve their difference, prepare for mediation, and prepare for arbitration or trial.

This book addresses the topic of construction delays, the resulting effects, and damages. This is a timely subject, since the failure to meet schedules can result in serious consequences with unprecedented cost implications. The financial significance of delays demands that the project owner, general contractor, construction manager, designer, subcontractors, and suppliers educate themselves regarding delays and the associated added costs. This book is designed to serve as a primer for that education process. All too many texts on this subject focus on the legal perspective, in legal language. This book is intended as a practical, hands-on guide to an area of construction that is not well understood.

All construction industry professionals should know the basic types of delays and understand the situations that give rise to entitlement to additional compensation for delay. Most importantly, they should understand how a project schedule and project documentation can be used to determine whether a delay occurred, quantify the delay, and assess the cause of the delay. They should also be able to assess the delay's effects on the project and quantify any costs or damages.

Many techniques are used to analyze delays. Some of these methods have inherent weaknesses and should be avoided. This book points out the shortcomings of these faulty methods and explains how a delay analysis should be performed. This book describes—specifically—how the analysis is done with Critical Path Method (CPM) schedules. The discussion will cover the subtleties of the process, such as shifts in the critical path, and noncritical delays.

The subject of damages is covered in detail, including the major categories of extended field overhead and unabsorbed home office overhead costs. Likewise, the damages suffered by the owner, either actual or liquidated, are also explained.

Chapters are devoted to managing the risk of delays and time extensions from the viewpoints of the various parties to a construction project and to common contract provisions governing time and the cost of time. A discussion of early completion schedules, concurrent delays, inefficiency, and constructive acceleration is also included.

The authors' substantial experience in analyzing delays and quantifying damages provides the readers with numerous benefits, including:

- A clear, concise definition of the major types of delays
- A simple, practical explanation of how delays must be analyzed
- A detailed explanation of how delays are defined and quantified for projects with CPM schedules, bar charts, or no schedule at all
- A glimpse of some of the less obvious problems associated with delays, such as delays to noncritical activities
- An understanding of the shortcomings of some delay analysis methods that may not provide reliable results
- A detailed understanding of the many areas where costs can increase and how to calculate these costs
- An understanding of the risks that delays present to various parties to the Project, and how each of those parties can manage those risks
- Examples of common contract provisions related to scheduling, delays, and delay costs

An explanation of delays and delay costs, presented in a straightforward, accessible manner, should be useful to public and private owners, construction managers, general contractors, subcontractors, designers, suppliers, and their attorneys.

CHAPTER ONE

Project Scheduling

THE PROJECT SCHEDULE

If we were to ask a contractor, a construction owner, or an architect or engineer if they plan their construction projects, undoubtedly they would respond in the affirmative. In the beginning, everyone has some form of a plan as to how the work will be executed.

Their plans might include some or all of the following elements:

- The tasks that must be performed
- The time required to perform each task
- The sequence that these tasks will be performed
- The subcontractors, trades, and numbers of workers that will perform each task
- The physical aspects of the project and the project site that might affect the performance of the tasks
- The availability of materials

While this list of typical plan components could be expanded, the concept is straightforward: each component represents an important consideration associated with the planning of a construction project.

A project schedule is a written, graphical, or computerized model of the project team's plan for completing a construction project. Schedules emphasize the elements of sequence and time. The plan will typically identify the major work items (activities), the time (duration) needed to perform these activities, and the sequence (logic) of construction of these activities to complete the project.

At its most basic level, a project schedule will visually depict the intended timing of the major work items necessary to demonstrate how and when the project team will construct the project.

The project schedule should include the significant elements of the project sequenced in a logical order from the beginning of the project through its completion. In addition, the schedule should define specific time periods for each activity in the schedule. The sequencing and

Construction Delays.
DOI: http://dx.doi.org/10.1016/B978-0-12-811244-1.00001-X

summation of the individual time elements will define the overall project duration.

The level of detail shown in a construction schedule will vary, depending on a number of different factors. To name just a few, those factors include the type of schedule used, the contract requirements, the nature of the work, and the contractor's practices.

Overall, the project schedule should portray in a clear way the construction tasks that must be performed, the time allocated to each task, and the sequence of the tasks.

THE PRIMARY PURPOSE OF A PROJECT SCHEDULE

Just as a bid is an estimate of the costs required to construct each piece of a project and the project as a whole, the project schedule is an estimate of the time required to construct each piece of a project and the project as a whole. A project schedule is a valuable project controls tool that is used by project managers to effectively manage construction projects. As noted earlier, the project schedule should include every major element of the construction project. In this manner, it should provide a complete picture of the project's planned construction sequence from start to finish, forecasting when the project will complete. Additionally, if the project schedule is properly developed and updated throughout the duration of the project, it will provide periodic snapshots of the plan to complete the project as the project progresses and as the plan changes over time.

When the project schedule is properly updated and revised to reflect the current construction plan, it enables the project participants to measure and control the pace of the work, provides the project participants with reliable information to make timely decisions, and serves as the primary tool to evaluate the effect of changes and other potential delays on the project plan as these events occur.

Effectively depicting and communicating the construction plan

Successful contractors use project schedules to depict and communicate the construction plan to the owner, the owner's representatives, the contractor's subcontractors and suppliers, and other project participants. The development of the construction plan should be a collaborative process that includes the contractor and its subcontractors, and the owner and the owner's team. Involving the subcontractors in the development of the construction schedule will significantly facilitate acceptance by the subcontractors of the overall approach to building the project. Additionally, incorporation of the subcontractors' means and methods will give greater credibility to the project schedule as a tool that accurately depicts the planned construction activities, durations, and sequencing.

The planning effort is the first essential step to successful execution of a construction project. This is because the development of the construction plan requires the project manager, superintendent, and other key team members to determine and identify how the project will be constructed. Although a project's planned sequence of construction may appear to be straightforward, there are many decisions that have to be made to develop a fully thought-out and comprehensive plan. Those decisions usually begin with identifying the project's work scope. That work scope is broken down into more detail by area, location, trade, and individual work item. To accomplish each work item, the team has to agree on the most efficient and cost effective use of the available labor and equipment resources. This, in turn, drives the decision regarding how much time must be allotted to each work item. Additional considerations are the coordination of the individual work items and the use of subcontractors. Often, a contractor's competitive advantage is derived from its ability to manage its resources and risk, and apply its means and methods in a manner that is more efficient and cost effective than its competitors.

Estimating the time needed to complete a specific operation or trade work element, such as foundations, steel erection, or roof installation, involves many considerations, including:

- Understanding the project and its unique elements
- Understanding the physical conditions under which the work has to be performed, such as location constraints and limitations, usage of the project during construction, climate, and the effect of these on the labor and equipment to be used
- Understanding the quantity and quality of the available labor resources
- Identifying the required materials, sources, and lead times
- Identifying the required and available equipment
- Understanding how the integration of these above factors affect the predicted productivity of an operation
- Incorporating predictable risks or events that could affect how long an operation or individual activities would take to complete

Once the contractor has a project schedule that it believes is an accurate representation of the construction plan, the contractor should share the schedule with the owner to demonstrate its plan and to inform the owner when it will need to perform its obligations, which may include the review and approval of shop drawings and submittals and inspection of the work. Effectively communicating the work plan to all parties involved is not only a sound project management practice, but it also promotes a culture of cooperation and partnering.

Additionally, a properly updated project schedule will also document changes in the contractor's plan to complete the project. Successful contractors and owners know that, as a project progresses, they may encounter unexpected problems or issues. In response to these, the contractor may need to alter portions of its construction plan, such as its work sequence, crew sizes, and operating hours. Project schedules should be periodically updated to reflect the contractor's then-current construction plan. These updates will provide snapshots of the contractor's plan as it changes during the course of the project.

Track and measure the work

A project schedule that is properly and periodically updated throughout the life of the project will enable the contractor and owner to accurately track and measure the project's progress. Using the project schedule for tracking and measuring occurs on at least two levels. In most instances, owners use the project schedule to track the contractor's progress, keep the project stakeholders informed of the project's status, and ensure that the contractor completes the project in accordance with the contract. Also, the project schedule includes activities representing the subcontractors' agreed-upon scopes of work. As such, the contractor is able to track and measure the progress of its subcontractors to ensure that their work is completed in accordance with their subcontract agreements.

Timely decisions

In addition to tracking and measuring the project's progress, a properly maintained project schedule will also enable the parties to identify and deal with unexpected issues as they arise. When a problem is encountered that may delay some element of the project, the project participants can use the project schedule as a tool to predict the effect of the delay on the completion of the overall project. In addition to predicting the effect of the problem, they can also decide on an appropriate course of action to deal with the problem, which may include accelerating the work, relaxing contract restrictions to more quickly advance the project, or deleting work items. This ability to predict and deal with a problem that may delay the project before it actually does so is perhaps a project schedule's most valuable attribute. Most project managers can see and deal with problems as they occur. However, good project managers can also predict how problems today will affect the project in a month, in 6 months, and even farther in the future. Relying on the project schedule as a planning, scheduling, and management tool will enable project managers to more competently and reliably control and manage their projects.

TYPES OF PROJECT SCHEDULES

A contractor can use many different types of schedules to depict its construction plan. Selection of the most appropriate scheduling technique depends on the size and complexity of the construction project, the preferences of the party preparing the schedule, and the scheduling requirements of the contract. The most common scheduling techniques used for construction projects are narrative schedules, Gantt Charts or bar charts, linear schedules, and Critical Path Method (CPM) schedules.

Narrative schedules

Narrative schedules are typically used on very small construction projects that have very few activities. A narrative schedule consists of a narrative description of the contractor's planned construction sequence and is typically submitted prior to the start of work. For example, a narrative schedule may tell the owner that the contractor plans to work across the project site from east to west. In addition, in the narrative the contractor should also identify the number of crews it plans to use and how long it believes the work will take. The degree of detail in a narrative schedule will vary from project to project, but most narrative schedules range between one paragraph and two pages. When a narrative schedule is requested, the contract scheduling specification should identify the level of detail required. An example of a narrative schedule is shown in Fig. 1.1. The physical project as described in the narrative schedule is shown in Fig. 1.2.

RE: Bridge Project
Narrative Schedule

Dear Engineer,

The bridge construction project consists of two abutments (Abutment #1 and Abutment #2), three pile-supported piers (labeled Pier #1, Pier #2, and Pier #3 from west to east), and four spans with steel girders and concrete decking (labeled Span 1, Span 2, Span 3, and Span 4 from west to east).

I will mobilize to the site on April 1, 2008, and plan to construct this bridge by moving from west to east. I plan to construct the bridge abutments and piers concurrently. One crew will construct Abutment #1 and then Abutment #2 and at the same time Piers #1 through #3 will be constructed starting at Pier #1 through Pier #3. The construction of the spans will follow the installation of the piers.

I expect the construction of each abutment to take approximately 6 weeks and the construction of each pier to take approximately 8 weeks. I expect that all of the abutments and piers will be in place and ready to accept the steel girders for the spans by the end of July.

From the end of July until the middle of October, I plan to erect the steel for the four spans, form and pour the decks, and form and pour the parapets and sidewalks. After completing any punchlist work, I plan to open the bridge to traffic on October 20, 2008.

Please feel free to contact me with any questions.

Sincerely,

Bob Smith

Project Manager

Figure 1.1 Narrative schedule.

Gantt charts

A Gantt chart, also called a bar chart schedule, visually depicts the project's major work activities in relation to time. A Gantt chart or bar chart schedule is a simple and straightforward depiction of the construction plan, showing the duration and timing of the work activities with the sequence of tasks implied. The major work items or activities are identified along the vertical axis, and time is tracked along the horizontal axis. The chart contains columns along the left-hand side of the page that identify the number and title of the major work activities, activity durations, and the activity start and finish dates. To the right of the columns are horizontal

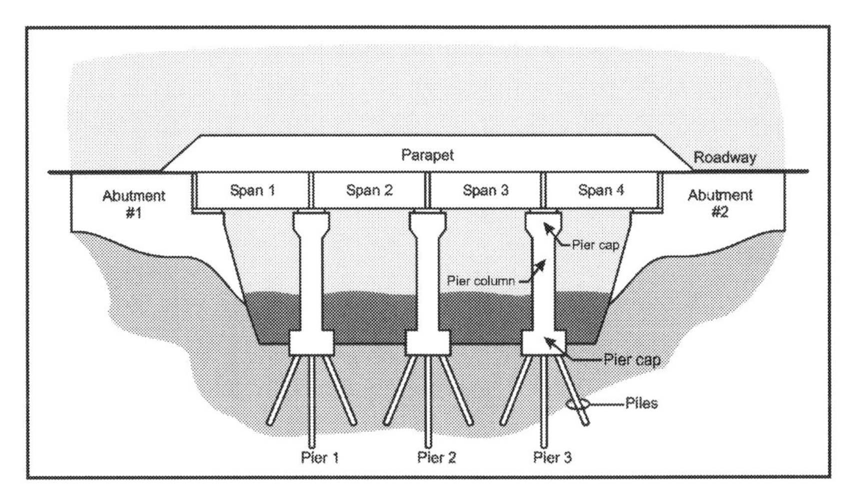

Figure 1.2 Example bridge project.

bars that represent the work activities described in the columns. The work activities are typically organized in descending chronological order, with the earliest work item in the first row, the next earliest work item in the second row, and so on. In addition to organizing the work items chronologically, they can also be grouped according to similar locations, phases, and so forth. A compilation of all of the horizontal bars should provide a visual representation of the work for the entire construction project. An example of a bar chart for a construction project is shown in Fig. 1.3.

Once work begins, the actual performance of the project can be tracked using the same chart. As the project progresses forward in time, the horizontal work bars that originally depicted when work was planned to begin can be compared to actual work bars that are drawn into the schedule as work progresses. These actual work bars can reflect progress and the actual start and finish dates of activities. If required, Gantt charts or bar chart schedules can be updated periodically throughout the project to show the status or progress of the work as of that time and be updated to forecast the timing and sequencing of the remaining contract work. The Gantt chart or bar chart schedule is typically used on smaller projects for which the relationships among the activities are obvious or easily recognized. A typical Gantt chart or bar chart schedule only summarizes the major work items and is usually only one to two pages long. It should be noted that bar charts can be very large and very detailed. If done correctly, they can be an acceptable method for scheduling the work, depending on the nature of the project.

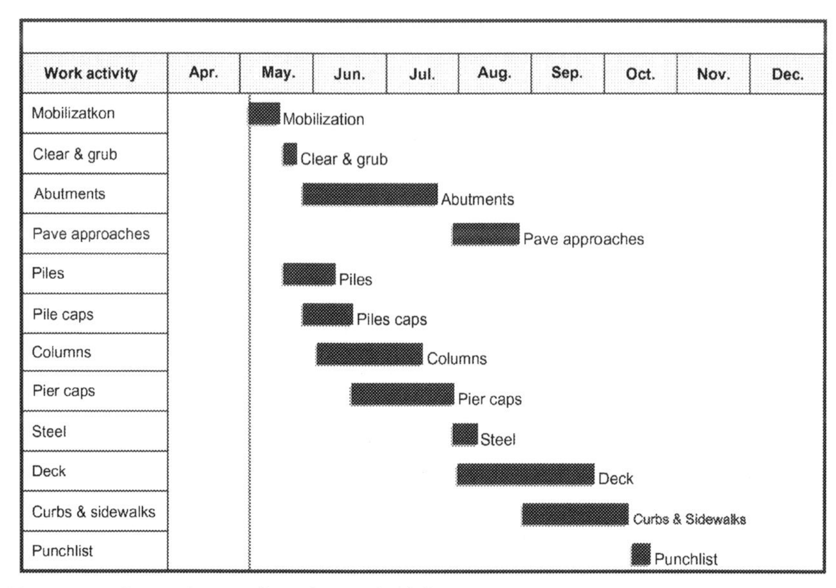

Figure 1.3 Gantt chart or bar chart schedule.

Linear schedules

The Linear Scheduling Method is also referred to as Line of Balance scheduling. It is most effectively used to plan and manage construction projects that are repetitive or linear in nature, such as highway construction and pipeline and power line construction. A linear construction schedule is usually depicted as a graph with an *X*- and *Y*-axis. The entire project duration (time) starting from Day 0 to the end of the project is plotted along the *X*-axis, and the project's scope of work is plotted along the *Y*-axis. Specific work locations along the length of the project are plotted from one end of the project to the opposite end, ascending from the lower limit of the *X*-axis to its upper limit of the *Y*-axis. The measure of production is typically represented as specific locations along the project and, depending on the type of construction project being depicted, the measure of production can be defined as stationing for highway and pipeline projects or floors for high-rise building construction. The schedule activities are usually represented as a line that starts at the *X*-axis and that extends upward toward the upper limit of the *Y*-axis that depicts performance of the schedule activity along the entire length of the project. The slope of the schedule activity line will represent the production rate of the schedule activity's operation. Fig. 1.4 is an example of a linear schedule for a roadway project.

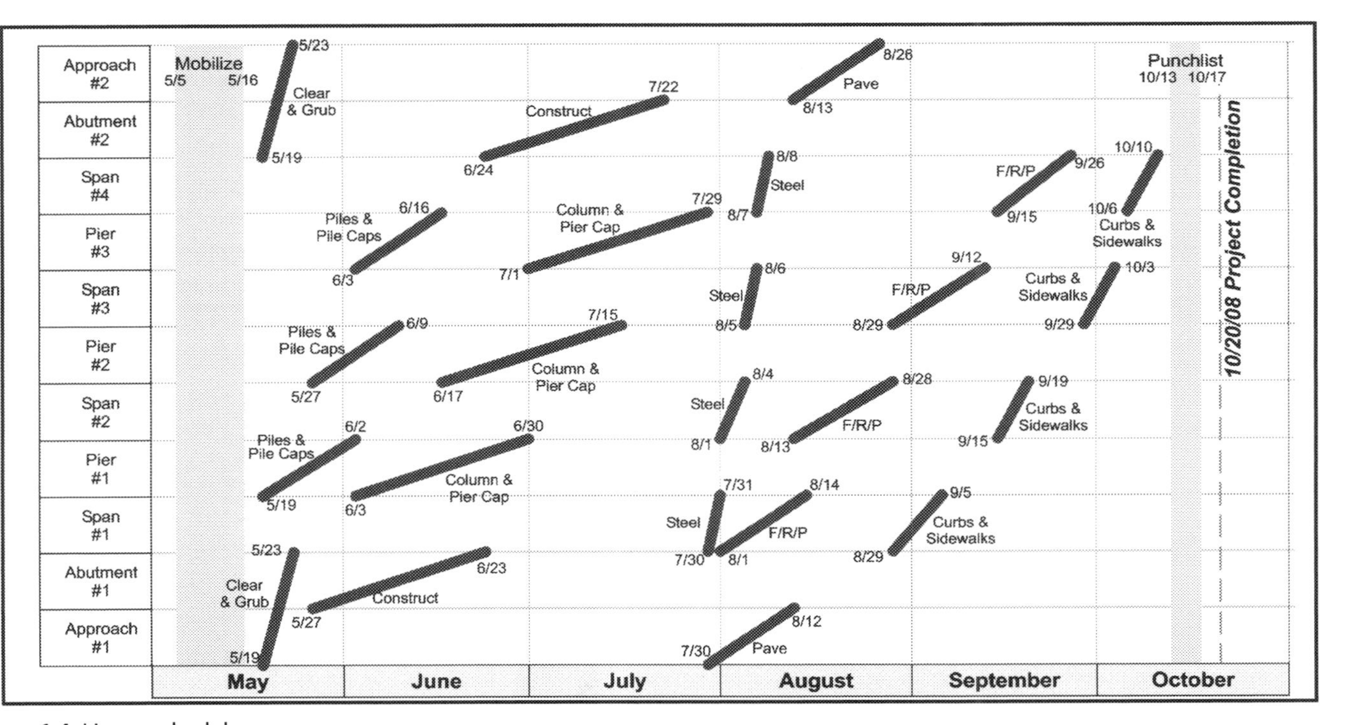

Figure 1.4 Linear schedule.

The major advantage that the linear scheduling method has over other scheduling techniques—e.g., a narrative schedule, Gantt chart, and CPM schedule—is that a linear schedule allows the user to easily track planned and actual production rates of individual schedule activities. When used on linear projects, such as roadway and pipeline projects, productivity is often an important factor in measuring efficiency, profitability, and, ultimately, success. However, the major weakness of this technique is that it does not identify the project's critical path.

It should be noted that whether or not a project has a schedule, the project will always have a critical path. The inability of the linear scheduling method to identify the project's critical path stems from the fact that the schedule activities are not linked to one another and, thus, the schedule cannot accurately correlate delays to the completion of the project to specific schedule activities in a cause-and-effect manner. One of the major reasons that many owners and contractors have chosen the CPM scheduling technique over the linear scheduling methods is the CPM scheduling technique's ability to identify the critical path of the project. With the critical path defined, the owner and contractor are better able to manage the project with respect to time. Along with this is the benefit of being able to assign project delays to specific schedule activities and determine whether the contractor is entitled to additional contract time.

The use of linear schedules is quite rare these days.

Critical path method schedules

A CPM schedule is similar to a bar chart schedule in that it contains the work activities necessary to construct the project. However, the CPM schedule usually contains all of the project work items and connects or links those work activities to one another according to their planned sequence. The result is a network of interrelated activities that defines the various paths of work necessary to construct the project. The linking or interdependency of the work items is a major strength of CPM scheduling because it enables the identification of the critical path or the longest path of work through the network. The critical path predicts the earliest date that the project can be completed.

CPM schedules are the most frequently used scheduling technique for the planning and scheduling of construction projects, both large and small. Whereas Gantt charts and bar chart schedules usually only depict the major items of work, properly developed and updated CPM schedules

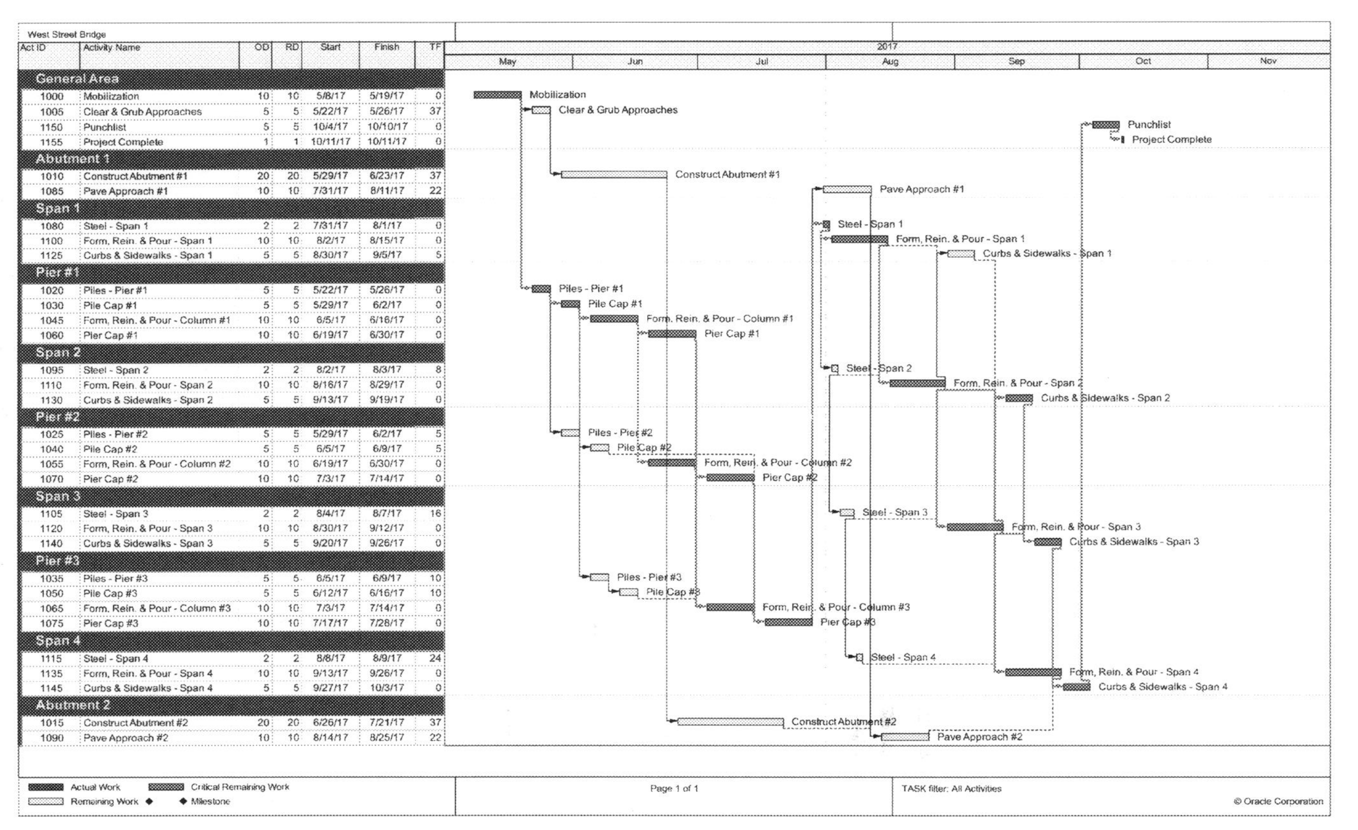

West Street Bridge

Act ID	Activity Name	OD	RD	Start	Finish	TF
General Area						
1000	Mobilization	10	10	5/8/17	5/19/17	0
1005	Clear & Grub Approaches	5	5	5/22/17	5/26/17	37
1150	Punchlist	5	5	10/4/17	10/10/17	0
1155	Project Complete	1	1	10/11/17	10/11/17	0
Abutment 1						
1010	Construct Abutment #1	20	20	5/29/17	6/23/17	37
1085	Pave Approach #1	10	10	7/31/17	8/11/17	22
Span 1						
1080	Steel - Span 1	2	2	7/31/17	8/1/17	0
1100	Form, Rein. & Pour - Span 1	10	10	8/2/17	8/15/17	0
1125	Curbs & Sidewalks - Span 1	5	5	8/30/17	9/5/17	5
Pier #1						
1020	Piles - Pier #1	5	5	5/22/17	5/26/17	0
1030	Pile Cap #1	5	5	5/29/17	6/2/17	0
1045	Form, Rein. & Pour - Column #1	10	10	6/5/17	6/16/17	0
1060	Pier Cap #1	10	10	6/19/17	6/30/17	0
Span 2						
1095	Steel - Span 2	2	2	8/2/17	8/3/17	8
1110	Form, Rein. & Pour - Span 2	10	10	8/16/17	8/29/17	0
1130	Curbs & Sidewalks - Span 2	5	5	9/13/17	9/19/17	0
Pier #2						
1025	Piles - Pier #2	5	5	5/29/17	6/2/17	5
1040	Pile Cap #2	5	5	6/5/17	6/9/17	5
1055	Form, Rein. & Pour - Column #2	10	10	6/19/17	6/30/17	0
1070	Pier Cap #2	10	10	7/3/17	7/14/17	0
Span 3						
1105	Steel - Span 3	2	2	8/4/17	8/7/17	16
1120	Form, Rein. & Pour - Span 3	10	10	8/30/17	9/12/17	0
1140	Curbs & Sidewalks - Span 3	5	5	9/20/17	9/26/17	0
Pier #3						
1035	Piles - Pier #3	5	5	6/5/17	6/9/17	10
1050	Pile Cap #3	5	5	6/12/17	6/16/17	10
1065	Form, Rein. & Pour - Column #3	10	10	7/3/17	7/14/17	0
1075	Pier Cap #3	10	10	7/17/17	7/28/17	0
Span 4						
1115	Steel - Span 4	2	2	8/8/17	8/9/17	24
1135	Form, Rein. & Pour - Span 4	10	10	9/13/17	9/26/17	0
1145	Curbs & Sidewalks - Span 4	5	5	9/27/17	10/3/17	0
Abutment 2						
1015	Construct Abutment #2	20	20	6/26/17	7/21/17	37
1090	Pave Approach #2	10	10	8/14/17	8/25/17	22

Figure 1.5 CPM schedule.

can include virtually every item of work in the contract and may contain as few as 10 or as many as 30,000 (or many more) interrelated work activities. CPM schedules have been and continue to be used as planning tools from simple to complex construction projects that require the integration of many components and incorporation of phasing and coordination. CPM schedules can be used on any size and type of construction project. If constructed properly and updated correctly, they are the most effective form of schedule for a construction project.

Most CPM schedules are updated with progress on a monthly basis, and in some instances the schedule may even be updated more frequently. When the schedule update records the project's progress, the timing and order of the remaining work items will automatically be calculated to reflect the actual performance of the completed work items. The strength of the CPM scheduling technique is that it is a dynamic modeling tool that can identify issues and problems *before* they arise. If problems arise, the CPM schedule will identify potential areas of delay and provide a measure of the magnitude of the delay. The CPM schedule's ability to accurately reflect project progress and the effect it has on the overall plan helps the project management staff to identify and deal with issues in a real-time setting. An example of a relatively simple CPM schedule is shown in Fig. 1.5.

SUMMARY

At its heart, the project schedule is a model of the project team's plan for the successful completion of the construction project. The strength of this model is a function of the care invested in its development and the consideration given to the many variables that can affect the success of the project.

It is the rare construction project that is executed precisely as planned. Consequently, in order to remain current, the schedule must be kept up to date.

Considered as a whole, the schedule and its updates constitute a history of the execution of the project's plan. Thus, not only is the project schedule a model of a successful plan for completion of the project, but the schedule and its updates also show how the plan failed, including how the project was delayed. Understanding this fact provides a key insight into the importance of the schedule as a tool for the measurement of project delays.

CHAPTER TWO

Float and the Critical Path

The explicit identification of float and the critical path are unique features of a Critical Path Method (CPM) schedule. These are the features that set CPM scheduling apart from other scheduling methods.

As discussed in Chapter 1, Project Scheduling, at its most basic level, a CPM schedule is a network consisting of activities that represent the project's scope of work with logic relationships connecting the activities to one another. These logic connections provide the sequence or order in which the activities will be completed. The work activities and their sequence should match the contractor's plan to complete the project.

A properly functioning CPM schedule will identify a period of time within which each activity can begin and must be completed so as to not delay the project. These periods of time are established by the activity's early and late dates, which will be discussed later in this chapter. Recognizing that this period may contain more time than is needed to perform the work associated with the activity is a first introduction to float.

WHAT IS FLOAT?

Float is often misunderstood. To resolve any confusion, float is best defined in two ways, which we will call the conceptual definition and the technical definition.

The conceptual definition of float is what most people mean or refer to when they use the term "float." This conceptual definition is "the amount of time that an activity can be delayed before it delays the project." This definition is linked to the guiding principle governing the analysis of delays, which is that "only delays to the project's critical path can delay the project's scheduled completion date"; this principle underlies the analysis of delays and will be discussed in more depth in this and subsequent chapters. The combination of the conceptual definition of float and the principle that the only way to delay the project is to delay work

Construction Delays.
DOI: http://dx.doi.org/10.1016/B978-0-12-811244-1.00002-1

on the critical path leads to the misconception that activities on the critical path cannot have float.

The technical definition of float is the total float calculation. Each activity's total float value is often one of the column headings in the tabular reports that are a common output of CPM scheduling software. It is usually labeled "Total Float" or "TF." An activity's total float value is the difference between its calculated early dates, which are the earliest dates an activity can start and finish according to the schedule, and its calculated late dates, the latest dates an activity can start and finish according to the schedule.

If an activity's early dates are planned to occur before its late dates, then the activity's total float value will be a positive total float value. Many times, but not always, the activity's positive total float value means that the activity can be delayed by the number of workdays equal to its total float value before the activity and its work path will become critical and begin delaying the project. Similarly, if an activity's early dates are the same as its late dates, then its total float value will be zero.

If an activity's early dates are planned to occur after its late dates, then the activity's total float value will be negative. Negative float can exist only when there is a constrained date in the schedule, usually a constraint on the end date. When an activity has negative float, it may still have float relative to the project's longest path (a concept that will be discussed at length throughout this book) and, thus, can still be delayed by the number of workdays equal to this "relative float" before the activity and its work path will become critical and potentially begin delaying the project.

Historically, the way to identify the critical path in a CPM schedule was to look for the "zero-float" work path. However, this is no longer the case. With advances in CPM scheduling software, particularly the ability to better model the plan through the use of multiple work calendars and activity date constraints that restrict when work can occur, float alone is no longer a reliable tool from which to identify the project's critical path.

As an aside, another calculated float value in a CPM schedule is free float. In contrast to total float values, which are calculated with respect to the project's end date, the activity's calendar, and constraints, free float is calculated with respect to an activity's successor activities. Free float is the amount of time that an activity can be delayed before delaying the start of its immediate successors (the early start of a successor activity, to use schedule parlance).

From this point on, all references to float refer to the technical definition, that being total float, not the conceptual definition and not free float.

The forward and backward passes

As introduced above, one of the more significant benefits of CPM scheduling over a bar chart or Gantt chart schedule is that a CPM schedule results in the calculation of a period of time within which an activity can be completed. That period of time is bookended by the earliest date an activity can start (early start) based on its position in the schedule's network, and the latest date it can finish (late finish) based on its position in the network. An activity's early dates (early start and early finish dates) and late dates (late start and late finish dates) are calculated by what are referred to as the forward pass and the backward pass. To illustrate how a CPM schedule forward and backward pass calculations are performed, we will use the following simple CPM schedule shown in Fig. 2.1.

Note that this Simple CPM Network example consists of four work activities (A, B, C, and D). In a construction schedule, rather than a letter designation, each activity would have a name, like Mobilize or Excavate Area A. Each activity has a duration in workdays. The calendar used for

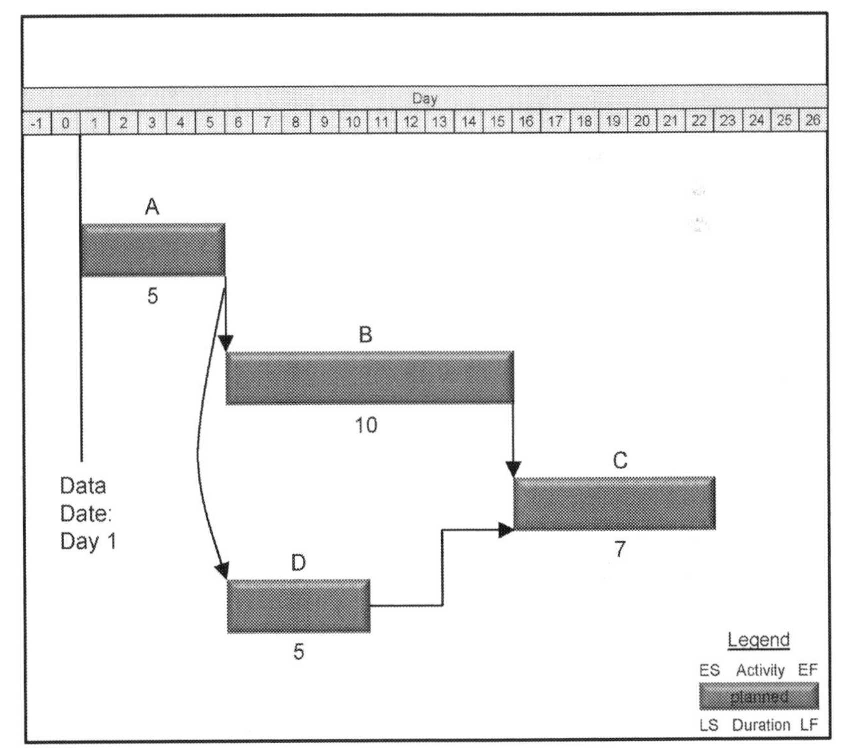

Figure 2.1 Simple CPM Network. *CPM*, Critical Path Method.

this schedule is an "ordinal" calendar, meaning that every day is a workday identified by sequential numbers. The use of an ordinal calendar will simplify the calculation of the activities' early dates, late dates, and total float values. Also, note that the Simple CPM Network's data date is the morning of Day 1.

The analysis begins with the forward pass. The forward pass calculation starts at the schedule's data date and adds the durations of incomplete activities in sequence, according to the network's logic relationships. This determines the earliest date that each activity can start and finish. In addition to calculating the early dates for every activity in the network that has not started, the forward pass also predicts the earliest date that the project can finish.

To illustrate the forward pass date calculations, see Fig. 2.2. (The calculated results of the forward pass are depicted above the bar. As indicated by the legend, the number at the top left of each bar is the

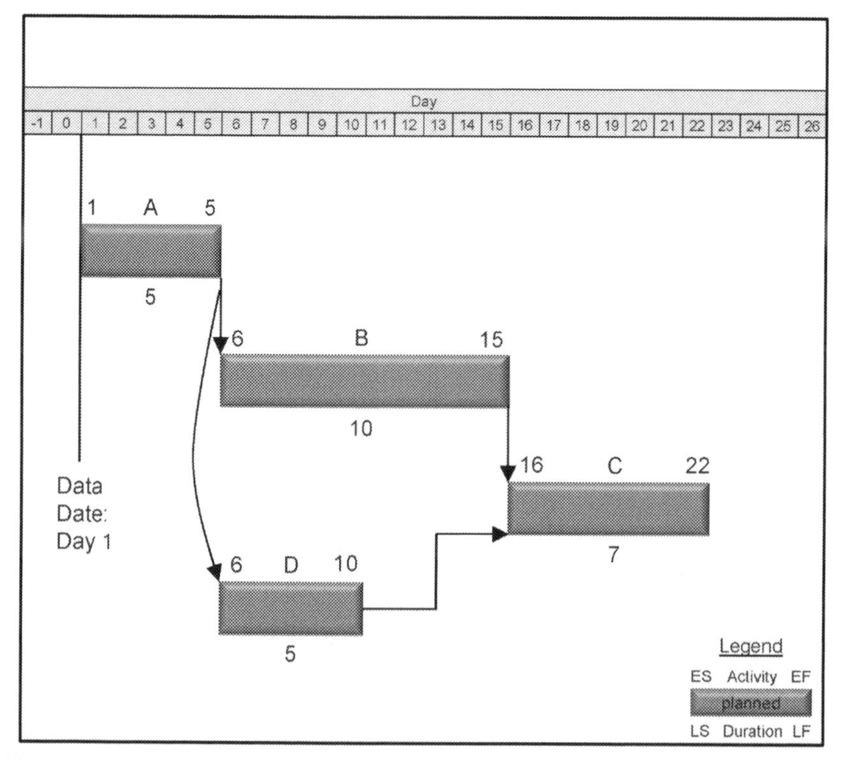

Figure 2.2 Forward pass calculation.

early start (ES) date. The number at the top right of each bar is the early finish (EF) date.)

The forward pass begins at the data date. Consequently, the ES date for Activity A is the data date, Day 1 (the morning of Day 1, to be precise). Activity A has a planned duration of 5 days. If Activity A starts the morning of Day 1 and takes 5 days to complete (where each day is a workday), Activity A's EF date is Day 5 (precisely, the evening of Day 5). These are the earliest dates that Activity A can start and finish.

Continuing the forward pass, the next step is to move sequentially to the next activities following the logic of the schedule. If the earliest date that Activity A can finish is Day 5, then the earliest date that Activities B and D can start is Day 6, the next workday.

Focusing on Activity B first, if the earliest date it can start is Day 6, then, given its 10-workday duration, the earliest date it can finish is Day 15. Looking at Activity D, given its ES date of Day 6 and its 5-workday duration, the earliest date it can finish is Day 10.

Following the logic of the schedule, the next step in the forward pass is to determine when the next activity, Activity C can start and finish. Both Activities B and D are logical predecessors to Activity C. The EF date of Activity B is Day 15, and the EF date of Activity D is Day 10. Because Activity C cannot start until both Activities B and D finish, the earliest date that Activity C can start is Day 16 (the day after Activity B finishes).

If the earliest date that Activity C can start is Day 16 and it has a planned duration of 7 days, then the earliest day it can finish is Day 22.

The backward pass is similar, but the opposite of the forward pass, and, in the simplest case, begins at the latest early finish date calculated by the forward pass. The backward pass is performed by subtracting the duration of each activity from its latest possible finish date, following the logic of the schedule backward from the completion date of the last activity. The results of the backward pass are illustrated in Fig. 2.3.

The math of the backward pass is identical to the math of the forward pass, just in reverse. Focusing an Activity D, because the latest date that Activity C can start is Day 16, then the latest date that each of its predecessors can finish is the day before—Day 15.

Similarly, if the latest date that Activity B can start is Day 6 and the latest date that Activity D can start is Day 11, then the latest date that Activity A can finish is the day before Activity C must start—Day 5; it cannot finish on Day 10 as that would delay Activity B.

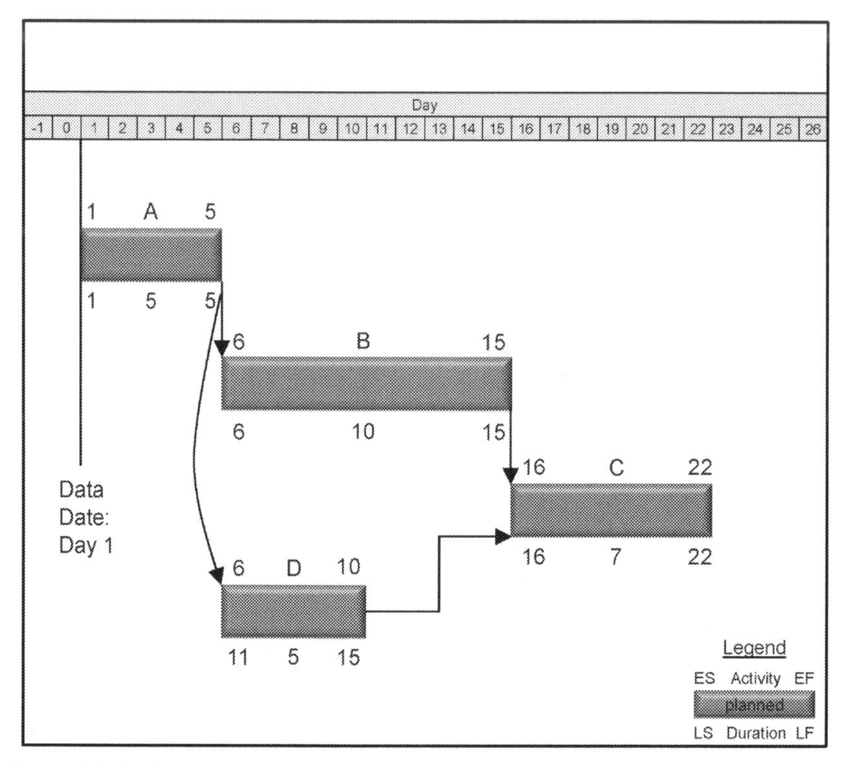

Figure 2.3 Backward pass calculation.

The difference is the total float

Each activity's total float value is calculated from the early and late dates determined by the forward and backward passes. As defined earlier in this chapter, total float is the difference between the early and late start dates or the early and late finish dates of each activity. The resulting value is the total float for each activity. The results of this calculation for each activity are shown in Fig. 2.4.

The critical path of work through the schedule shown in Fig. 2.4 consists of Activities A, B, and C. Please note that each of these activities has a total float value equal to zero. Based on this observation, it would be tempting to conclude that the critical path is always the path of zero total float. Fight this temptation. Float is affected by multiple calendars, activity constraints, and relationship ties. There is only one calendar in this schedule, and none of the activities have constraints. Also, all activity relationships are "finish-to-start," meaning that no activity can start before its predecessor finishes. As a consequence, the critical path is the path of zero total float.

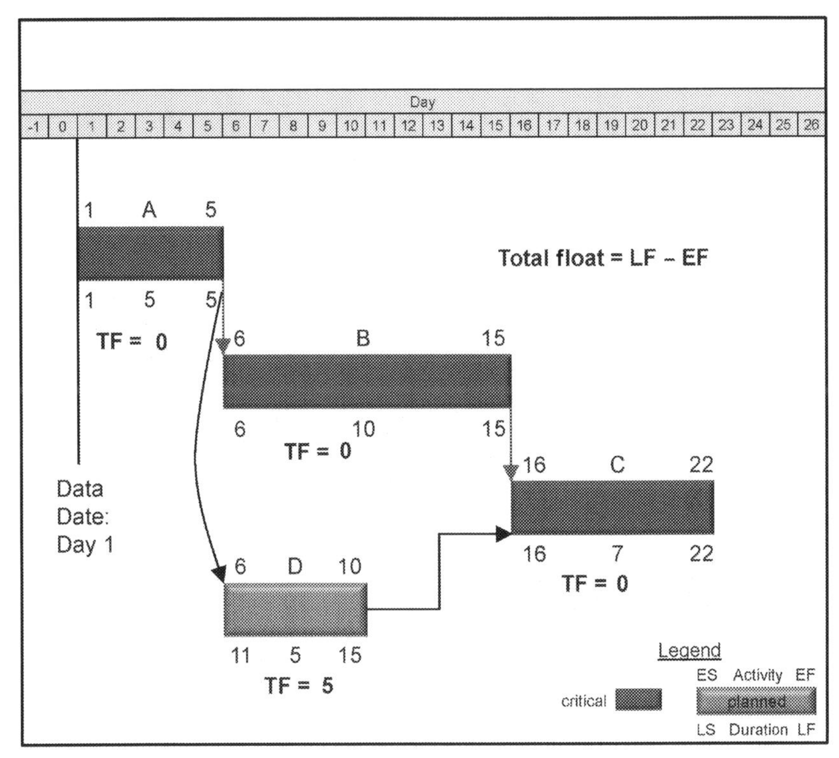

Figure 2.4 Identification of the critical path.

If a schedule utilizes multiple calendars, constraints, and more complex relationships, the critical path may not be the path of zero total float. For these schedules, the more reliable way of identifying the critical path is to identify the longest path of work. For the schedule in Fig. 2.4, not only is the path defined by Activities A, B, and C the path of zero total float, it is also the longest path of work.

Note that in the schedule shown in Fig. 2.4, any delay to Activities A, B, or C will result in a day-for-day delay to the project's completion date. Note, also, that Activity D would have to be delayed at least 5 days before it could begin to delay the project's scheduled completion date.

NEGATIVE FLOAT

Now that we have got a better feel for float, let us direct our attention to negative float. First of all, where does it come from? Negative

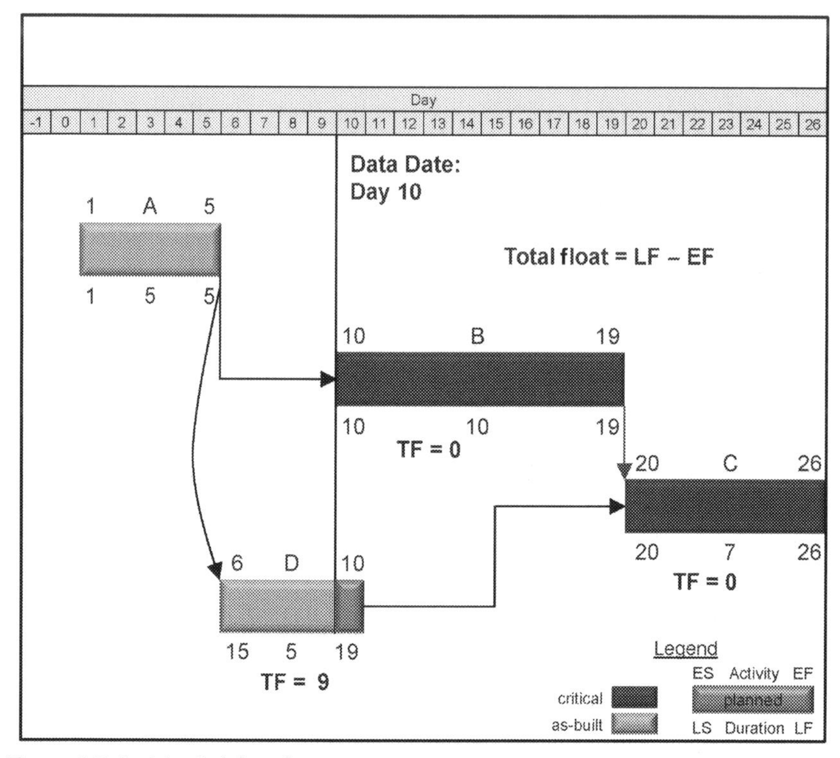

Figure 2.5 Activity B delayed start.

float can result from the application of a constraint to a specific activity or to the network as a whole. Recall from the prior example that the critical path had zero float and ran through Activities A, B, and C. Recall, also, that Activity C had a completion date of Day 22. Let us assume that the start of Activity B is delayed 4 days from Day 6 to Day 10, and that Activities A and D make progress as expected. The result is depicted in Fig. 2.5.

Because the start of Activity B was delayed 4 days to Day 10 and it was on the project's critical path, the project's completion date experienced the same 4-day delay from Day 22 to Day 26. Note that Activities B and C still have Total Float values of 0 workdays.

If we add a "Finish On or Before" constraint (using the terminology used by Oracle's Primavera Project Management (P6) software; other software packages use different terminology to name this constraint) to

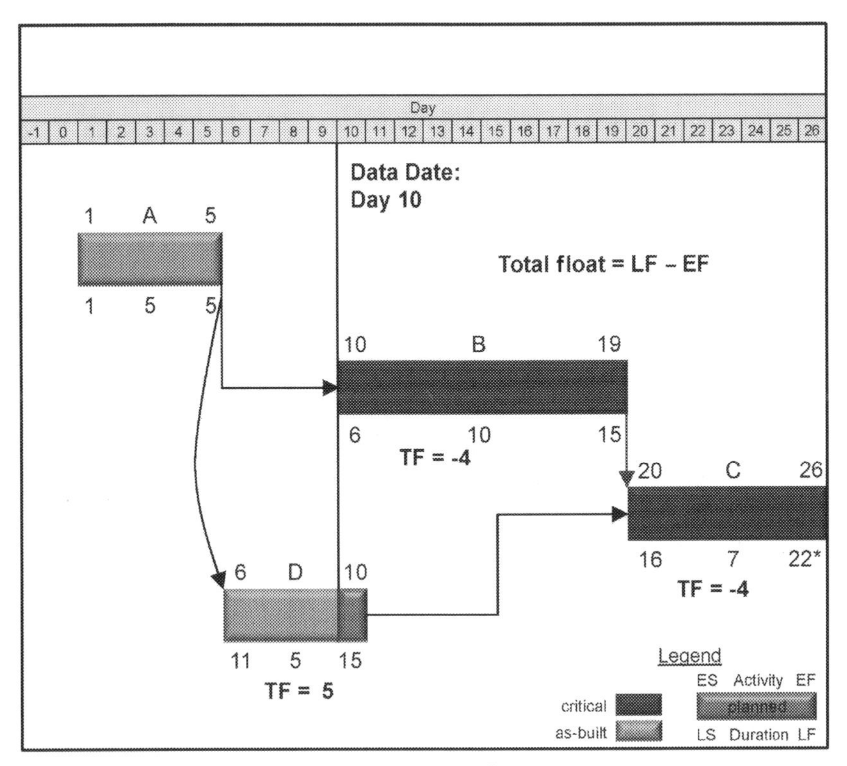

Figure 2.6 Activity B delayed start and negative float.

Activity C of Day 22, the result is depicted in Fig. 2.6. This "Finish On or Before" constraint is represented by an *asterisk* in Fig. 2.6.

When a Day 22 "Finish On or Before" constraint is applied to Activity C, the result is that Activities B and C now have total float values of −4 workdays, which is based on the fact that total float is calculated as the difference between an activity's late finish date and early finish date. The Day 22 constraint causes the backward pass to be calculated backward from Day 22, not Day 26. The result is that, for the critical activities, the late dates are earlier than the early dates. What this really means is that the work cannot be completed in time to meet a Day 22 completion date. In essence, for this simple schedule, negative float is a measure of delay. The project is 4 days behind and the critical path has 4 days of negative float.

One might argue that negative float is useful because it indicates that there is a delay. That is a reasonable position, in that negative float calculates how far behind, or late, a particular activity is forecast to finish with respect to a constraint. But negative float in and of itself does not establish that an activity or path of activities is critical. We are most interested in delays that affect the project's completion date, which means delays to the critical path. Although negative float is a useful indication that an activity is forecast to finish late with respect to a constraint, the first step in evaluating project delay begins with identifying the critical path.

Before discussing how to identify the critical path, let us continue our discussion of float and try to answer the question, "Who Owns the Float?"

WHO OWNS THE FLOAT?

Many construction contracts go beyond merely defining float. They may include provisions that both define float and assign ownership. When project-specific questions arise regarding the ownership of float, the project's contract documents should always be the first place to look for guidance. The statements made in this chapter regarding float ownership do not take precedence to the language in your contract regarding the definition and determination of float ownership. However, if your contract is silent with regard to the definition of float and its ownership, then the discussions in this chapter may be a valuable guide.

Absent contract language to the contrary, the "project" is said to own the float. In other words, float is a commodity shared by the parties to the contract—usually just the contractor and the owner. It is available to both parties as needed until it is fully consumed. A more direct way to say it is to say that float is available on a first-come, first-served basis until it is gone.

Many contracts have adopted this industry standard approach to float ownership. For example, here is what the 2016 Minnesota Department of Transportation Standard Specifications for Construction says about float ownership:

> *The contractor acknowledges that all float (including Total Float, Free Float, and Sequestered Float) is a shared commodity available to the Project and is not for the exclusive benefit of any party; float is an expiring resource available to accommodate changes in the Work, however originated, or to mitigate the effect of events that may delay performance or completion of all or part of the Work.*

In contracts between contractors and their subcontractors and suppliers, it is more common for the general contractor to restrict the availability of float. For example, the subcontract might dictate specific dates for delivery of materials or performance of work. In such contracts, the general contractor has retained ownership of float and is not sharing it with the subcontractors. The subcontract or purchase order might also contain language that states that the subcontractor or supplier has to complete its work by the early dates shown in the schedule. If the subcontractors and suppliers have to finish their work by the early dates in the schedule, then, technically, float is not available to them.

Some contracts provide for a more complicated accounting of float. In such contracts, the parties can create float for their own use. For example, if a contractor mobilizes additional crews or equipment and completes some aspect of the work more quickly, then the rest of the work on the path (if it is not on the critical path) will gain float. Subject to such contract provisions, this becomes the contractor's float and is not available to the owner. Similarly, if the owner returns a submittal more quickly than planned, the path of work associated with the submittal might also pick up float. Again, subject to such contract provisions, this added float would belong to the owner and not be available to the contractor for use.

WHAT IS THE CRITICAL PATH?

It is essential to both clearly understand what the "critical path" is and to be able to properly define it. First and foremost, the critical path is the "defining" feature of the CPM scheduling method. The development of a properly constructed schedule network is necessary to perform the forward and backward passes, which in turn is necessary to identifying the "critical path" of the schedule.

It is important to note that even when construction projects do not have an accompanying CPM schedule to identify the project's critical path, the critical path still exists. Whether you are traveling from location A to location B, cooking Thanksgiving dinner, or constructing a physical project, the critical path is the sequence of work items that forecasts when your "project" will be complete.

Recognizing this fact, the Oracle Primavera P6 Professional Help website defines the critical path as follows:

> *The critical path is a series of activities that determines a project's completion time. The duration of the activities on the critical path controls the duration of the entire project; a delay to any of these activities will delay the finish date of the entire project. Critical activities are defined by either the total float or the longest path in the project network.*

Note that this quote definitively states that the critical path is the work path or "... series of activities that determines a project's completion..." and that the "... duration of the activities on the critical path controls the duration of the entire project." This is the critical path's unique and defining feature—the critical path is the path of work that determines when a project can be completed. The critical path can or, at least, should be defined by this attribute.

Note that P6 also states that critical activities and, therefore, the critical path can be "defined by either total float or the longest path in the project network." But a word of caution is necessary here. In some schedules, the critical path can only be defined as the longest path.

REDEFINING THE CRITICAL PATH AS THE LONGEST PATH

In the past, when project schedules used only one work calendar and all activity relationships were finish-to-start with no activity constraints, the critical path could be quickly and simply identified as the path of zero float. But "we're not in Kansas anymore, Toto."

CPM scheduling software packages now give users the ability to model construction project plans with more precision than ever before. These software packages allow us to precisely model the physical and contractual restrictions common to construction projects by using, among other things, multiple calendars, activity constraints, and a variety of relationship ties. All of these facets of modern schedules affect the calculation of float.

The use of multiple calendars enables the scheduler to account for the many work calendars associated with construction project work activities. These work calendars might be based on environmental limitations (no work in the river during fish spawning season), seasonal limitations (no paving during the winter), or the type of work (concrete cures every hour of every day). To model these situations, modern scheduling software enables users to create work calendars for specific work activities, such as paving, that do not allow these activities to be scheduled during the winter months.

There are many examples of contractual work restrictions that dictate when and where contractors can perform their work. Modern scheduling software incorporates the ability to create multiple work calendars that match the planned means and methods. Such calendars may include calendars that depict 4-day, 10-hour work weeks; 5-day, 8-hour work weeks; 6-day work weeks; 7-day, no holiday work weeks; and no-work weeks or months.

A consequence of using multiple calendars is that each work calendar contains a different set of available workdays. By having multiple calendars with different available workdays, the calculation of total float values of activities on the same work path can vary if some of those activities are assigned to different calendars. This is why we say that in some schedules, the critical path can only be defined as the longest path.

To address the problem with identification of the critical path when float is not sufficient, P6 incorporates a "Longest Path" filter. This filter allows the identification of the critical path without relying on the use of float. In fact, the Oracle Primavera P6 Professional Help website recognizes the difficulty of using float to identify the critical path:

> *If your project uses multiple calendars, defining critical activities based on the longest path in the project provides an alternative to viewing critical activities based on float. Defining float in a multicalendar project is more complicated, since P6 Professional bases float calculations on calendar definitions, including workperiods, holidays, and exceptions. Using float to identify critical activities may prove misleading, since some activities have large float values due to their calendar assignments but are still critical to the completion of the project.*

The Oracle Primavera P6 Professional Help website goes on to describe the longest path's importance in the section identified as "Define critical activities," as follows:

> *In a multicalendar project, the longest path is calculated by identifying the activities that have an early finish equal to the latest calculated early finish for the project and tracing all driving relationships for those activities back to the project start date.*

Said another way, because the longest path, through driving relationships, determines the latest calculated early finish for the project, the longest path will forecast when the project will finish and, thus, is the project's critical path. The critical path should simply be defined as the project's longest path to completion.

How multiple calendars affect total float on the critical path?

To illustrate a common reason that positive float appears on the critical path, consider the example of a schedule containing multiple calendars and a critical path activity assigned to a "work calendar" that contains a nonwork period. The classic example is temperature-sensitive work, like the placement of hot mix asphalt, which cannot be performed when ambient air temperatures fall below a specific temperature.

An example of how positive float would occur on the critical path is depicted in Fig. 2.7.

As depicted in Fig. 2.7, the critical activity "Install HMA Base Course" cannot be performed during the winter months and has been assigned to a work calendar that classifies the winter months as nonworkdays. As such, the work activity for the installation of HMA Base does not allow this work to occur during the winter. Additionally, the first

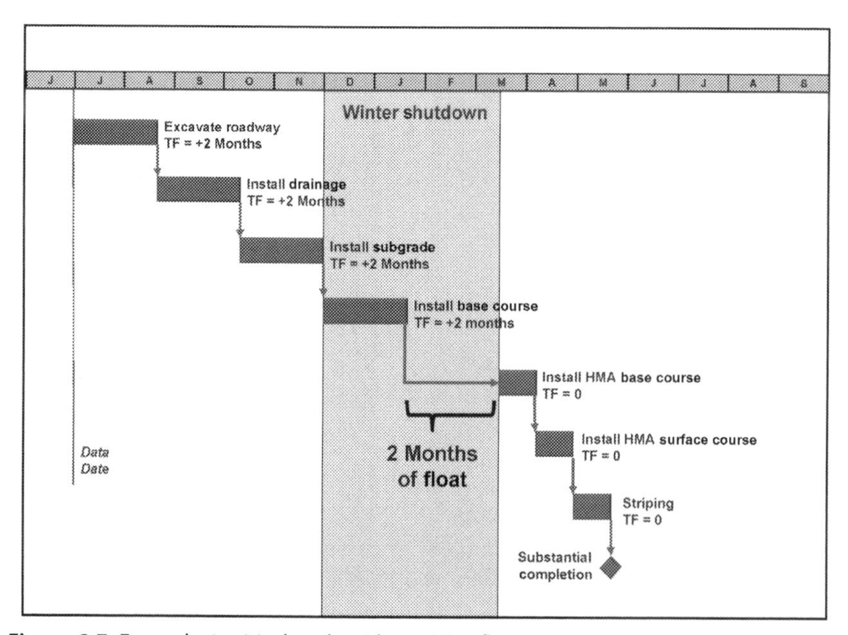

Figure 2.7 Example 1 critical path with positive float.

four activities on the critical path depicted in Fig. 2.7, which can all occur during the winter months, all have a total float value of 2 months. These two months of float are caused by a "calendar effect."

This calendar effect results in the first four activities in Fig. 2.7 having positive float. When the "Install Base Course" activity, which is the driving predecessor to the "Install HMA Base Course" activity, is scheduled to occur into the winter, because of the different calendar it pushes the "Install HMA Base Course" activity through the winter nonwork period to the following spring. The gap of time that results between the finish of the "Install Base Course" activity and the start of the "Install Hot Mix Asphalt Base" activity is positive float on the critical path. Thus, delays to the "Excavate Roadway" activity, in the example above, will delay the completion of the "Install Base Course" activity and consume the available positive float on the first four activities, but will not delay the "Install HMA Base" activity until the finish of the "Install Base Course" activity is delayed through the winter and begins to delay the start of the placement of the HMA Base in the following spring. This example also demonstrates the axiom that *Only delays to the critical path can delay the project, but not all delays to the critical path will delay the project.*

You may be tempted to believe that activities on the longest path cannot be critical activities if they have positive total float values and that the first four activities in Fig. 2.7 are not critical because they have positive float, despite the fact that they are on the project's longest path and responsible for pushing the paving activity into the spring. This dilemma begs the question: Are all activities on the project's longest path said to be critical activities? To answer this, you must consider the term "critical" in the context of Critical Path Method scheduling. In that case, the term "critical" becomes a term of the art for CPM schedules. Thus, the only logical answer to the question is yes, by definition, critical activities are those activities on the project's longest path.

Consider another example, a project that consists of the construction of a concrete, rigid-pavement roadway. Consider a schedule update that was submitted toward the end of the project during which time the initial activity on the critical path is the placing of the concrete roadway. It has a 2-day duration on a 5-day/week calendar and its successor is a 10-day activity representing the curing of the concrete, which is assigned to 7-day/week calendar. The curing activity is followed by the striping of the roadway, removal of traffic control devices, and the Open-to-Traffic project milestone. This work path is the project's critical path and is depicted in Fig. 2.8.

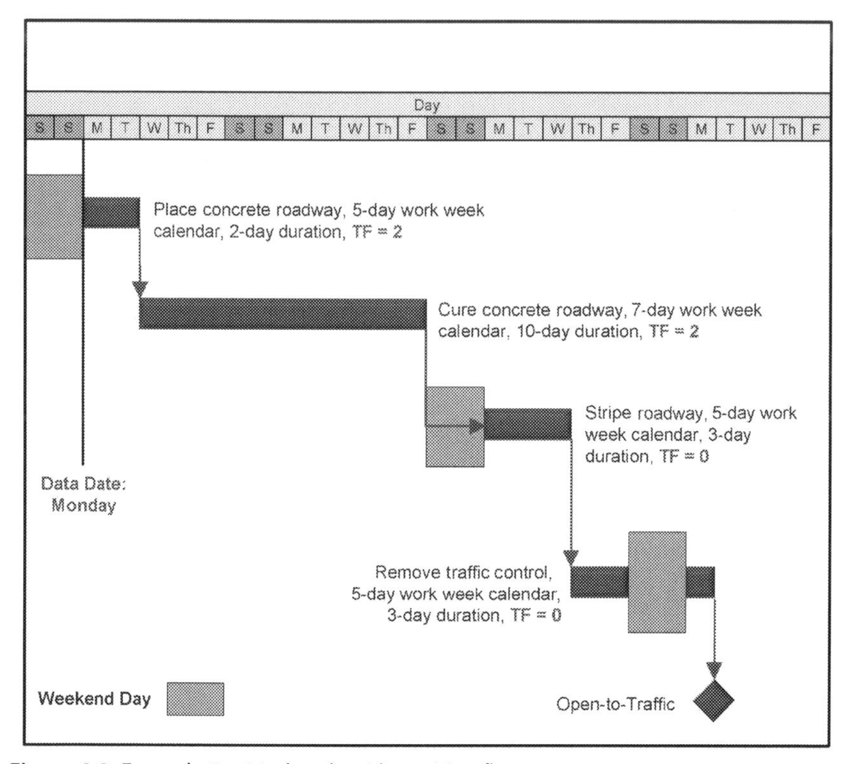

Figure 2.8 Example 2 critical path with positive float.

Fig. 2.8 shows that the concrete placement is planned to finish on Tuesday, the 10-day concrete cure activity will begin on Wednesday and finish on the following Friday, with striping starting the following Monday. Note that because the concrete cure activity, which is planned to finish on Friday, is assigned to a 7-day work week calendar, it and the concrete placing activity both have a total float value of +2 workdays. This positive float is created by the fact that the concrete placing activity and the concrete curing activity can be delayed up to 2 days, which would push the concrete cure activity to Saturday and Sunday, but would not delay the start of the striping activity or, thus, the project. In this example, all of the activities depicted in Fig. 2.8, regardless of their respective total float values, are on the project's longest path and, thus, the critical path. All of these activities are responsible for determining when the project will finish and, thus, all are "critical activities." Float does not determine whether an activity is on the longest and critical path. Rather, the length of the work paths and when the work can occur according to their respective work calendars will determine the critical path and, thus, the critical activities.

Note, also, that the path of work with positive float on each of these paths is the path of work that drives when the zero-float paths can start. The spring in the case of the paving work and Monday in case of the striping work.

Simply put, the longest path is the critical path. Activities on the longest path are critical activities regardless of their total float value.

How constraints affect activities and their total float?

Modern scheduling software allows the scheduler to employ many different types of constraints that affect when activities are planned to occur and how activity total float values are calculated. To demonstrate the effects of using different activity constraints, the four-activity example schedule from earlier in this chapter was entered into Oracle P6 scheduling software and updated through the end of Day 5. (The data date is Day 6, which is February 6, 2016 in this example.) Fig. 2.9 is the example represented in P6 without any constraints.

- *"Finish On or Before" constraint* in P6 ("Late Finish" constraint in P3): Fig. 2.10 depicts the start dates, finish dates, and total float values of the example schedule activities when a "Finish On or Before" constraint of Day 20, or February 20, 2016, is applied to Activity C.

  When a "Finish On or Before" constraint of Day 20 is applied to Activity C, the total float values of the critical path activities (Activities B and C) are reduced from 0 workdays to −2 workdays and the total float value of Activity D is reduced from 5 workdays to 3 workdays. This constraint is the correct constraint to use to model a specified completion date, because the constraint is applied to the

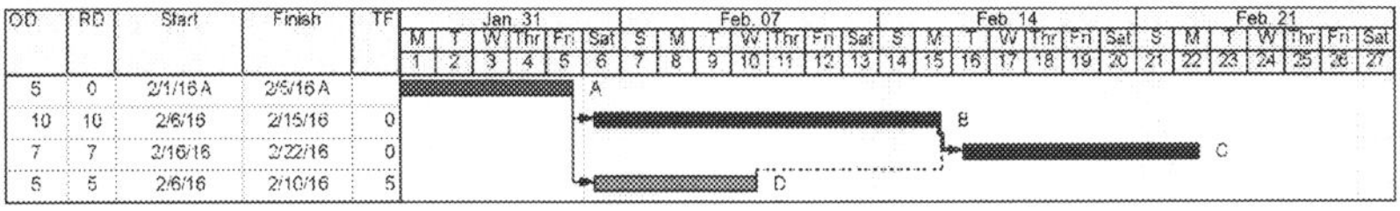
OD RD Start Finish TF
5 0 2/1/16 A 2/5/16 A
10 10 2/6/16 2/15/16 0
7 7 2/16/16 2/22/16 0
5 5 2/6/16 2/10/16 5
Jan 31 Feb 07 Feb 14 Feb 21
A
B
C
D

Figure 2.9 Example represented in P6 without any constraints.

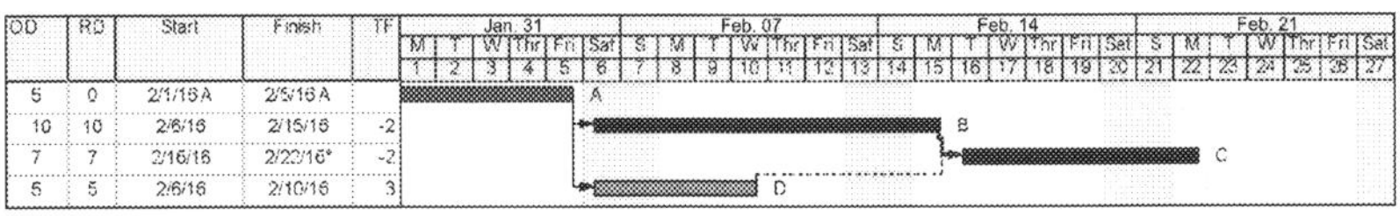
OD RD Start Finish TF
5 0 2/1/16 A 2/5/16 A
10 10 2/6/16 2/15/16 -2
7 7 2/16/16 2/22/16* -2
5 5 2/6/16 2/10/16 3
Jan 31 Feb 07 Feb 14 Feb 21
A
B
C
D

Figure 2.10 "Finish On or Before" constraint on Activity C in P6 ("Late Finish" constraint in P3).

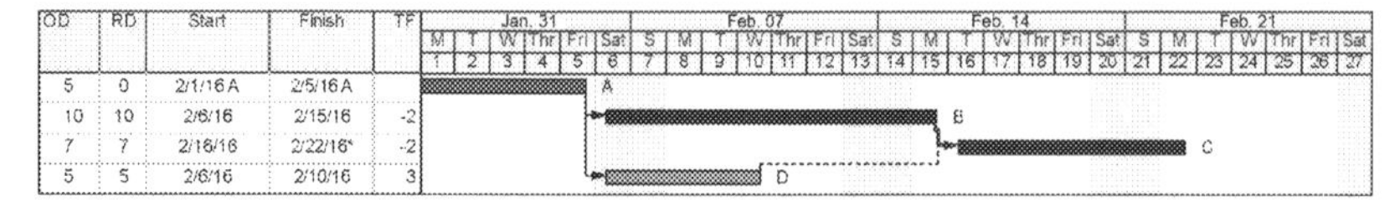

OD	RD	Start	Finish	TF
5	0	2/1/16 A	2/5/16 A	
10	10	2/6/16	2/15/16	-2
7	7	2/16/16	2/22/16*	-2
5	5	2/6/16	2/10/16	3

Figure 2.11 "Finish On" constraint on Activity C in P6.

OD	RD	Start	Finish	TF
5	0	2/1/16 A	2/5/16 A	
10	10	2/6/16	2/15/16	-2
7	7	2/14/16	2/20/16*	0
5	5	2/6/16	2/10/16	3

Figure 2.12 "Mandatory Finish" constraint on Activity C in P6 (and P3).

activity's late finish date. The constraint only affects the backward pass by setting Activity C's late finish date to Day 20 or February 20, 2016. The backward pass is started from this date. Because this constraint only affects the late finish date, the forward pass properly calculates the early dates of the remaining, uncompleted activities and the project completion date.

- *"Finish On" constraint* in P6: Fig. 2.11 depicts the start dates, finish dates, and total float values of the example schedule activities when a "Finish On" constraint of Day 20, or February 20, 2016, is applied to Activity C.

  This constraint affects the float of the example schedule activities the same way as the "Finish On or Before" constraint.
- *"Mandatory Finish" constraint* in P6 (and P3): Fig. 2.12 depicts the start dates, finish dates, and total float values of the example schedule activities when a "Mandatory Finish" constraint of Day 20 or February 20, 2016, is applied to Activity C.

  When a "Mandatory Finish" constraint of Day 20, or February 20, 2016, is applied to Activity C, there are two significant changes to the example schedule activities that affect the activities' early dates and total float values. First, note that because Activity C is assigned a Mandatory Finish of Day 20, the software forces the activity to finish on Day 20. Additionally, the software also changes the activity's total float value to zero. However, the most significant change is that because Activity C is forced to finish on Day 20, the software changes the early dates of Activity C, forcing work on Activities B and C to be performed concurrently for 2 days—Day 14 and 15. The finish-to-start relationship between Activity B and C is essentially ignored.

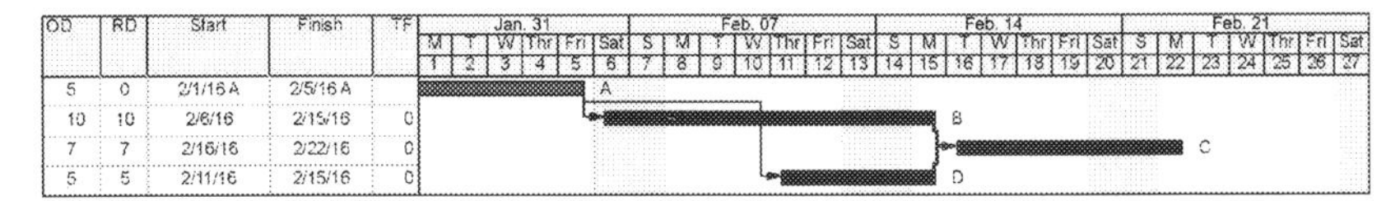

Figure 2.13 "As Late As Possible" constraint on Activity D in P6 ("Zero Free Float" in P3).

Mandatory constraints should be used with great caution given their effect on schedule logic and total float values.

- *"As Late As Possible" constraint* in P6 ("Zero Free Float" in P3): Fig. 2.13 depicts the start dates, finish dates, and total float values of the example schedule activities when an "As Late As Possible" constraint is applied to Activity D.

  When an "As Late As Possible" constraint is applied to Activity D, the result is that Activity D is now shown to finish as late as possible without delaying the project. The use of this constraint results in the activity's early dates being changed to its late dates and the activity's total float value being changed to 0 workdays. Additionally, when this constraint is applied to Activity D in this example, P6 places this constraint on the critical path, in the longest path sort.

  Despite the fact that P6 places Activity D on the longest path, should it be considered a critical path activity? This constraint may be used to model "just-in-time delivery." For example, it may be necessary to deliver steel to the site and then pick the steel directly off the truck because there may be no place to store it if it were delivered early. This constraint makes it possible to model such a scenario. It is not inappropriate or unrealistic. But, is Activity D a critical activity? The pragmatic answer is yes. Once the scheduler decides to model the delivery to start just in time, it is critical to the completion of the project. If the truck breaks down or the delivery is otherwise delayed, it will delay the project. In effect, the use of this constraint placed the activity on the longest path. Admittedly, the use of the "As Late As Possible" constraint is tricky. Generally, the start of a critical activity is driven by its predecessor. However, in the case of a "As Late As Possible" constraint, the start of the activity is driven by its successors. Because of this, the "As Late As Possible" constraint should be used with extreme caution.
- *"Zero Total Float" constraint* in P3: Fig. 2.14 depicts the start dates, finish dates, and total float values of the example schedule activities when a "Zero Total Float" constraint is applied to Activity D.

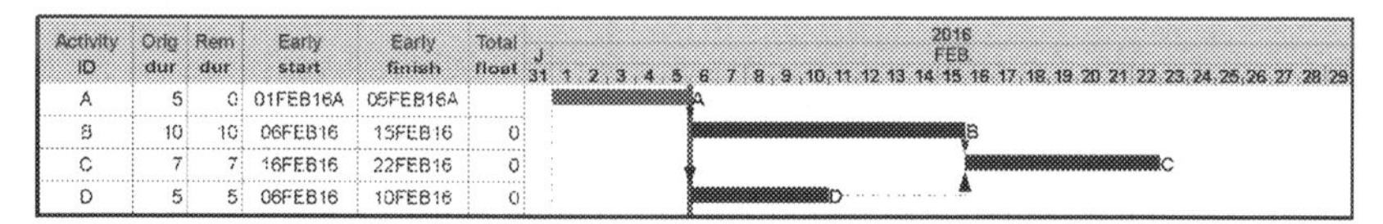

Activity ID	Orig dur	Rem dur	Early start	Early finish	Total float
A	5	0	01FEB16A	05FEB16A	
B	10	10	06FEB16	15FEB16	0
C	7	7	16FEB16	22FEB16	0
D	5	5	06FEB16	10FEB16	0

Figure 2.14 "Zero Total Float" constraint on Activity D in P3.

When a "Zero Total Float" constraint is applied to Activity D, the result is that Activity D's total float value is now shown as 0 workdays. The use of this constraint results in the activity's late finish dates being changed to its early dates and the activity's total float value being changed to 0 workdays. The fact that Activity D has a total float value of 0 workdays suggests it is critical, but, if this activity is delayed, it would not result in a delay to the project. The reason that Activity D is not critical has nothing to do with its total float value, it has to do with the fact that it is not on the project's longest path. The use of the "zero total float" constraint artificially makes this activity's calculated total float value of zero; however, Activity D actually has a positive total float value of 5 workdays, which means it is not responsible for determining when Activity C will start. This constraint has no real purpose and some may argue that its use in an instance like this distorts the meaning of float by attempting to suggest an activity is critical when it is not.

As demonstrated above, activity constraints can affect total float values and override the schedule network's logic relationships. Because of this effect, particularly when combined with the effect on float of multiple calendars, float alone cannot be used reliably to identify the critical path.

MULTIPLE CRITICAL PATHS

Multiple critical paths can be a confusing term and concept. It can be used to mean concurrent critical paths, which will be discussed in a later chapter in this book, or it can be used to describe an instance when a project has separate project milestone or completion dates each with its own critical path. For the purposes of this discussion, we will address the latter case.

For multiphase or multistage projects, the contract-prescribed sequence of the phases or stages usually dictates whether the work in the phases or stages is independent or sequential.

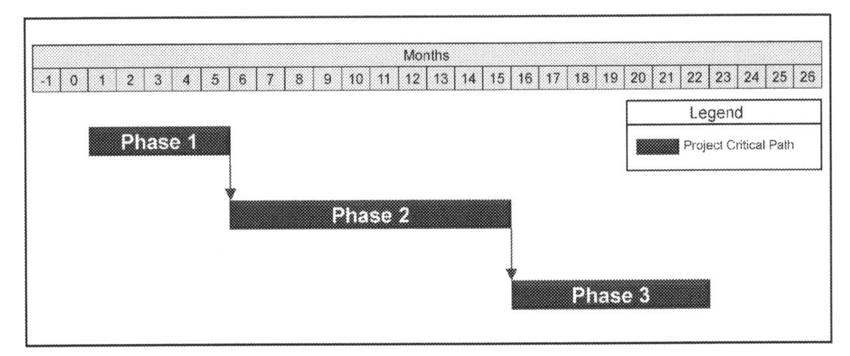

Figure 2.15 Critical path with sequential phases.

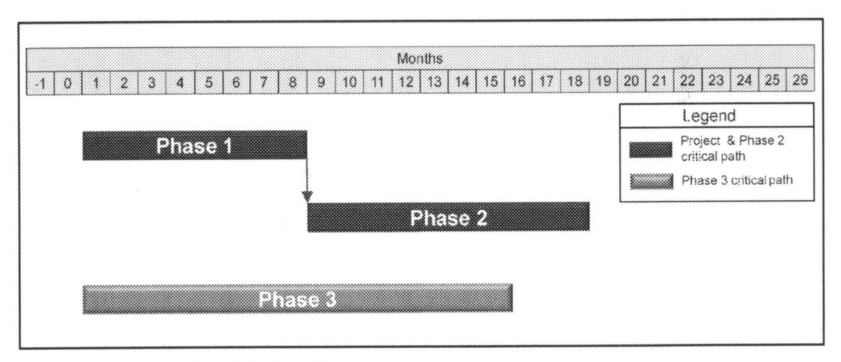

Figure 2.16 Multiple critical paths.

On multiphased projects in which the phases have to be completed in sequence and there is only one contract-specified completion date, then it would be reasonable to expect the project's critical path to be continuous from the start to the finish, as depicted in Fig. 2.15.

Phase 1 would start on Month 1 and finish at the end of Month 5. Upon the completion of Phase 1, Phase 2 would start at the beginning of month 6 and finish at the end of Month 15. Upon the completion of Phase 2, Phase 3 would start at the beginning of Month 16 and finish in Month 22. Delays to Phase 1 would delay both Phases 2 and 3, and the project.

On multiphased projects in which not all of the phased work is required to be completed in sequence and some of the phases have independent, contract-specified completion dates, the project may have more than one critical path, as depicted in Fig. 2.16.

In this example, Phases 1 and 2 have to be completed in sequence and Phase 3 does not. Phase 1 starts in month 1 and finishes at the end of

Month 8. Upon the completion of Phase 1, Phase 2 starts at the beginning of Month 9 and finishes in Month 18. In this example, the project's critical path consists of Phases 1 and 2.

Delays to Phase 1 will delay Phase 2 and the project. As such, this path of work is both the longest path in the schedule network and is the project's critical path.

Phase 3 work is performed independently from the work of Phases 1 and 2. Phase 3 work must finish by its own contract-specified completion date. Phase 3 would have its own critical path to its own completion date. The result is that the project has multiple critical paths—one critical path to the completion of Phase 2 and another critical path to the completion of Phase 3.

CHAPTER THREE

Reviewing the Project Schedule

The project schedule is best prepared by those who are going to execute the work. That is because a Critical Path Method (CPM) schedule is most effective when it properly models the project plan. Who better to model the project plan than the contractor who will be performing the work? The CPM schedule should reflect the contractor's means and methods for performing the work. In this regard, it is said that the contractor "owns" the project schedule. But, if that is the case, what is the owner's role?

Most construction contracts require the contractor to develop a CPM schedule for submission to the owner. The contract may also require the contractor to submit periodic schedule updates. So, while the schedule "belongs" to the contractor, the owner has an interest in making sure that it exists. The submission of the baseline schedule provides an owner with the opportunity to understand the contractor's plan for the project. This should include the contractor's assumptions regarding the owner's role and the roles of other parties.

Upon submission, the owner should review the baseline schedule to ensure that the contractor's plan, complies with the contract requirements, represents a reasonable plan, and follows CPM scheduling best practices.

Similar to the baseline schedule submission, the submissions of periodic schedule updates are also an important consideration for the owner. Because the CPM schedule is the only project management tool that identifies the project's critical path and, thus, forecasts when the project will finish, it provides the most reliable measure of project delay through the life of the project.

Thus, it is recommended that the owner require the contractor to prepare and submit a project schedule that meets the contract requirements to ensure that both parties know the plan and have a reliable tool to evaluate progress and facilitate timely completion of the project. To accomplish this, owners should perform a thorough and knowledgeable review of the project schedule. This submittal should be treated similar to every other submittal received from the contractor. For example, just as an owner relies on a qualified structural engineer to review the contractor's steel shop drawings, the owner should have a qualified and experienced scheduling professional to evaluate the project schedules.

Construction Delays.
DOI: http://dx.doi.org/10.1016/B978-0-12-811244-1.00003-3

REVIEWING THE BASELINE SCHEDULE

A project's baseline schedule should represent the contractor's initial plan for performing the project work. As the initial plan, the baseline schedule should not contain any actual start or finish dates. This is an important point that should not be ignored. If the reviewer decides for some reason that showing some progress in the baseline is acceptable, be sure to confirm that the program is set to calculate using Retained Logic. CPM software allows the scheduler to select a variety of calculation methods, the most common being Retained Logic, Progress Override, and Actual Dates. These calculation methods are described in detail later in this chapter as they are more relevant to schedule updates that contain actual progress.

Because the baseline schedule is a model of the contractor's initial plan, it should represent the contractor's plan for completing the project based on the contract requirements and project conditions known at the time of bid, not on any information that only became known after the contract was awarded. Such new information may trigger a change to the contractor's plan. It is important that the baseline schedule represent the plan at bid so that the effect of any change can be assessed by comparison to the baseline schedule.

Because the baseline schedule is used to track progress, quantify project delays, and mitigate delays, both the contractor and owner should work together to develop and approve a baseline schedule that complies with the contract, represents a reasonable construction plan, and complies with CPM scheduling best practices. That said, owners should understand that when reviewing a baseline schedule, perfection is not the criteria for an acceptable schedule. Owners should also recognize that just as important as having a baseline schedule that is acceptable to all parties is the need to have the baseline schedule accepted as quickly as possible.

As stated above, the key aspects of a baseline schedule review are:

- Confirming that the schedule complies with the contract.
- Determining that the schedule represents a reasonable plan.
- Verifying that the schedule utilizes good CPM scheduling practices.

Does the baseline schedule comply with the contract?

The first question to answer when reviewing the contractor's baseline schedule submission is: Does the baseline schedule comply with the contract? The first step to answering this question begins with the submittal. The submission must be complete. It should include all of the required components and printouts described in the contract's scheduling specification. If there are missing reports or charts, then the submission should be considered incomplete and, potentially, nonresponsive. However, rejecting the submission on this basis may not be the correct action. If the submission includes essential elements, such as the native CPM schedule file and a baseline schedule narrative, it may be possible to complete the review without the required printouts. The basis for rejection of a baseline schedule is always a judgment call that should be balanced against the need to obtain timely acceptance of the baseline schedule.

It is advisable that the project's preconstruction meeting be used as a forum to discuss the contractor's schedule. The preconstruction meeting is a good opportunity for the contractor to present the schedule to the project stakeholders and to explain how it intends to use the schedule and other project controls to manage and monitor the progress of the work. This meeting is also a good opportunity for the owner to discuss how it intends to use the project schedule during the course of the project, such as to manage the time impact of changes on the project, to evaluate time extension requests, and to mitigate delays during the project.

When reviewing a baseline schedule, it is first necessary to have a thorough understanding of the project's scope of work and contract requirements, which includes understanding and enforcing the requirements of the project's scheduling specification. Armed with this knowledge, the reviewer should evaluate the baseline schedule's compliance with the following items. This list is not intended to represent a complete list of all contract requirements for every construction project; however, it is recommended that the reviewer, at a minimum, ensure that the baseline schedule complies with these common contract requirements.

- Meet contractual completion dates and duration(s). The baseline schedule should not forecast achievement of a contract completion date beyond the contract-required date or in excess of the contract duration. Examples of contract completion dates include, but are not limited to:
  - Substantial Completion Dates
  - Final Completion Dates
  - Maintenance-of-Traffic Milestone Dates
  - Interim Milestone Completion Dates
  - Phase Start and Completion Dates
  - Incentive Dates

- Include the entire project work scope. Examples of categories of activities that should be incorporated in the baseline schedule should include, but are not limited to:
  - Preparation, submission, and review of permit applications
  - Issuance of permits
  - Preparation and submission of submittals
  - Initial review, possibly second review, and approval of submittals
  - Fabrication and delivery of materials
  - Right-of-way acquisitions
  - Owner-supplied materials
  - Utility work
  - All on-site construction work
  - All off-site construction work
  - Testing, balancing, and commissioning of systems
- Include work of all responsible parties. Examples include schedule activities for the owner, subcontractors, sub-subcontractors, suppliers and material-men, all third parties (architect, engineers, utilities, permitting agencies, railroads, etc.); every party or entity who is responsible for completing work, has decision-making responsibility, or is responsible for a deliverable on the project.
- Comply with all contractual work restrictions. Examples include, but are not limited to:
  - *Environmental work restrictions.* Examples include a restriction on how close the contractor's operation can get to the nest of a particular species of bird during mating or fledgling season, or an in-water work restriction, which would not allow the contractor to perform any in-water work during a fish-spawning season.
  - *Navigable water work restrictions.* Examples include restrictions on the hours during the day or days of the week that a contractor's barge can be located within the navigable portion of a waterway.
  - *Storm water management requirements.* Examples include requiring the contractor to install temporary storm water control measures before physical construction work begins.
- Comply with contract lane closure, phasing, or traffic staging requirements. Examples include:
  - Compliance with the contract phasing or staging.
  - Requirements to maintain a specified number of lanes to be open to traffic during construction.

- Requirements that specify the number of lanes that need to be open to traffic during a.m. and p.m. rush hours.
- Requirements for the contractor to perform road work at night and have the road open to traffic during daytime hours.

- Include contract-specified durations for activities such as owner submittal reviews, delivery of owner-supplied materials, etc.
- Comply with the contract scheduling specification requirements, such as:
  - Maximum activity durations
  - Types of logic relationships used
  - The use of relationship lags or leads
  - Specified level of detail
  - Consideration of inclement weather, as required
  - Inclusion of contract-stipulated milestones
  - Inclusion of activity coding or WBS coding
  - Assignment of party responsible for the activity
  - Use of the longest path, not float, to identify the critical path
  - Retained Logic, Progress Override, and Actual Dates options
  - The use of activity constraints

The baseline schedule submission's noncompliance with any contract-mandated requirement typically constitutes adequate grounds for rejecting the submission and requiring the contractor to revise and resubmit the baseline schedule. However, it is appropriate to use some judgment when gauging the significance of the deficiency. Many deficiencies can be noted with a request for the deficiency to be corrected and the baseline schedule be resubmitted prior to submission of the first schedule update. Remember that, while it is desirable to have a baseline that adequately models the contractor's plan in the context of the contract requirements, both parties will benefit from having an accepted baseline schedule as early in the project as possible.

Does the baseline schedule represent a reasonable plan for completion?

Determining whether or not the baseline schedule represents a reasonable plan for completion is more subjective and requires more judgment than determining whether the baseline schedule complies with the contract requirements. The best way to evaluate the reasonableness of the baseline schedule is to begin by asking some of the following questions:

- Does the critical path make sense? As stated earlier, the critical path is the defining feature of a CPM schedule. It is the work path that determines when the project will finish. As a result, only delays to the critical path will delay the project. Therefore, the baseline schedule's critical path has to make sense with regard to the project's scope of work. Knowledge of the project's scope of work should enable the reviewer to predict the work path or type of work that he or she should expect to see as responsible for determining the project's completion date even before reviewing the contractor's baseline schedule submission. For example, if the project is a multiyear bridge replacement, the reviewer would not expect to see landscaping as the initial work item on the critical path. Plus, on a multiyear project, the critical path shouldn't begin at a moment of time in year two or three without a thorough explanation.
- Does the sequence of work in the baseline schedule violate mandatory construction sequencing? This concern is not usually addressed in the contract's scheduling specification. However, an experienced reviewer who is knowledgeable about the project's scope of work should walk through the baseline schedule's sequence of work and verify whether or not the sequence of work violates mandatory construction sequencing. For example, does the baseline schedule show drywall completing on a particular floor before in-wall MEP rough-in is complete? The reviewer should keep in mind that there is a difference between "mandatory" construction sequencing and "preferential" construction sequencing. Mandatory sequencing is the order of the work that all contractors have to abide by, and is generally dictated by the physical conditions that exist regardless of the contractor's means and methods. An example of mandatory logic would be the need to place rebar in a slab before the concrete can be placed. Preferential sequencing is used to depict work sequences that the contractor prefers. These may be a function of the contractor's selected means and methods, or they may be a function of resource limitations or pricing restrictions. For example, the contractor may prefer to build the project using only one crane and this may dictate a certain construction sequence. Note that rejecting the baseline schedule based on the contractor's preferential construction sequencing might be viewed as directing the contractor's means and methods.
- Do the near-critical paths make sense? Near-critical paths are work paths with low total float values that, if delayed before the next update,

might consume their available float, become critical, and, potentially delay the project. These work paths are usually identified by their total float values. For example, the reviewer could define near-critical work paths as work paths with total float values that range between 0 and 20 workdays (roughly 1 month, which is the time between updates). This range should be determined both by the project's complexity and its duration. Complex, short-duration projects might have more near-critical paths of work than projects that have a greater duration.

Based on the project's scope of work, just as the reviewer of the baseline schedule should be able to predict the work path that is the critical path, the reviewer should also be able to predict the work paths that he or she would expect to be one of the near-critical paths. When evaluating whether the near-critical paths make sense, the reviewer should look for work paths that should be included, but are not, as well as work paths that are included, but should not be.

When a near-critical path is unexpected, the reviewer should trace the logic along that work path to determine if the path contains unnecessary logic revisions or inflated activity durations that would suggest the contractor is attempting to sequester or hide float along the work path.

When a work path that would be expected to be near critical has more float than expected, the reviewer should check to see if an activity on the path is forecast to be performed earlier than it could be due to missing mandatory logic. If a logic relationship is missing, then the reviewer should be able to explain this deficiency in its review comments.

- Does the baseline schedule identify night work, shift work, or overtime hours? Based on discussions with the contractor, the reviewer should find out if the contractor's bid is based on working beyond normal-workday hours. This is also something that should be discussed in the narrative report, which will be discussed later in this chapter.
- Does the baseline schedule properly consider and depict the contractor's obligation to meet the requirements of other entities, such as Federal, State, Local, Municipal permitting-related work limitations, and safety-requirements? For example, does the baseline schedule consider and properly account for the time needed to prepare, submit, review, and approve the crane permit application or roadway shutdowns needed for construction?
- Are the activity durations reasonable? The contractor's activity durations should either be consistent with typical production rates for the

respective work activities or operations, or the contractor should be prepared to commit to extra resources to achieve short durations. The contractor's primary resources used to perform the work are its selection of labor and equipment. Just as different equipment has different capacities, not all work crews produce at the same rate. Therefore, the contractor's selection of the equipment and labor that are available to complete the work will affect its planned production rates. Keep in mind that the contractor's activity durations may also be influenced by the:

- Timing of the work and work hours. Shift work is often considered to be less productive and could result in longer work activity durations.
- Project working conditions. Factors such as seasonal weather, temperatures, and where the project is located may also result in longer work durations.
- Time to mobilize and demobilize labor and equipment. Depending on the needs of each work activity, mobilizing for a particular work activity may result in longer activity durations.
- Safety considerations.
- Site access.
- Size and proximity of laydown areas.
- Predicted downtime or inefficiencies.
- Additional time for performance risk. The contractor has the risk of performance of the work. Thus, it may include activity durations to provide a cushion or contingency to account for likely risks.

Unlike instances when nonconformance to contract requirements may provide an immediate basis for rejection of the baseline schedule, when a reviewer questions whether the baseline schedule represents a reasonable plan for completion of the project for the reasons discussed above, the reviewer must use judgment when determining the right response to the contractor. The reason for this subjectivity is that questions and concerns related to the items discussed above do not automatically justify rejection of the baseline schedule; some might and some might not. For example, if the reviewer believes and can show that the critical path depicted in the baseline schedule does not make sense based on the project's scope of work, then this deficiency may be considered as an adequate reason alone to reject the baseline schedule. However, when the reviewer has concerns related to, e.g., the reasonableness of activity durations and the contractor's level and utilization of labor and equipment, the reviewer should request the contractor to provide more information or schedule a meeting to address the reviewer's concerns. Again, the overriding concern should be to achieve an acceptable baseline schedule as early in the project as possible. For this reason, the parties should look for ways to directly communicate to expedite the review and revision process.

Does the baseline schedule violate good CPM scheduling practices?

If a baseline schedule violates good CPM scheduling practices, the result could be a baseline schedule network that does not respond properly to progress, the lack of progress, or schedule changes. Examples of issues that violate good CPM scheduling practices include are:

- Improper or overuse of constraints. As discussed in Chapter 2, Float and the Critical Path, constraints can alter activity total float values and create instances that may violate mandatory construction sequencing. Constraints are not a suitable substitute for appropriate network logic relationships. Constraints should only be used to model actual or contractual work requirements or restrictions. For example, later access to a portion of the work may be represented using a Start-On-or-After constraint, also called an Early-Start constraint. Similarly, a contractual completion date, such as the Substantial Completion date or an interim milestone date may be represented using a Finish-On-or-Before constraint, also called a Late-Finish constraint. One way to limit the use of constraints is to limit their use to modeling only contract-mandated dates. All other constraints must be by mutual agreement.
- Proper use of work calendars. As discussed in Chapter 2, Float and the Critical Path, the schedule's work calendars identify the workdays and nonworkdays that determine when activities can occur. Calendars may be used to model contractual limits, environmental restrictions, and seasonal limitations. They may also be used to model contractor preferences, such as longer work weeks and multiple shifts. The calendar definitions and assignments should be reviewed to ensure that the work calendars properly represent the work limitations in the contract, but that they do not overly restrict when work can occur.
- Overuse of lags. A lag is the amount of time inserted between logic relationships. A lag is most commonly used when two activities are linked to one another with a Start-to-Start (SS) relationship. For

example, by connecting two activities with an SS relationship with a 2-day lag, the scheduler is instructing the software to forecast the start of the successor activity 2 days after its predecessor is planned to start. Fig. 3.1 depicts this SS, 2-day lag relationship between these two activities.

It is important to note that lags do not represent an activity's progress, they just represent the passage of time. For example, as depicted in Fig. 3.1, if two activities are connected with an SS relationship with a 2-day lag, the 2-day lag may be intended to represent the amount of progress made by Activity A that may be necessary to enable Activity B to begin. This use of a lag enables the scheduler to connect these two activities in a way that better models how the work will be completed. However, as depicted in Fig. 3.1, the forecast start of Activity B is determined by two things. First, the SS, 2-day lag relationship links the start of Activity B to the start of Activity A. In other words, this relationship directs the software to assume that Activity B can only start after Activity A starts. Second, the 2-day lag instructs the software that the earliest that Activity B can be forecast to start is 2 workdays after Activity A starts. As stated above, lags only represent the passage of time. Therefore, when Activity A actually starts, the software will show that Activity B can start 2 workdays later regardless of the progress made on Activity A.

The use of lags in and of themselves is not bad scheduling practice. However, a problem arises when lags are overused. Their overuse can result in instances where the work paths, even the critical path, are not being driving by the actual progress of activities, but rather by the lags themselves, in other words, by the passage of time itself and not by progress. Fig. 3.2 depicts an instance of the overuse of lags and how this can mask or misrepresent the true status of the individual work activities and, thus, the project.

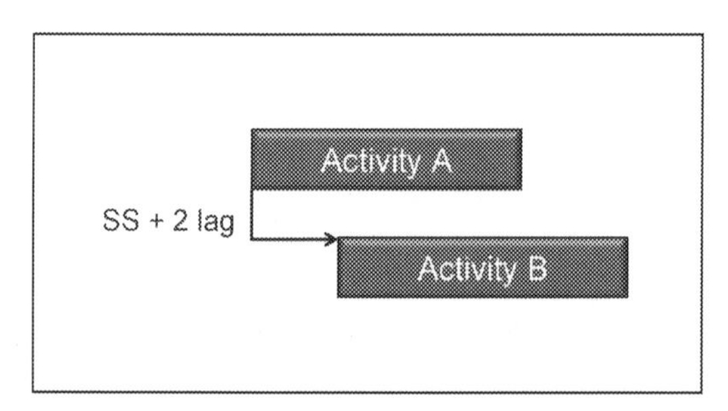

Figure 3.1 Example of start-to-start, 2-day lag logic relationship.

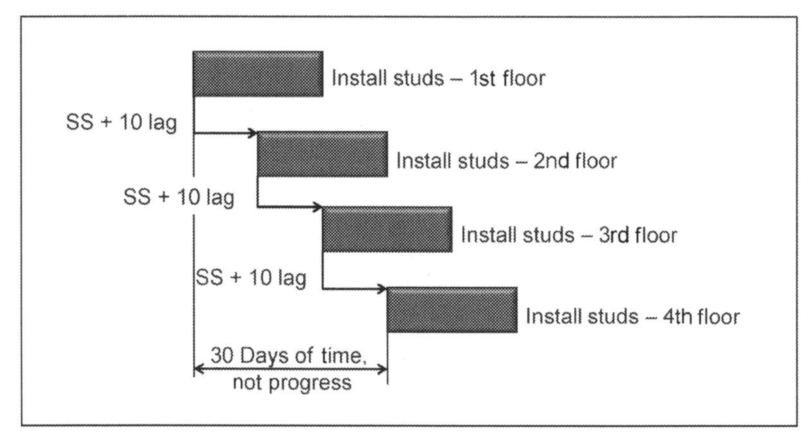

Figure 3.2 Example of overuse of lags.

Fig. 3.2 represents a way that contractors often link work activities using SS lag relationships in their project schedules. Let us assume that the four activities in Fig. 3.2 are the four initial activities on the project's critical path. The use of successive lags, as depicted in Fig. 3.2, results in a situation where the actual progress on the first three activities on the critical path is not driving when their successor activities can start. This is not in itself bad, but it does place a burden on both the contractor and the owner to understand this limitation in the driving relationships. As a result, extra care will need to be taken in both preparing and reviewing the schedule updates. The ability of a CPM schedule update to incorporate the "actual progress achieved" coupled with the time needed to complete the remaining work will determine when the schedule will forecast project completion. So, while these lags may adequately define the original plan, it may be necessary to modify or change this logic in the update.

- Use of negative lags. Contractor's sometimes use negative lags to force activities to start on or finish on specific dates. In most instances, the contractor's use of negative lags in its CPM schedule should be considered an indication that shortcuts are being taken and the scheduler is not following good CPM scheduling practices.

A negative lag is the opposite of a positive lag. A positive lag represents the amount of time separating the starts of two activities linked with a SS relationship or the amount of time separating the finish of one activity and start of another activity linked with a Finish-to-Start (FS) relationship. Conversely, a negative lag allows the successor activity to begin

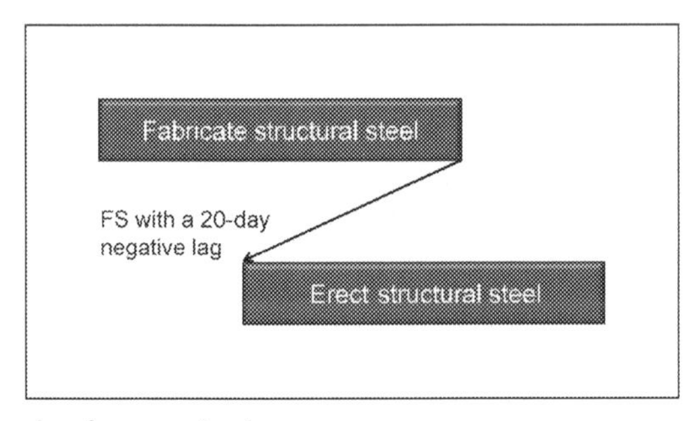

Figure 3.3 Example of a negative lag.

before the predecessor starts or finishes depending on the relationship linking the activities.

An example of the use of a negative lag is depicted in Fig. 3.3.

Fig. 3.3 shows two activities, the fabrication of structural steel and the erection of structural steel, and how they are connected to one another with an FS relationship with a 20-day negative lag. This FS relationship with a negative lag between these two activities may appear to show that not all of the structural steel fabrication needs to be complete to begin structural steel erection.

However, planning to start steel erection 20 days before fabrication completes is counterintuitive to the typical thinking behind planning and scheduling. Steel erection typically requires expensive resources, such skilled labor and cranes, and involves considerable risk during erection until the structure is complete and stable. To make the process as efficient as possible, steel erection would be planned to start once a sufficient amount of steel has been fabricated at a proven production rate. Therefore, a better model would be to breakdown the structural steel fabrication activity into preerection and posterection fabrication, allowing those activities to be connected with FS relationships without lags. By breaking down the fabrication and erection activities into even smaller, more defined component parts, the model may be improved even more. A reviewer should question the use of negative lags and an owner should consider using a scheduling specification that prohibits the use of negative lags.

- Use of Start-to-Finish (SF) Relationships. The four types of logic relationships that can be used in CPM schedules are:
  - Finish-to-Start—This logic relationship connects two activities by linking the finish of the predecessor activity to the start of its successor activity. This logic relationship option is the one most

commonly used and does not allow a successor activity to begin until its predecessor activity has completed.

- Start-to-Start—This logic relationship connects two activities by linking the start of the predecessor activity to the start of its successor activity.
- Finish-to-Finish—This logic relationship connects two activities by linking the finish of the predecessor activity to the finish of its successor activity.
- Start-to-Finish—This logic relationship connects two activities by linking the start of the predecessor activity to the finish of its successor activity.

Fig. 3.4 provides graphical examples of these four logic relationships.

The SF relationship is as counterintuitive as the FS relationship is intuitive and logical. Similar to the use of negative lags, a reviewer should consider the use of SF relationships as a potential red flag that may indicate that the scheduler is inexperienced and using software features improperly. At a minimum, this logic should be questioned by the reviewer.

- Misuse of Scheduling Features. Other problems arise when the contractor combines the improper use of lags and constraints to overcome a lack of detail. Fig. 3.5 depicts a situation where the use of a 10-day lag artificially forces the bridge steel submission and procurement work path to be the project's critical path.

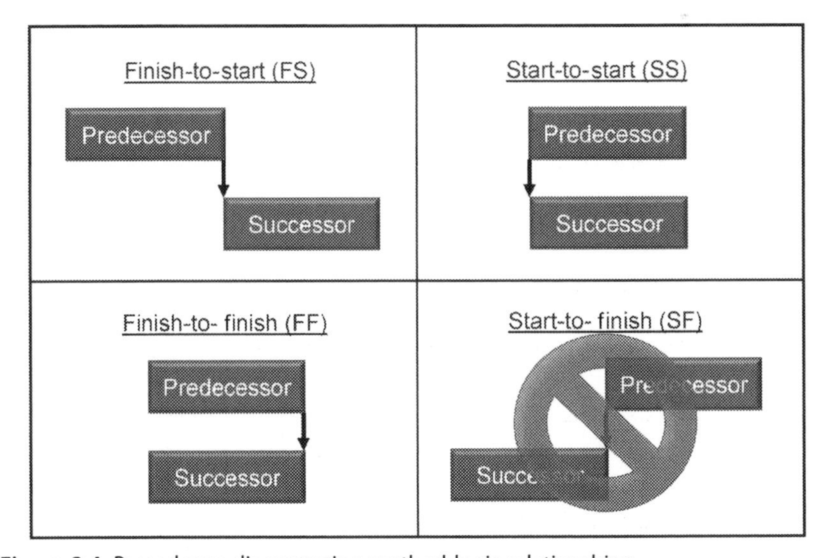

Figure 3.4 Precedence diagramming method logic relationships.

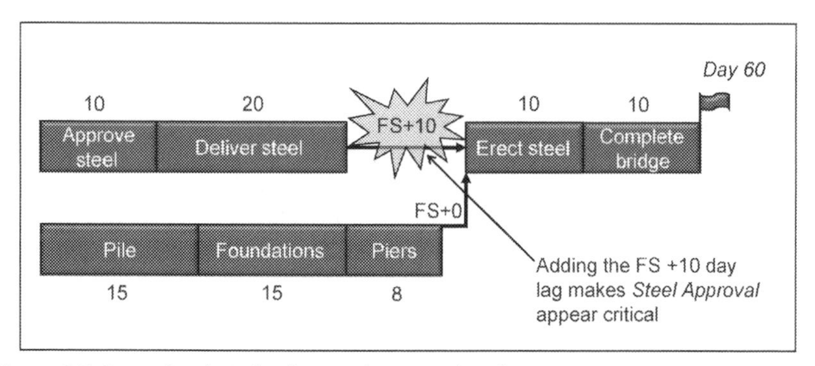

Figure 3.5 Example of 10-day lag on the critical path.

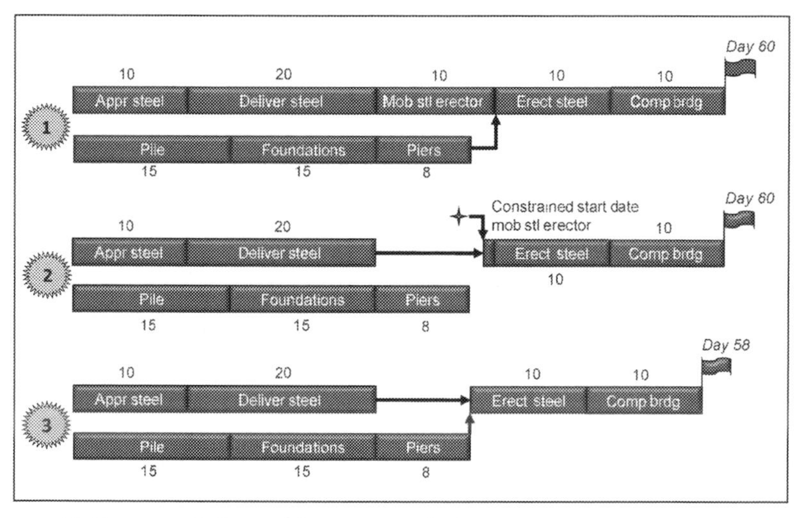

Figure 3.6 3 representations of the same project.

Fig. 3.5 shows that the piers are ready to accept the placement of structural steel on Day 39. However, the 10-day duration between the delivery of the steel and erection of the steel is forcing the steel erection to begin on Day 41 and, thus, forecasting the completion of the project on Day 60.

Fig. 3.6 depicts the same project using three different logic theories. So, which option is the best way to model the intended sequence of the work?

Option 1 forecasts that the project will finish on Day 60 and the critical path starts with the approval of the steel shop drawings. That activity is followed by the delivery of steel, mobilization of the steel erector, the erection of the steel, and finally the completion of the bridge on Day 60.

Additionally, the substructure work path, which consists of the installation of the piles, foundations, and piers, has 2 workdays of float relative to the critical path.

However, when digging a little deeper, the reviewer should ask two questions. First, will it take 10 days for the steel erector to mobilize? Second, does the contractor plan to wait until all of the steel is delivered until the steel erector mobilizes? On a typical project, the answer to these questions may be no. In that case, similar to the logic in Fig. 3.8, the schedule logic is overly restrictive.

Option 2 also forecasts that the project will finish on Day 60. However, due to the use of the constrained start of the mobilization of the steel erector on Day 40, the critical path starts on Day 40, not Day 0. In fact, the steel approval and deliver has 9 days of float and the substructure work has 1 day of float. From a practical perspective and barring extenuating circumstances, a predefined start date to mobilize the steel erector should not determine when the erection of the steel would begin and the contractor should plan for the steel erector to be mobilized and ready to begin erecting steel when the piers are ready to accept steel and the steel is on site.

Option 3 forecasts that the project will finish on Day 58, 2 days earlier than the other options. In Option 3, the critical path begins with the substructure work and as soon as the piers are complete and can accept the steel, steel erection is scheduled to begin.

Of the three options, Option 3 represents the best way to model the example plan without using unnecessary constraints, lags, or activities.

Fig. 3.7 depicts two ways to depict a plan for completing a road construction project.

If you were asked to decide which of the two options above provides a more useful depiction of the project's scope of work and plan for completion, the obvious answer is Option 2. The problem with Option 1 is that the lack of detail and the use of non-FS logic relationships with lags has created a situation where all of the project work is on the critical path. Option 2 depicts the work in smaller and more measurable work packages. The added detail in Option 2, combined with the use of FS relationships, will enable the project team to identify which portions of the work are planned to occur each week and how the actual progress of the work activities will affect the critical path and the end date of the project.

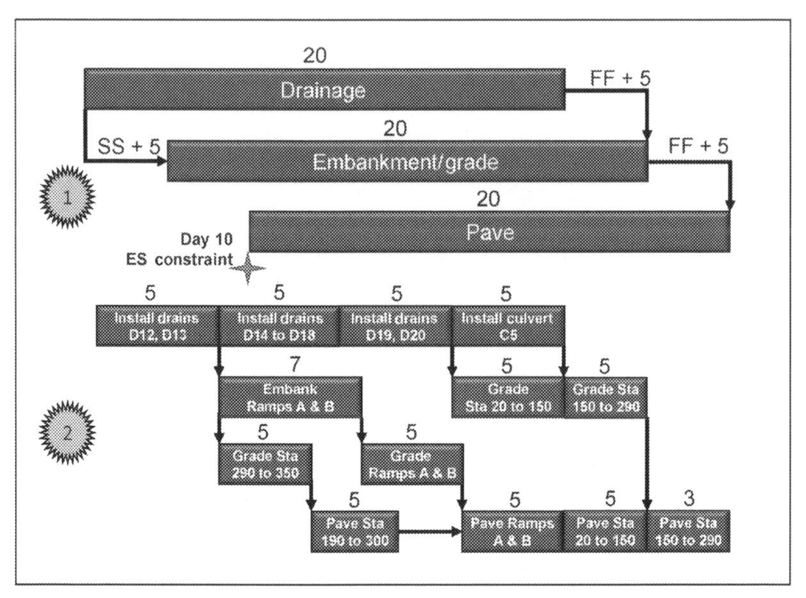

Figure 3.7 2 depictions of the same project.

- Sequestering Float. The sequestration of float is the elimination or "hiding" of float in the schedule by inflating activity durations or by the use of unnecessary preferential logic relationships to reduce total float values. By hiding float in this manner, contractors improperly reserve float for their own use. It is important to note that the sequestering of float conflicts with the idea that float is a shared commodity owned by the project and not for the sole use and benefit of one party or another. Therefore, a clear understanding of how the contract addresses float ownership is essential.

The effect of sequestering float can range from merely reducing total float values along a single work path to hiding so much float that the scheduler makes work paths critical that should not be. Identifying instances of float sequestration is often difficult, because they can be explained away as the contractor's intended means and methods.

While contractor's may be tempted to consider the sequestration of float to be beneficial, doing so has a large downside for the contractor in that it essentially masks potential delays caused by the owner. In such cases, where the reasonable duration would have demonstrated delay from the owner's actions, the expanded duration will appear to absorb the additional time and the contractor will be unable to demonstrate the impact caused by the owner. The bottom line is that there are few games that can be played with float. More importantly, knowledgeable contractors understand that, without a crystal ball, such games may as easily be detrimental as helpful.

The primary reason that both the owner and the contractor should want the baseline schedule to be developed using these CPM scheduling best practices is that both parties ultimately benefit from having a project schedule that responds properly to progress and to changes that are incorporated into the schedule. The schedule should respond to actual progress by showing the effect the actual progress has had on the other activities in the schedule.

For example, when an activity on the critical path makes slower-than-expected progress, the schedule should show the corresponding delay to the completion date. If the activity making slower-than-expected progress is on a noncritical work path with float relative to the critical path, the schedule should reflect the consumption or reduction of that float. If the work was sufficiently delayed such that all of this float was consumed, the critical path may shift to this delayed work path whereby further delays will delay the completion date.

Conversely, if the contractor is able to perform some element of the project critical path in less time than anticipated, then the schedule should show the resulting savings to the completion date. And, if the better-than-expected progress was made on a work path that was not on the critical path, then the schedule should show an increase in the total float values of activities on that work path.

In a similar manner, when changed work, added work, or unanticipated events are inserted into the schedule as they occur, the schedule should reliably predict the effect of these on the end date of the project and on the total float values of the schedule activities.

When a reviewer believes it has identified a departure from CPM scheduling best practice in its review of the baseline schedule, it should ask the contractor to explain the necessity for that particular feature and, if the contractor cannot provide a sufficient explanation, the owner should request that the contractor revise the schedule accordingly.

The best way to increase the chances that the baseline schedule is developed in accordance with good CPM scheduling practices is to include language and provide guidance on acceptable and unacceptable CPM scheduling practices in the contract's scheduling specification. Alternatively, the contract could reference a relevant standard. An example of a good standard concerning CPM scheduling practices is the AGC's "Construction Planning and Scheduling Manual."

Responding to the contractor's baseline schedule submission

Again, the contractor's baseline schedule submission should be treated like every other submission the contractor makes during the course of the project. As such, it requires an adequate review and, just as importantly, the correct response.

Most construction contracts expect the owner to respond to all contractor submissions. Absent contract language stating that the lack of a response from the owner is considered a rejection of the submission, when an owner does not respond to the contractor's submission, it is sometimes viewed as tacit approval or acceptance. This means that the owner's inaction can equate to acceptance or approval of the baseline schedule submission. Because there is never a guarantee that the baseline schedule submission will comply with the contract, the owner should provide a timely and complete response to the contractor's baseline schedule submission. Additionally, because the project schedule will be used to manage the project, to address instances of delay, and to develop delay mitigation efforts, the owner should, at the very least, review and accept or reject the contractor's baseline schedule submission.

The determination of either accepting or rejecting the baseline schedule submission should consider the need to have an acceptable or approvable baseline schedule as quickly as possible. Both the owner and the contractor will derive a maximum benefit from the schedule by having it in place as quickly as possible. This is necessary so that this key time and risk management tool is available as the project begins. Deviations from the contractor's plan are common early in the life of a project. This is because the early days of most projects carry considerable risks. These risks can be the responsibility of either party; e.g., site conditions on the part of the owner and start-up and learning curve challenges on the part of the contractor. Many disputes have erupted because of the lack of an agreed-to schedule in these early days of the project. By having this key time and risk management tool available for use by the parties during this key time period, many of these disputes can be more easily resolved or avoided.

So, what should the owner's response to the contractor's baseline schedule submission be? When should the owner accept the contractor's baseline schedule, perhaps with comments, and when should it simply be rejected?

First, if the baseline schedule submission satisfactorily meets all of the elements that were discussed above—complying with the contract requirements, forecasting an on-time completion, representing a reasonable plan for completing the entire work scope, etc., then the response should be simple—approval, or approval as noted to address minor concerns.

However, the owner should consider rejecting the baseline schedule submission when it does not comply with the contract requirements such as, the contract completion dates, required phasing or staging, contract-stipulated work restrictions, or when the schedule improperly depicts owner review times, deliverables, and obligations. Other reasons to consider rejecting the schedule may include instances when the work sequencing depicted in the baseline schedule violates mandatory construction sequencing or the critical path does not make sense.

However, rejection of the schedule will necessarily result in a delay to obtaining an acceptable schedule. For this reason, judgment should be used in determining the degree of noncompliance and the effect of such noncompliance on the early aspects of the scheduled work. If, e.g., the noncompliant aspect of the schedule can be corrected in the first schedule update without adversely affecting the use of the schedule prior to the submission of that update, it may make sense to accept the schedule "as-noted." In some cases, it may be advisable to provide an explanation of any limitations on the use of the schedule until it is resubmitted. Keep in mind that perfection is not the measure of a good schedule. Rather, having a reliable model that can be improved or corrected in the next update may provide more benefit to both parties than an outright rejection.

When responding to the contractor with either a rejection, or acceptance as-noted, the owner should provide a complete list of the deficient items and detailed questions that will clarify the contractor's plan. For example, the owner should ask the contractor to clarify unclear logic relationships, lags, and questionable activity durations. However, the owner should be careful not to include comments that the contractor could interpret as direction to modify its means and methods.

Unfortunately, the review cycle for the baseline schedule submission can take more than one round. However, both parties should strive to agree on an approved baseline schedule as quickly as possible.

APPROVAL VERSUS ACCEPTANCE

Many owners state clearly in their scheduling specification that they accept the contractor's project schedule submissions, but that they do not approve them. To many, this distinction appears essential. However, it is really a distinction without a difference and, perhaps, a dangerous distinction to make. Most construction contracts have language that alerts the contactor that the owner's approval of a submittal does not relieve the contractor of its obligations to meet all of the terms of the contract. Many contracts go on to describe the scope of the owner's review to be limited to certain aspects of the submittal, but state that the owner's acceptance or approval still does not waive the contract requirements related to that scope of review. Limitations to such caveats may relate to items for which approval is required to proceed, such as with the approval of a mock-up for aesthetic purposes, or the approval of trench bedding before backfill.

When a contractor submits its bid for a project, it acknowledges that it will complete the project in accordance with the contract requirements. The project schedule is a depiction of the contractor's plan to fulfill those contract requirements. Therefore, in the case of a schedule submission, it is important for both parties to act in a way that preserves the schedule as the contractor's plan. This means that the owner, in its review, should not dictate alterations to that plan. Rather, it should identify aspects of the plan that do not meet the contract and question any aspects of the plan that it does not understand or that it does not believe are feasible or that are otherwise potentially in conflict with the contract. In a similar vein, the contractor should not expect that the owner's review and acceptance of the plan would exonerate the contractor from aspects of the plan that do not meet the contract. To avoid confusion, the owner's project scheduling specifications should define the scope of its review and the limits of its acceptance.

EARLY COMPLETION SCHEDULES

Another important aspect of baseline schedule reviews is the submission of an early completion schedule. An early completion schedule typically consists of a baseline schedule submission that forecasts the project finishing earlier than the specified contract completion date or in less time than the specified contract duration.

Many owners are reluctant to accept an early completion schedule as the baseline schedule for the project. One reason is because they view it as the contractor's way to sequester float. In other words, owners object to the fact that if the owner delays the project and prevents the contractor from finishing early, the owner may be responsible to pay the contractor's delay costs. But if the contractor delays the project and subsequently fails to meet the early completion date, the contractor does not have to pay the owner liquidated damages. This looks like float sequestration because the time between the scheduled early completion date and the contract completion date can be consumed by the contractor without apparent consequence, but the owner has to pay to use this time. To explore this issue further, consider the following example.

Consider the contractor who submitted a baseline schedule showing that it planned to finish two months early. Then, during the project, the owner delays the work on the critical path resulting in a delay to the project's forecast early completion date of 1 month. Note that the owner-caused delay did not delay the project's forecast completion date beyond the project's contract completion date and the contractor is still on track to finish the project 1 month before the project's contract completion date. As a result, the contractor requests a time extension and corresponding delay costs.

In this case, there is often a tendency for the owner to deny the contractor's request for a time extension reasoning that the contractor is not entitled to a time extension because the contractor did not complete the project later than the required contract completion date. Conversely, the contractor would argue that is entitled to a compensable time extension because it has a right to complete the project early and the owner cannot interfere with that right without compensating the contractor for the additional costs incurred to maintain the project work for an additional month. The contractor's argument will likely include the assertion that it bid the project to finish early and the owner received a lower price as a result. If true, the contract amount did not include field office and home office overhead costs for the 2 months of time between when the contractor planned to complete the project and the project's contract completion date.

The first test of both the owner's and contractor's positions is determined by how the contract defines when the project must be completed.

Most construction contracts state that the project must be completed within a specified duration or by a specified completion date and, thus, provide the contractor with the opportunity to finish early. The terms "within" or "by" are used because most owners prefer to have their projects finish early. Such language also provides an incentive to the innovative contractor who can devise ways to complete early and, thus, provide a lower bid. However, despite this, many owners have devised ways to deal with early completion schedules that appear to be in conflict with this desire by:

1. Including contract language that states that the contractor can submit a baseline schedule that forecasts an early completion, and if the owner accepts the early completion schedule, then the contract completion date will be changed from the original contract completion date to the early completion date forecast in the baseline schedule. For owners, this approach appears to be a fair way to ensure that if the owner delays the project, the contractor is able to recover its delay damages and that if the contractor delays the project, it is responsible to pay the owner liquidated damages. However, this approach changes the contractor's risk of completing "on-time" from that defined by the bid documents. In other words, in its proposed schedule, the contractor's risk of completing later than planned was limited to its own cost. The change in the completion date adds the liquidated damages to the contractor's risk. As a result, owners using such language should expect to get few if any bids based on early completion schedules.
2. Including contract language that states that the contractor can submit a baseline schedule that forecasts an early completion and, if the owner accepts the early completion schedule, then the time between the forecast early completion date and contract completion date will be considered float available to all project participants and the contractor will not be entitled to recover any compensation for delay costs for an owner-caused delay until such delay consumes that float and delays the forecast completion date beyond the contract completion date. However, this approach changes the contractor's risk of owner-caused delays from that defined by the bid documents. In other words, in its proposed schedule, the risk of owner-caused delays would entitle the contractor to compensation for time-related costs. By defining the period as float, the owner prevents the contractor from recovering these costs. As a result, owners using such language should expect to get few if any bids based on early completion schedules.

3. Including contract language that requires the contractor to submit a resource-loaded baseline schedule whenever it plans to submit a baseline schedule forecasting early completion. This allows the owner to better validate the reasonableness of the contractor's early completion schedule. In such cases, the contract would not modify the treatment of the early completion period. In the absence of such language, the contractor's position that it is entitled to be compensated for time-related costs incurred solely due to an owner delay that prevented it from completing early will likely be found to have merit. Of course, such findings will be dependent upon many factors related to the project and the prevailing applicable law.

If the owner does not want the project to be completed early, then the contract completion should be defined in the contract to occur at the completion of a specified duration or on a certain date, rather than "within" or "by." Such provisions may be necessary when the owner is unable to take possession of the project earlier than it has planned as in the case of a planned transfer of staff and resources from an existing production facility to a new one.

The submission of an early completion schedule may be a benefit to both the owner and the contractor. However, the acceptance of an early completion schedule places a great burden on both parties to diligently manage time. This means that the owner must understand the resources that the contractor intends to utilize to achieve the early completion and must closely monitor the progress of the work so that it can identify when the contractor is falling behind the schedule and why. Similarly, the contractor must also closely monitor the progress of the work and timely notify the owner when it believes that the owner is preventing it from achieving the planned early completion.

REVIEWING A SCHEDULE UPDATE

The project participants are only able to take full advantage of the capabilities of the project schedule when it is updated. A construction project is rarely completed precisely as planned, particularly when changes or unanticipated events or conditions are encountered. Because a contractor's plan is likely to change during the course of the project due to these unanticipated events or conditions, or simply due to its own performance, the contractor's schedule update submission should represent the contractor's plan to complete the project at the time of the update. That plan should evolve in response to actual project events and conditions. For this reason, it is essential that the project schedule be updated periodically (monthly at a minimum) to record actual progress and to reflect any changes in the plan, including new or added work, changes in sequence, and other changes from the original plan. A properly updated schedule will enable the project participants to:

- Identify critical and near-critical activities accurately
- Identify and possibly mitigate problems early
- Have a reasonable forecast of activity and project completion dates
- Have an accurate record of when activities started and finished
- Make informed decisions about the effect of changes on the project
- Maintain an understanding of the resources needed to accomplish the work
- Identify third-party responsibilities
- Plan for the work in the field using short-term look-ahead reports
- Evaluate pay applications based on reported progress
- Identify areas that will require special attention
- Improve the chances that a project will be completed on time and profitably

Because a schedule update includes actual progress information, it can provide a significantly different portrayal of the project than the baseline schedule. For example, although a project's baseline schedule and its schedule updates should depict the contractor's plan for completion as of a particular moment of time, the schedule update includes additional information that provides the participants with the ability to compare actual progress to the plan in order to identify how progress, the lack of progress, changes, and unanticipated events or conditions affect the ability to meet the project completion and milestone dates. Therefore, the steps involved in reviewing a schedule update include more than just validating the plan to complete the remaining work.

Instead of revalidating the contractor's entire plan for completion when reviewing a schedule update submission, the objective of a schedule update review is to determine how the following factors effected the project completion and milestone dates during the update period.

- The effect of progress or lack of progress
- The effect of minor revisions to the schedule logic

- The effect of added or deleted activities representing changes
- The effect of unanticipated events or conditions

After determining the effect of these factors on the project's completion and milestone dates, the next step is to determine the party responsible for the delay or savings resulting from each of these factors during the update period. Then the owner can decide on the best option to address the delay, which may involve granting a time extension, directing acceleration, or requiring the contractor to resolve the delay by submitting a recovery schedule.

STEPS TO REVIEW A SCHEDULE UPDATE

The first step in reviewing a schedule update is similar to the first step in reviewing a baseline schedule, which is to verify that the submission is complete and in accordance with the scheduling specification. For example, if the schedule update submission is missing the required narrative, then the owner should notify the contractor that the narrative is missing and that the review of the schedule update will not commence until the narrative is submitted. However, the owner should also be reasonable when determining whether the submission is responsive. Similar to the baseline schedule review, the basis for rejection of a schedule update is a judgment call that should be balanced against the need to obtain timely acceptance of the schedule update.

The review of the schedule update should involve an evaluation of the following components:

- Confirm that the schedule is being calculated using Retained Logic
- Identify the status of the contract completion and milestone dates
- Confirm that the as-built information recorded in the update submission is correct
- Review the narrative
- Identify and measure the delay experienced during the update period
  - Evaluate the effect that progress had on the forecast project completion date and other contract milestone dates, if applicable
  - Evaluate the effect of schedule revisions on the update and the resulting delay or savings
- Properly respond to the schedule update submission

The remainder of this chapter addresses each of these components.

Confirm that the schedule is calculated using retained logic

Primavera P6 Project Management scheduling software includes different user options, known as Retained Logic, Progress Override, and Actual Dates, to instruct the software as to how to calculate the network when there are activities that begin and make progress earlier than expected based on their position in the network. Such early progress is known as "out-of-sequence" progress. Figs. 3.8, 3.9, and 3.10 illustrate the differences between selecting the Retained Logic, Progress Override, and Actual Dates when an activity begins earlier than expected and makes out-of-sequence progress. The figures depict a simple, four-activity schedule in which the activities are connected to one another with FS relationships in sequence. Fig. 3.8 compares how the Retained Logic and Progress Override options deal with an activity that starts early and progresses out of sequence.

For example, when the Retained Logic option is selected and Activity C starts before its predecessor Activity B finishes, the Retained Logic

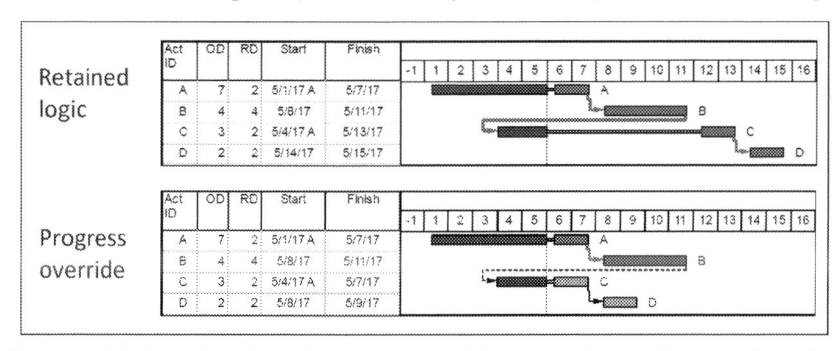

Figure 3.8 Retained logic and progress override comparison with an activity that started early.

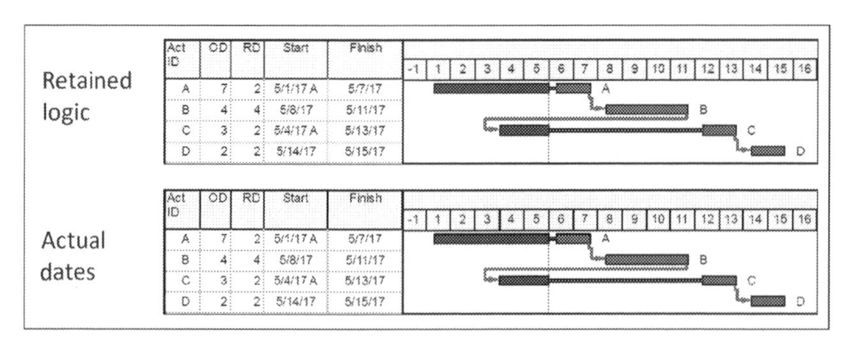

Figure 3.9 Retained logic and actual dates comparison with an activity that started early.

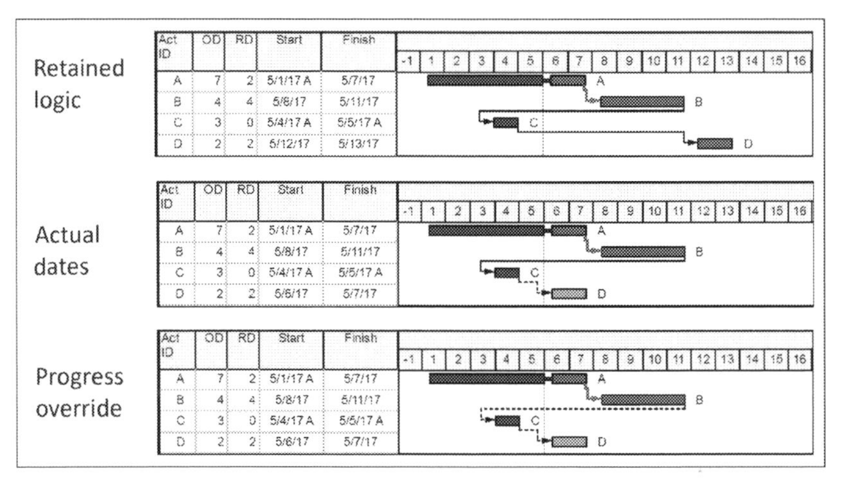

Figure 3.10 Retained logic, actual dates, and progress override comparison with an activity that started and finished early.

option respects or "retains" the logic relationship between the finish of Activity B and the remainder of progress on Activity C. In contrast, using the same schedule and progress, the Progress Override option ignores the logic relationship between the finish of Activity B and the start of Activity C and assumes that work on Activity C may continue to be performed as early as possible, with no relationship to the completion of work on Activities A or B.

The use of Progress Override is a powerful feature that directs the software to ignore precedent logic relationships when an activity starts out of sequence. In fact, Fig. 3.1 shows that when using Progress Override, the software allows Activities A and C to occur at the time same regardless of the nature of the work. More to the point, the software does not know or consider the type of work depicted by each of the schedule activities, it merely performs the forward and backward pass calculations based on the parameters set.

The use of Retained Logic is favored over using Progress Override under most scheduling situations. And while its use is really only relevant to schedule updates because the baseline should not contain progress, it is recommended that the reviewer confirm that the contractor did, in fact, select Retained Logic when it calculated the baseline schedule. This is to ensure that, if the contractor has not selected Retained Logic, the owner's preference for it can be made clear, either in the enforcement of a contract requirement or as good scheduling practice. The use of the Progress

Override or Actual Dates options is not considered to be good scheduling practice because these methods do not distinguish between logic connections that can be broken and those that cannot.

The Actual Dates option addresses out-of-sequence progress in a less severe manner than the Progress Override option. Fig. 3.9 compares how the Retained Logic and Actual Dates options deal with an activity that starts early and progresses out of sequence.

In contrast to the Progress Override option, the Actual Dates option deals with an activity that starts early and out of sequence in the same way as the Retained Logic option. The Actual Dates option does not direct the software to ignore the logic relationship between Activity B and Activity C just because Activity C starts early.

However, Fig. 3.10 compares how the Retained Logic, Actual Dates, and Progress Override options deal with an activity that completes early.

Similar to Figs. 3.1 and 3.2, the Retained Logic option used in Fig. 3.3 again respects or retains the logic relationship between Activity B and Activity C, despite Activity C completing early, and does not allow Activity D to begin until all of its predecessor activities are complete.

However, notice that the Actual Dates option allows Activity D to begin as soon as possible after its direct predecessor activity, Activity C, finished early. Also note that the Actual Dates and the Progress Override options deal with the completion of Activity C in the same manner, which is to allow Activity D to start as soon as possible.

Interestingly, when comparing Figs. 3.2 and 3.3, it becomes apparent that the Actual Dates option calculates the start date of activities whose predecessor activities finished early differently from activities whose predecessor just started early but have not finished. As such, the Actual Date option appears to recognize that when an activity "completes" out of sequence its predecessor activities do not hold up the ability of its successor activities to begin, but if an activity only "starts" out of sequence, but does not finish, it still retains the logic relationships between its predecessors and successors.

Note that both the Progress Override and Actual Dates options assume that the logic of the schedule is incorrect or no longer valid, at least as it relates to the interpretation of the significance of out-of-sequence progress. When used, these options actually modify the logic of the schedule, whether such a modification is appropriate or realistic. For this reason, these options are typically prohibited by the project's scheduling specification.

Status of contract completion and milestone dates

Most construction projects only have one contract completion date. However, other more complex projects with multiple phases or stages, or interim milestones may have more than one contract completion date. When beginning a detailed review of a contractor's schedule update submission, the first question an owner must answer is: Are the project completion and milestone dates forecasted to complete on time, early, or late?

The review should detail the status of the forecast contract completion and milestone dates. In doing so, the reviewer should also compare these forecasted dates to the current contract dates, record any differences, and report whether the forecast dates are in conformance with the contract. Additionally, the reviewer should also compare the forecast completion dates in the current update to the same dates from the previous update. In other words, if the forecast completion and milestone dates are late, then the reviewer should report both the number of calendar days that the project is forecast to finish beyond the contract-required dates and additional delay incurred during the update period.

This first item is key, because most scheduling provisions state that the owner will not approve or accept a schedule update submission that forecasts late completion. Therefore, a forecast late completion alone may be grounds for rejection. If the schedule update forecasts a late completion, then the contractor should at least provide an explanation for the forecast late completion in its accompanying narrative.

Confirm that the as-built information is correct

The as-built information recorded in the schedule update consists of the actual start dates for activities that have started, the actual finish dates for activities that have finished, and reduced remaining durations or increased percentages of completion for activities that had begun, but not finished during the update period.

This as-built information identifies the progress or lack of progress that was achieved on the project during the update period. The parties need to ensure that this information is correct so that it can be relied upon to compare the planned timing of the work from the previous schedule submission to determine the effect that the progress had on the forecast project completion and milestone dates during the update period.

It is recommended that this as-built information be agreed upon by both parties before the submission of the schedule update to avoid the owner rejecting the schedule update submission based on incorrect actual performance information. On larger projects, the project team may have a separate monthly scheduling meeting that occurs a day or so before the submission of the schedule update. It is recommended that the project participants discuss the project schedule's status as part of the regularly scheduled weekly or bi-weekly progress meeting or Owner/Architect/Contractor meeting.

Review the narrative

The update narrative should complement the submitted schedule update file. At the very least, the reviewer should ensure that the narrative is contract compliant. It is recommended that the contractor's update narrative address the following items:

- Identify the status of the project completion and milestone dates. If these dates are forecasted to complete late, then the contractor should explain why these dates are late by identifying the responsible activities and explain who is responsible for the delay. If the delay was the contractor's responsibility, the narrative should explain how the contractor plans to mitigate the delay.
- Describe the work completed during the update period. The contractor should be able to specifically identify the work it completed during the update period including which submittals were submitted and approved, the materials delivered and their quantities, and the actual work elements completed or in progress. This description should mirror the work completed during the update period in the schedule update file.
- Identify current and upcoming, potential issues or concerns that, if not addressed in a timely manner, may delay the project. The narrative is the vehicle by which the contractor should be able to keep the owner informed of potential problems. In most instances, inclusion of these potential problems in the update narrative does not constitute written notice in accordance with the contract. However, the continued reporting of potential problems in the update narratives may, at least, represent instances of the contractor informing the owner of the existence of potential problems contemporaneously.
- If the project is experiencing delays or emerging problems might delay the project in the near future, then the contractor should suggest potential mitigation efforts. All parties should be reminded that they all have an obligation to mitigate delays and additional costs.

- If the contractor makes revisions to the schedule update, which might include the addition or deletion of activities, the addition or deletion of logic relationships, the addition or deletion of constraints, changes to calendar assignments, changes to the workdays depicted in calendars, and changes in original durations of activities that have not started, the contractor should identify the changes it makes to the schedule and explain the reasoning behind each. Many contracts are written in such a way that revisions must be submitted and approved by the owner before they can be included in the schedule update. The inclusion of such language is not recommended; however, if such language exists, the parties should discuss ways to expedite the process.

Identifying and measuring the delay or savings experienced during the update period

The reviewer should identify the reasons for any changes that occurred in the projected dates and work sequences from the previous update or baseline schedule. These will have occurred either due to the actual progress reported in the update or due to logic changes that were made to the update. In doing so, the reviewer should determine and report the effect that progress or the lack of progress during the update period had on the critical path and, thus, the project's forecast completion date. This evaluation is essential to properly utilize the project schedule to plan and manage the project. Because the project schedule is the only tool available to owners and contractors that is able to reasonably forecast when the project will finish, owners and contractors do not have to wait until the project is complete to determine why the project finished late. Contractors and owners should use the contemporaneously submitted schedule updates to their fullest and determine whether the project was delayed during the update period, what caused the delay, and which party was responsible, in order to identify and implement measures to address the delay.

It is important to note that when the schedule update forecasts an on-time completion it is still possible that the project was delayed during the update period. For example, during the update period the project could have been delayed by slow progress or the lack of progress caused by the owner, the contractor, a third party, or a force majeure event and the contractor may have elected to revise the schedule logic associated with future, uncompleted work to offset or eliminate the resulting forecasted delay. Said another way, contractors may adjust their plan in order to show an on-time completion. In fact, this is what they are obligated to do under most contracts. It is for this reason that the contractor should show the impact of progress separately from any mitigation efforts that it plans to undertake.

Revising the schedule to offset or recover delay is a common practice that occurs more often than most people realize. Making changes to recover time is not necessarily a problem if it occurs in a planned and attainable manner. However, it could be more problematic when it occurs in every update or when the revisions to the logic are impractical or unrealistic. If a project is continually delayed, not achieving the expected level of progress, it could be an indicator of a number of different issues, including:

- The contractor is not constructing the project in accordance with the plan depicted in the schedule.
- The contractor is unable to make reasonably expected progress.
- The contractor's planned progress was overly optimistic.

Note that not all of these reasons for progress-related delays are always the contractor's responsibility. In fact, owners should recognize that contractors may make revisions to the update to eliminate delays caused by the owner in order to "keep the client happy." If the contractor is offsetting or mitigating the owner's delay, then it should notify the owner immediately to preserve its rights under the contract to request a time extension and seek additional compensation.

As part of every schedule update review, the reviewer should identify and measure the delay that occurred during the update period. As alluded to above, there are two ways that project delays can affect the project schedule: (1) through the progress achieved during the update period and (2) through revisions made to the schedule logic during the preparation of the schedule update. Properly identifying how both progress and schedule revisions effect the project's forecast completion date during the update period allows the owner to identify individual delays and the party responsible for each and to determine the right approach to mitigate the delay so as to finish the project on time.

Evaluate the effect of progress on the completion and milestone dates during an update period

The best way to determine the effect that progress has had on the project during an update period is to perform a schedule analysis using the currently submitted schedule update and previous schedule update submission. The only schedule delay analysis method available that separately identifies and measures the project delay and savings resulting from progress and the project delay and savings resulting from schedule revisions is the Contemporaneous Schedule Analysis. The Contemporaneous Schedule Analysis method will be discussed in more detail in later chapters.

A simple explanation of how the Contemporaneous Schedule Analysis method is performed is that it closely mirrors the process that contractors typically follow when preparing a schedule update. For example, consider the review of the contractor's July 31 schedule update for which the previous update had been submitted as of June 30. The first step in this review analysis consists of measuring the effect that the actual progress achieved during the update period had on the project schedule. In this example, the reviewer would begin by identifying the actual progress achieved during the update period, which consists of the actual start dates, actual finish dates, and reduced remaining durations of activities that progressed between the two updates. Next, the reviewer would track the actual progress of the work, comparing planned progress to actual progress, on a daily basis from the data date of the June update to the data date of the July update. This tracking of progress on a daily basis from update to update allows the reviewer to identify the critical path activities that were responsible for the delays to the project completion date during the update period.

This step closely mirrors the typical process that the contractor follows when preparing an update. When preparing its July update, the contractor would first copy the June update file and then insert all of the actual progress achieved during the update period into the copied file. The contractor would then change the data date of the copied June update file, with the actual progress, to the new data date and run or calculate the schedule. After running this schedule, this copy of the June update, which contains the progress achieved up to the data date of the July update, represents the status of the project based solely on the progress achieved during the month. It is essentially the progress-only version of the new July update.

This progress-only version of the new July update indicates the status of the project based on actual progress. It will reveal whether the progress achieved during the month was sufficient to keep the project on schedule. If the contractor's progress-only version of the July update shows that it made better-than-expected progress and forecasts an improved completion date or shows that it made expected progress and forecasts an on-time completion, then, the absence of any added work or changes issued during the update period, the contractor will probably submit the update as a progress-only version to the owner.

If, on the other hand, the progress-only update does not forecast an early or on-time completion, then the contractor will likely need to change its plan and revise the schedule to recover the delay shown by the progress-only update. If the contractor believes that it was not responsible for the project delay experienced in the update period, then it should consider submitting the update in two phases: progress only, showing the delay, and the completed update, showing the planned mitigation of the owner-caused delay.

Evaluating the effect of schedule revisions

Before we begin discussing the effect of schedule revisions, we need to understand what schedule revisions are. In a general sense, schedule revisions, which may also be referred to as logic changes, are changes that the contractor makes to the schedule for the incomplete or remaining project work. Schedule revisions consist of adding or deleting activities, adding and deleting logic relationships, changing activity original durations, changing activity constraints, changing workdays in the schedule's work calendars, and the reassignment of activity work calendars, all of which may also be characterized as logic changes. Such changes would also include changes in the way the schedule is calculated and presented.

A contractor's schedule revisions really fall into two categories. The first category is best described as "refinements to the plan." As the project progresses, the contractor's original construction plan will evolve and change in response to the availability of labor, equipment, material, and subcontractors, and to the actual progress of the work. These refinements will include minor changes to activity durations and sequencing, but usually will not cause changes to the project's critical path, near-critical paths, or completion date.

The second category of schedule revisions consist of the types of changes that most people think of as schedule revisions. These types of changes are major changes or full-scale resequencing of the work, the addition of activities representing new work, or the deletion of original work scope. These types of changes are often the kinds of changes that cause delays to the project's completion date or result in the recovery of forecasted delay.

When reviewing the contractor's schedule update submissions, it is essential that the reviewer not only report delays and savings to the completion date resulting from the progress achieved during the update period, but it is also essential to identify and report the delays and savings that result from any logic changes made by the contractor.

With this knowledge, determinations may also be made regarding which party or parties were responsible for the project delay and savings during the update period, and how delays should be resolved or mitigated. For example, if the owner added work or suspended the project during the update period, then it may decide that the right way to address this delay is to grant a time extension or, if late completion of the project is not acceptable, then the owner may decide to pay for acceleration measures that will recover or mitigate the delay.

Conversely, if the contractor is continually falling behind schedule and recovering its own delays through logic changes, then the owner needs to monitor these revisions closely. Too often these repeated revisions are like tightening a watch spring. If overtightened, the spring breaks; similarly, if the contractor borrows too much time from the future, it will simply run out of time to complete all of the work. To combat this situation, when evaluating the schedule changes that contractors make to schedule updates to eliminate or mitigate their delays, owners should review the changes and determine whether they are realistic or just kicking the can down the road. If the latter, it may be prudent to challenge the contractor to demonstrate that it has the planned resources sufficient to complete the work in accordance with the revised plan.

Properly responding to the contractor's schedule update submissions

By now, we should have conveyed the point that more than most project submittals, which simply need to be reviewed and accepted or approved in time to allow the project work to remain on schedule, the submission of schedules and particularly schedule updates need to be accepted as quickly as possible. This is because the schedule is the time management tool itself. The further the schedule updates lag the project work, the less useful they become. Also, because the schedule "belongs" to the contractor, meaning that it is the model of the contractor's plan to build the project in accordance with the contract, owners can give the contractor some latitude with respect to its "accuracy." For this reason, owners are urged to include contract language that indicates that the contractor's schedule update submissions are considered accepted upon submission unless and until otherwise indicated by the owner. If an owner is uncomfortable with this approach, as a minimum the owner should include contract language that requires the contractor to submit the subsequent month's update even if the current update has not been accepted. In this manner, the schedule update process will never be bogged down and delayed by a prolonged review and approval process.

Even with such language, owners should strive to provide a proper and timely response to the contractor's schedule update submission. If the contractor's update forecasts an on-time completion and it does not include schedule revisions, which means that the contractor's actual progress during the update period was sufficient to keep the project on schedule, then the response is an easy one.

However, if the contractor's update submission forecasts late completion, then the reviewer should perform a schedule delay analysis to determine the cause of the delay. For example, was the delay caused by slow progress or a schedule revision? When the analysis of delays is completed, the reviewer should discuss the results with the project team to determine the party responsible for the delay. If the delay was determined to be the contractor's responsibility, then the owner should require the contractor to recover the delay or submit a time extension request demonstrating why a time extension should be granted. If the delay was the owner's responsibility, then the owner will need to decide on the appropriate course of action. Options for the owner may include granting a time extension, requesting a proposal to affect a particular change to the schedule that the owner believes will mitigate the delay, or requesting that the contractor propose a priced mitigation plan.

Additionally, and as discussed earlier in this chapter, even if the schedule update submission forecasts an on-time completion, the reviewer should perform a schedule delay analysis to determine whether the project was delayed during the update period. Depending upon what delays are found and how they have been mitigated, the owner can decide on appropriate courses of action in responding to the submission.

Lastly, if the contract requires the owner to accept or approve each update, we recommend that the review cycle for a single schedule update not be extended beyond one round. Again, this is to avoid having the schedule updates lag the project work by an extended period. Because each schedule update builds on the previous update, when approvals are required, the next update cannot be prepared until the previous update is approved. Limiting the review cycle to one round ensures that the contractor's schedule update submission does not get caught up in a never-ending review loop.

Ultimately, the parties need a time-management tool that they can use to manage the project and evaluate delays on a real-time basis. If the schedule update lags too far behind the project work, the schedule will no longer be a useful tool to timely manage the project work. As with the baseline review, it is important to recognize that an imperfect schedule update is better than having no schedule update at all. Most concerns and deficiencies can be noted with a request for the issue to be addressed and corrected in the next schedule update. Remember that, while it is desirable to have a schedule that appropriately models the contractor's plan to complete the remaining work, both parties will benefit from having an up-to-date schedule update at all times during the project work.

CHAPTER FOUR

Types of Construction Delays

WHAT IS A DELAY?

There are a number of definitions for delay:
- Something that happens later than expected
- Something that is performed later than planned
- An action that is not timely

Each of these definitions can describe a delay to an activity of work in a schedule. On construction projects, it is not uncommon for delays to occur. It is what is being delayed that determines if a project or some other deadline, such as a milestone, will be completed late. Before any discussion of delay analysis can begin, a clear understanding of the general types of delays is necessary. There are four basic ways to categorize delays:
- Critical or noncritical
- Excusable or nonexcusable
- Compensable or noncompensable
- Concurrent or nonconcurrent

The chart shown in Fig. 4.1 presents a general overview of how the excusable and nonexcusable categories of delay can be viewed. Note that this figure represents the interpretation of a typical construction contract. However, the circumstances that may be characterized as excusable or compensable can vary significantly, depending upon the contract. The discussion that follows elaborates on this simple summary chart.

When determining the effect of a delay on a project, the analyst must determine whether the delay is critical or noncritical. The analyst must also determine if delays are concurrent. All delays that are identified in the analysis will be either excusable or nonexcusable. Excusable delays can be further divided into compensable or noncompensable delays. This chapter provides basic definitions of these types of delays. Concurrent delays will be addressed in a later chapter.

Construction Delays.
DOI: http://dx.doi.org/10.1016/B978-0-12-811244-1.00004-5

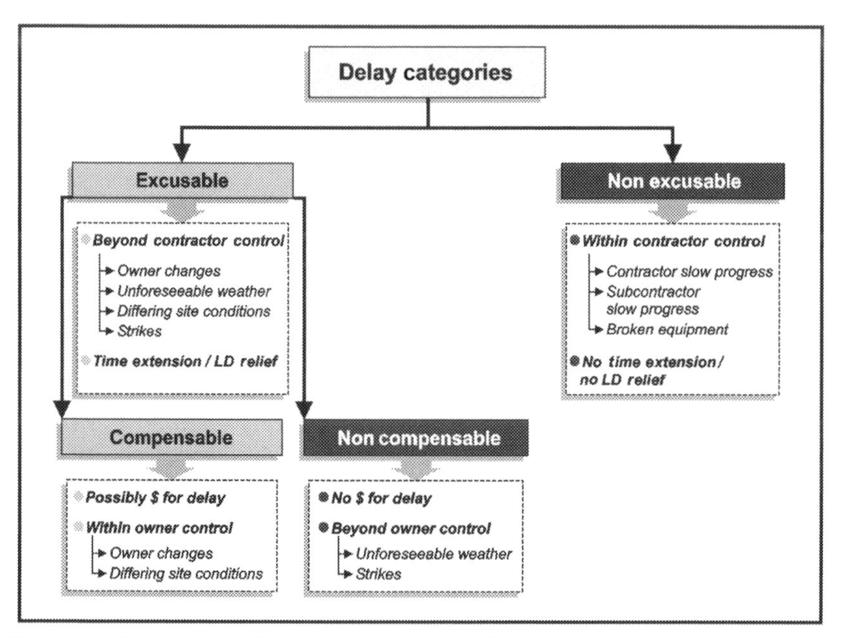

Figure 4.1 Descriptions of excusable, nonexcusable, compensable, and nonexcusable delays.

CRITICAL VERSUS NONCRITICAL DELAYS

In any analysis of delays to a project, the primary focus is on delays that are critical to some element of the work, those are delays that will affect the completion of that work element. In most cases, the completion of the project is the focus of the analysis. For simplicity, we use the project completion as the reference point for our delay analysis discussions in this book. In some cases, the completion of a particular event, often referred to as a milestone, will be the focus of an analysis. But, the same concepts that apply to the analysis of delays to the project completion will apply to the analysis of delays to a milestone.

Critical delays are those activity delays that affect the progress of the project in such a way that result in a predicted delay to the project completion date. However, many delays occur that do not delay the project completion date. Delays that affect the project completion are considered critical delays, and delays that do not affect the project completion are considered noncritical delays. The concept of "critical" delays emanates from Critical Path Method (CPM) scheduling. While the determination of the critical path and the identification of critical activities is a major feature of CPM scheduling, all projects, regardless of the type of schedule, have "critical" activities.

A key concept of CPM scheduling is that only delays to the critical path result in a delay to the scheduled project completion date. This is because the critical path is the longest path through the schedule network and, as such, is the path that determines the length of the project and the date upon which the project is predicted to be complete. Thus, the delay to the completion date is a predicted delay based on the then-current project plan. A change in that plan may either mitigate or exacerbate that delay.

Determining which activities truly control the project completion date depends on the following:

- The project itself
- The contractor's plan and schedule (particularly the critical path)
- The requirements of the contract for sequence and phasing
- The physical constraints of the project—how to build the job from a practical perspective

Regardless of how one analyzes a project and the schedule to find the delays, there is one overriding criterion: The analysis must accurately consider the contemporaneous information when the delays were occurring. "Contemporaneous information" refers to the daily reports, the schedule in effect, and any other job data available that show the circumstances at the time of the delays. Proper research and documentation eliminates the "but-fors" and any other hypotheses contrived to advance predisposed conclusions or desired results.

EXCUSABLE VERSUS NONEXCUSABLE DELAYS

Excusable delays

All delays are either excusable or nonexcusable. These categories are typically defined by the contract. Generally, an excusable delay is a delay that is due to an unforeseeable event beyond the contractor's control. Normally, based on common general provisions in public agency specifications, delays resulting from the following events would be considered excusable:

- General labor strikes
- Fires

- Floods
- Acts of God
- Owner-directed changes
- Errors and omissions in the plans and specifications
- Differing site conditions or concealed conditions
- Unusually severe weather
- Intervention by outside agencies (such as the EPA)
- Lack of action by government bodies, such as building inspection
- Constructive changes

These conditions may be reasonably unforeseeable, not within the contractor's control, and not the contractor's fault or responsibility. When a delay is determined to be excusable, the contractor will be entitled to an extension of the time to complete the project work.

The characterization of a delay as excusable must be made within the context of the specific contract. The contract should clearly define the factors that might justify entitlement to a time extension to the contract completion date. For example, some contracts may not allow for time extensions caused by weather conditions, regardless of how unusual, unexpected, or severe, even though such delays would be beyond the control of the contractor.

Nonexcusable delays

Nonexcusable delays are events that are within the contractor's control, are the contractor's responsibility, or that are foreseeable. These are some examples of nonexcusable delays:

- Late performance of subcontractors
- Untimely performance by suppliers
- Faulty workmanship by the contractor or subcontractors
- A project-specific labor strike caused by either the contractor's unwillingness to meet with labor representatives or by unfair labor practices

Again, the contract is the controlling document that determines if a delay would be considered nonexcusable. For example, some contracts consider supplier delays to be excusable if the contractor can prove that the materials were requisitioned or ordered in a timely manner, but the material could not be delivered due to circumstances beyond the contractor's control, such as national or worldwide material shortage. Other contracts may not consider such delays to be excusable. Therefore, both owners and contractors should recognize the importance of clear and unambiguous contract documents when defining excusable and nonexcusable delays.

Compensable versus noncompensable delays

A compensable delay is a delay for which the contractor is entitled to both a time extension and additional delay-related compensation. Relating back to excusable and nonexcusable delays, only excusable delays can be compensable. A noncompensable delay means that the contractor is not entitled to additional delay-related compensation resulting from the delay. Some excusable delays may be compensable. All nonexcusable delays are noncompensable.

Whether or not a delay is compensable depends primarily on the terms of the contract. In many cases, the contract specifically defines the kinds of delays that are excusable, noncompensable, for which the contractor does not receive any additional money but may be allowed a time extension. Contracts distinguish between compensable and noncompensable delays in many ways, some of which are described in the following paragraphs.

Federal contracts

Federal government contracts normally define strikes, floods, fires, acts of God, and unusually severe weather as excusable but noncompensable delays. These are delays that are outside the control of both the contractor and the owner. Other forms of excusable delays may be compensable, such as differing site conditions or owner-directed changes.

ACTIVITY DELAY VERSUS PROJECT DELAY

A common mistake as it pertains to measuring project delay is to calculate project delay by merely comparing an activity's planned and actual dates, regardless of whether the activity was on the critical path when the delay occurred. For example, if the start of the building's excavation operation was planned to begin on Friday, September 15, but it actually started on Wednesday, September 20, one might be tempted to conclude that the contractor is entitled to a time extension of 5 workdays.

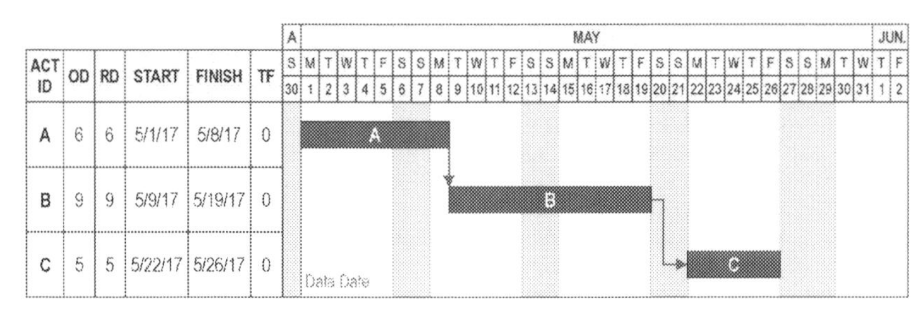

ACT ID	OD	RD	START	FINISH	TF
A	6	6	5/1/17	5/8/17	0
B	9	9	5/9/17	5/19/17	0
C	5	5	5/22/17	5/26/17	0

Figure 4.2 Example before late start of Activity A.

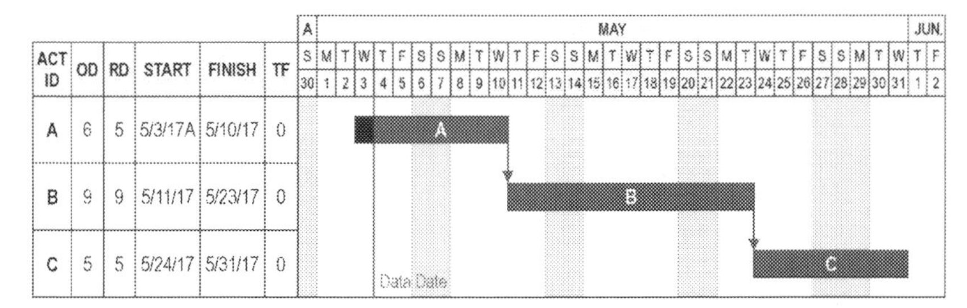

ACT ID	OD	RD	START	FINISH	TF
A	6	5	5/3/17A	5/10/17	0
B	9	9	5/11/17	5/23/17	0
C	5	5	5/24/17	5/31/17	0

Figure 4.3 Example after late start of Activity A.

However, a project can only be delayed when the critical path is delayed. So, the first thing that an analyst must do when attempting to measure project delay is to identify the critical path; remember, only delays to the critical path will cause project delay.

Figs. 4.2 and 4.3 illustrate how activity delay produces project delay and how project delay should be measured on a calendar day basis. The schedule example depicted in Fig. 4.2 consists of three activities connected in sequence. The data date of the schedule is May 1, 2017, and the project's forecast completion date is May 26, 2017. In Fig. 4.2, the initial critical activity is Activity A, which is planned to start on May 1, 2017, and the shaded portions of the bar chart indicate nonworking days.

Fig. 4.3 depicts the same schedule, but it demonstrates the consequence of the late start of Activity A, which is the initial activity on the critical path. The start of this activity is delayed from May 1, 2017, to May 3, 2017, which is a 2-workday delay and the project schedule's data date is May 4, 2017.

If the delay experienced by Activity A was used as the only basis for identifying and measuring the project delay, then a comparison of the May 1, 2017, planned start date to the May 3, 2017, actual start of Activity A would result in project delay of 2 workdays. However, project delay is not calculated by simply comparing the planned and actual performance of a critical activity. It is calculated by measuring the effect that the actual performance of a critical activity has on the scheduled project completion date. When Figs. 4.2 and 4.3 are compared on this basis, it should be clear that the 2-day delay (May 1 and 2) to the start of Activity A resulted in a 5-calendar-day delay to the scheduled completion date of the project from May 26, 2017, to May 31, 2017.

Said another way, if Activity A had actually started on May 1, 2017, and if all three activities had progressed as expected, then the project would have finished on Friday, May 26, 2017. However, Activity A started 2 workdays late on May 3, 2017, and, as a result, this 2-workday delay delayed the completion of the project 5 calendar days from Friday, May 26, 2017, to Wednesday, May 31, 2017. The difference between the 2-workday late start of Activity A and the 5-calendar-day delay to the project was due to the fact that the next available workdays after the planned completion date of Friday, May 26, 2017, were Tuesday, May 30, 2017, and Wednesday, May 31, 2017. This was because Saturday, May 27, 2017, through Monday, May 29, 2017, were nonworkdays due to the weekend and the observance of the Memorial Day holiday.

Therefore, when evaluating the magnitude of a delay, the analysis should consider not only the delay itself, but resulting delay to the scheduled project completion date.

NO-DAMAGE-FOR-DELAY CLAUSES

Some contracts are more restrictive in defining compensable delays. It is also not uncommon for a contract to use exculpatory language concerning delays. Exculpatory language is language that exculpates, or excuses, a party from some liability. One approach that might be used to limit compensation for delay is a broad no-damage-for-delay clause. The wording in this clause can take many forms, but in general the clause states that for any excusable delay the contractor may be granted a time extension, but no additional compensation will be paid. The time extension is the sole remedy for the contractor for any type of excusable delay. An example of such a clause is presented in Fig. 4.4.

If the contractor is delayed in completion of the work under the contract by any act or neglect of the owner or of any other contractor employed by the owner, or by changes in the work, or by any priority or allocation order duly issued by the federal government, or by any unforeseeable cause beyond the control and without the fault or negligence of the contractor, including but not restricted to acts of God or of the public enemy, fires, floods, epidemics, quarantine restrictions, strikes, freight embargoes, and abnormally severe weather, or by delays of subcontractors or suppliers occasioned by any of the causes described above, or by delay authorized by the engineer for any cause which the engineer shall deem justifiable, then:

For each day of delay in completion of the work so caused the contractor shall be allowed one day additional to the time limitation specified in the contract, it being understood and agreed that the allowance of same shall be solely at the discretion and approval of the owner.

No claim for any damage or any claim other than for extensions of time as herein provided shall be made or asserted against the owner by reason of any delays caused by the reasons hereinabove mentioned.

Figure 4.4 Example 1 no-damage-for-delay clause.

It is understood and agreed that the contractor has considered in his bid all of the permanent and temporary utility appurtenances in their present or relocated positions and that additional compensation will not be allowed for delays, inconvenience, or damage sustained by him due to any interference from the said utility appurtenances or the operation of moving them.

Figure 4.5 Example 2 no-damage-for-delay clause.

It should be noted that enforcement of these types of clauses has been questioned in the courts. Some courts have been reluctant to strictly enforce these clauses and, in cases where such exculpatory language may be enforced, it is often strictly construed. According to legal views researched, courts have often ruled narrowly on no-damage-for-delay clauses, thus limiting enforceability. Contractors, though, should not assume that these provisions will not be enforced.

There are many variations in contract clauses that address the compensability of delays. However, the broader the clause, the less likely it is to be enforceable. More specific clauses are more readily upheld by the courts. Public contracts at the state and municipal level often contain specific no-damage-for-delay clauses. An example of a no-damage-for-delay clause pertaining to work by utilities is shown in Fig. 4.5.

Similarly, the paragraph in Fig. 4.6 shows a no-cost-for-delay clause covering work by other contractors.

A no-damage-for-delay clause that specifically covers the review and return of shop drawings is shown in Fig. 4.7.

All parties to a project should clearly understand the clauses of the contract concerning delays and time extensions. If a contractor is

The contractor shall assume all liability, financial or otherwise, in connection with its contract, and hereby waives any and all claims against the Department for additional compensation that may arise because of inconvenience, delay, or loss experienced by it because of the presence and operations of other contractors working within the limits of or adjacent to the Project.

Figure 4.6 Example 3 no-damage-for-delay clause.

The contractor should allow thirty calendar days for the review of any shop drawing, samples, catalog cuts, etc. which are required to be submitted in accordance with the contract. This thirty day time period will begin on the date the submission is received by the Architect/Engineer (A/E) and terminated on the date it is returned by the A/E. The contractor should further allow thirty calendar days for each resubmission of any rejected submission. Should any submission not be returned within the thirty calendar days specified, it is understood and agreed that the sole remedy to the contractor is an extension of the contract time. For each day of delay in completion of the overall project caused by the late return of submissions, the contractor shall be allowed one day additional to the time limitation specified in the contract. No claim for any damages or any claim other than for extensions of time as herein provided shall be made or asserted against the owner by reason of any delays caused by the reasons hereinabove mentioned.

Figure 4.7 Example 4 no-damage-for-delay clause.

considering signing a contract with such language, it should consult qualified counsel familiar with construction litigation and the laws of the jurisdiction in which the clause will be enforced or adjudicated.

When a contract identifies specific items in a contract as being noncompensable, it should clearly define each one. For example, if unusually severe weather is a noncompensable delay, the contract should clearly state the restriction. The contract may define unusually severe weather as weather not ordinarily expected for the specific time of year and region. The definition in the contract may further clarify unusual weather as that which exceeds the historical weather conditions recorded by the National Oceanic and Atmospheric Administration at a specific location. The Corps of Engineers has taken this one step further by specifying in their contracts the exact number of days of rain greater than 0.01 in that the contractor can expect during each month of the project.

While the extent of detail provided by the Corps of Engineers may not be absolutely necessary, the owner should be sure that the contract does not have ambiguous wording. Some contracts will list "inclement weather" as an excusable, noncompensable delay, but "inclement" can have many definitions. It is also possible that inclement weather may occur but may not delay the project. Therefore, all parties to the contract must carefully read and clearly understand the compensable and noncompensable delays recognized by the contract.

CONCURRENT DELAYS

The concept of concurrent delay is a very important aspect of delay analysis. Concurrency is relevant, not just to the determination of critical delays, but also to the assignment of responsibility for delay-related costs. Owners may cite concurrent delays by the contractor as a reason for issuing a time extension without additional compensation. Contractors may cite concurrent delays by the owner as a reason why liquidated damages should not be assessed for its delays. Unfortunately, few contract specifications include a definition of "concurrent delay" or define how concurrent delays affect a contractor's entitlement to additional compensation for time extensions or responsibility for liquidated damages. To complicate matters further, there is a lack of consistent understanding in the industry concerning the concept of concurrent delay. So as not to diverge at this point, concurrent delays are discussed in detail in Chapter 7, Delay Analysis Using Critical Path Method Schedules.

CHAPTER FIVE

Measuring Delays—The Basics

The method used to measure or quantify delays on a construction project will generally be a function of and dependent upon the type and quality of documentation that is available for analysis. The majority of this book discusses delay analysis in the context of CPM schedules that have been determined to reliably model the plan and progress of construction. However, regardless of the method chosen, there are five essential principles that should be used to guide every analysis of delays.

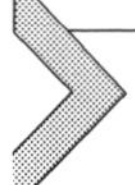

DELAY ANALYSIS PRINCIPLES

Delay analysis principle no. 1—only delays to the project critical path can delay the project

This principle has already been introduced in earlier chapters, because it is the basic principle upon which the analysis of delays is based. The proper application of this principle relies on the correct identification of the critical path. This is something that is not only easy to do correctly but also easy to do incorrectly.

As discussed in Chapter 2, Float and the Critical Path, and also below, the project critical path is the longest path of work through the schedule network and, as such, forecasts when the project will finish.

Delay analysis principle no. 2—not every delay to the critical path will delay the project

Although only delays to the critical path can delay the project, not every delay to the critical path will delay the project. As discussed in Chapter 2, Float and the Critical Path, the critical path can include critical activities that have positive total float values. If this circumstance exists, then these critical path activities can be delayed without delaying the project. The number of days of delay that can be absorbed will depend on the number

Construction Delays.
DOI: http://dx.doi.org/10.1016/B978-0-12-811244-1.00005-7

of days of total float. This situation is more common on construction projects that are subject to seasonal or environmental work restrictions.

Delay analysis principle no. 3—the critical path is the longest path

As demonstrated in Chapter 2, Float and the Critical Path, the critical path is the longest path. Why is this principle essential? As explained in Chapter 2, Float and the Critical Path, the use of multiple calendars and activity constraints will affect the total float values of activities in the schedule. As such, total float values alone cannot be used to identify the project's critical path in every schedule.

CPM scheduling software packages acknowledge this complexity and have embedded a feature called the longest path filter that allows users to identify the longest path of work in the schedule network. In fact, as described in Chapter 2, Float and the Critical Path, Oracle's Primavera Project Management (P6) scheduling software manual states that if a CPM schedule uses multiple calendars, then "using total float values to identify critical activities may prove misleading, since some activities have large float values due to their calendar assignments but are still critical to the completion of the project." In these circumstances, the longest path filter becomes the most reliable tool for identifying the critical path.

Delay analysis principle no. 4—the critical path can and does shift

The critical path is dynamic in nature and may change, or shift, during the project. Shifts to the critical path depend on how the project is planned; how the project work progresses or does not progress; and changes made to the schedule logic. Because critical path shifts occur while the project is progressing and when changes are made to the schedule, delay analysts should identify and account for shifts in the critical path when identifying and measuring project delay.

Delay analysis principle no. 5—activity delay and project delay are not the same

It is important to understand the difference between activity delay and project delay. While the delay to an activity may be important, in that there may be consequences when an activity starts or finishes later than it was planned to start or finish, each day of delay to an activity will either consume float or delay the project, but not both. Understanding the activity's position in the network and its total float value relative to that of the critical path is essential to determining when the delay to an activity has delayed the project. This, of course, relates back to Delay Analysis Principle No. 1.

Evaluating delays prospectively and retrospectively, what is the difference?

The first step in answering this question involves defining these two terms—prospective and retrospective.

The Merriam-Webster dictionary defines the term prospective as "relating to or effective in the future." The term retrospective is defined as "of or relating to the past or something that happened in the past."

These are the definitions, but what is the distinction between these two terms as they relate to the analysis of delays? As their definitions suggest, a prospective analysis of delays estimates or forecasts the delay caused by an event or a change in the future, before the affected or changed work is performed. In contrast, a retrospective analysis of delay identifies delays that have occurred in the past, after the affected or changed work is completed. A good analogy of these methods is the pricing of a change. A prospective analysis is like the forward pricing of a change in which an owner and a contractor agree to an estimated value of the work, prior to the work being performed; a retrospective analysis is like the T&M or "force account" pricing methods, which determine the actual cost of the work after it is performed.

With regard to prospective delay analysis, there is nearly universal industry agreement that the Prospective Time Impact Analysis method, known simply as Time Impact Analysis, or TIA, is the best method to forecast the delay resulting from an event or change before the affected or changed work is performed. The TIA method consists of creating a fragmentary network, or "fragnet," which consists of creating a network of activities that represent the event or change, and inserting that network into the current version of the project's schedule update. The schedule update file containing the fragnet for the change order work is recalculated or rescheduled to determine whether the added activities forecast a delay to the project.

Unfortunately, with regard to retrospective delay analysis methods, there is no consensus or agreement on the best method. In the remainder of this chapter, we explore some of the more commonly used approaches.

THE IMPORTANCE OF PERSPECTIVE

Reality is a question of perspective; the further you get from the past, the more concrete and plausible it seems.—Salman Rushdie, Midnight's Children

The identification of a critical delay is often a question of perspective. Every analyst has a way of illustrating this point, but the classic example is the "ribbon-cutting" story. Consider a project where, in addition to all its other responsibilities, the contractor must also provide the scissors for the mayor's ribbon cutting at the conclusion of the project. The architect rejected the contractor's original scissor submittal (the Contract specified something larger and grander). The project manager for the contractor shoved the rejected submittal to the bottom of the "to-do" pile, where it languished and was eventually lost. The project ultimately finished late due to an error in the design of the structural steel. The error necessitated refabrication of the steel, which was on the project's critical path. This delayed the start of the structural steel erection work that, in turn, delayed the project.

At the ribbon-cutting ceremony, it quickly became apparent that the scissors had not been purchased. The project manager, at the last minute, ran to the local office supply store and bought the biggest, brightest pair of scissors in stock and returned to the project site just as the mayor was about to cut the ribbon. As a result, the proceedings were held up only a few seconds as the project manager ran up to the entrance of the new water treatment plant.

After the ribbon-cutting ceremony, the project manager met with the architect to close out the project. The contractor sought a time extension due to the steel design error. The architect rejected the contractor's request, stating that even without the steel design error, the formal opening of the project would have been delayed by the lack of scissors to cut the ribbon.

The ribbon-cutting story points out the importance of perspective. Viewed solely from the end of the project, which we may call an after-the-fact or as-built perspective, the lack of a pair of scissors, and the flawed procurement process that caused them to be delivered just in time, could be considered critical to opening the project. Given these facts, most of us quickly see the error in the architect's logic. But what if the scissors are changed to aluminum tank covers? In response to the steel design error, assume that the project manager called the fabricator of the aluminum tank covers to let them know that the project would be a little late and that the delivery of the tank covers should be postponed. If the tanks were not ready when the covers were delivered, they would have to sit before they could be installed and might be damaged. As the project manager recommended, the tank covers were delivered later than originally scheduled, but they finally arrived and were installed as the delayed tanks were completed.

In this alternate story, the contractor and the architect again meet after the ribbon cutting to close out the project. Again, the contractor asks for a time extension, and, again, the architect refuses the request. This time, however, the architect denies the time extension because the "aluminum tank covers were late." We know all the facts, in that the aluminum tank covers were intentionally delivered when they could be immediately installed. So we, again, see the error in the architect's logic. But what if the facts were not known? What if there was no written record of the project manager's conversation with the tank cover fabricator? Absent verifiable facts, is the architect correct? Is the as-built perspective a relevant and valid way to view the project events and evaluate the critical project delays?

Perhaps it is only the as-built perspective that is problematic. What about the view from the beginning of project, or the as-planned perspective? Consider the same project. As required by the contract, the contractor prepared a CPM schedule. The first schedule prepared on the project is called the baseline, or "as-planned" schedule. It should only depict the contractor's initial plan for completing the project and should not include "as-built" or actual performance information. The critical path of the project as depicted in the contractor's as-planned schedule proceeded through the erection of structural steel. During the close-out meeting, the architect requires the contractor to prepare an analysis that demonstrated that the steel design error delayed the project. The contractor concluded that the best way to evaluate or "measure" the delay associated with the steel design error would be to simply "insert" this delay into its as-planned schedule. In other words, the contractor chose to use a TIA, this time done retrospectively or after the fact, to analyze the delay. The contractor believed that by inserting a fragnet representing the steel design error into the as-planned schedule, the recalculated schedule would show both that the error caused a critical delay and the magnitude of the delay. If we did not know anything else, this approach might be acceptable.

But we do know something else. We know that a dispute developed between the contractor and its steel erector. In fact, the steel erector abandoned the project. The contractor was not able to get another erector on site until after the fabricated steel was delivered to the site. But the contractor's analysis does not consider this problem. The only fragnet inserted into the schedule is the fragnet for the steel error, and this causes a critical project delay. Is the contractor entitled to a time extension for the steel design bust regardless of what else might be going on at the project site when the delay occurred? Is the as-planned perspective a valid and reliable way to view the project events and evaluate the critical project delays?

In addition to evaluating the critical project delays using an as-built perspective or an as-planned perspective, another option would be to evaluate the critical project delays as they occur—in other words, evaluate delays to the project at the time the delay is experienced. This would avoid both the ribbon-cutting error and the flawed approach of inserting only the structural steel design fragnet into the as-planned schedule; rather it would force the analyst to consider everything that is happening on the project as delays occur. But what if the analyst is not brought in until long after the project has been completed? Is the view from the time when the delay actually occurred still relevant and valid, even though the analyst knows what ultimately happens?

The answer to the questions of perspective are at the heart of many of the disagreements among analysts regarding the best way to analyze delays on a construction project. Does the analyst evaluate the delay from the perspective of the beginning of the project, adding delays to the as-planned schedule, or from the end of the project, evaluating only those delays that appear to ultimately hold up the project's completion (the ribbon-cutting example)? Or should analysts try to put themselves in the shoes of the project team at the time the delay occurs? It would be disingenuous to suggest that analysts are united in their answers to these questions. There is, however, an emerging consensus supported not only by many analysts but by case law, as well.

Perspectives—forward looking and backward looking

Though rarer now, there was a time when delays were sometimes analyzed by "impacting" the as-planned schedule. The as-planned schedule is usually defined as the earliest complete and owner-approved project schedule. It represents the contractor's plan for completion of the project before work begins. If delays are analyzed using an "impacted as-planned" approach, the delay (or impact) is inserted into the as-planned schedule, and the schedule is then recalculated. The difference between the originally scheduled completion date and the completion date that results from impacting the as-planned schedule is the project delay attributable to the impact. This type of analysis takes the position that delays should be measured from the as-planned perspective that considers only the project team's original plan and the delay being analyzed. The problems with this analytical approach will be discussed in more detail in another chapter, but here's what a judge had to say about this approach in Haney *v.* United States [30 CCF ¶ 70, 1891], 676F. 2d 584 (Ct. Cl. 1982).

> *We have found that [the contractor's] analysis systematically excluded all delays and disruptions except those allegedly caused by the Government.... We conclude that [his] analysis was inherently biased, and could lead to but one predictable outcome.... To be credible, a contractor's CPM analysis ought to take into account, and give appropriate credit for all of the delays which were alleged to have occurred.*

Essentially, the judge's criticism was that the outcome of an impacted as-planned analysis, because it ignores everything other than the as-planned schedule and the delay the analyst is evaluating, was predetermined. It would overstate the delay, if any, associated with the inserted delay. Years of experience analyzing impacted as-planned analyses have confirmed this judgment. They very nearly always overstate the project delay, predicting project delays well beyond the actual project completion date. On this basis, an analysis of delays based solely on the as-planned perspective that employs an impacted as-planned analytical technique is flawed and should be avoided.

The logical opposite of an impacted as-planned analysis is the "collapsed as-built." Again, the problems with this analytical approach are discussed in another chapter, but a discussion concerning perspective is appropriate here. Stripped to its essentials, a collapsed as-built analysis is performed by first creating the "as-built schedule" for the project. This is essentially a schedule showing how a project was actually constructed. It is not a schedule that ever existed on the project, though it is theoretically composed of actual project events. The analyst creates the "as-built schedule" after the project is completed. The next step is to identify the delay or delays to be analyzed. Note that this approach is a little like the tail wagging the dog. The delay must first be identified before it can be analyzed. The analysis is performed by removing the delay from the as-built schedule and then rerunning the schedule to see what happens. If the collapsed schedule shows an earlier project completion date, then the conclusion would be that the delay that was removed was responsible for a project delay equivalent to the improvement in the project completion date. This analysis presumes that delays are best analyzed from the perspective of the end of the project.

Setting aside the questions concerning the mechanics of a collapsed as-built analysis, consider what it means. Essentially, the collapsed as-built approach is based on the assumption that all that matters is what happened, not what was planned. To understand the problems with this assumption, consider the following example. A contractor is tasked with excavating a 100-foot rock face and then lining the face with concrete. Excavation began, and the contractor immediately encountered a problem. It turns out that a fault zone ran through the area of construction. This fault zone was oriented in such a way that as the contractor removed rock, the rock face that was left tended to slip into the excavated area. This was not only dangerous, but it prevented the contractor from excavating the planned 100-foot rock face. The owner and the contractor met to discuss the problem, and they decided to pin the rock face with rock anchors as the face was excavated in 10-foot lifts. Also, the owner decided that the concrete lining had to be constructed before the next 10 ft of the rock face could be excavated.

At the conclusion of the project, the contractor asked for a time extension to cover the additional time it had expended excavating and lining the rock face in 10-foot lifts as opposed to all at once, as planned. The owner responded with a collapsed as-built analysis showing that the only delay was the time required to install the rock anchors, which had not been contemplated in the original design. The rock excavation and concrete liner were not "delays," since this work had always been required.

The fallacy of the owner's analysis was that in addition to the rock-anchor delay, the contractor was also delayed because it was required to build the project in 10-foot lifts rather than all at once, as planned. Because the owner's delay analysis considered only what happened (the as-built schedule), it could not quantify delays associated with deviations from the contractor's original plan. And this is the essential failure of any analysis based solely on what happened.

If the perspectives from the beginning of the project and the end of the project are flawed as shown in the preceding examples, the only perspective remaining is to analyze the project at the point where the delay actually occurred. An analysis based on this at-the-time perspective has a name. It is called a contemporaneous analysis. Before discussing how such an analysis might be performed, consider this judge's decision.

> *Mr. Maurer, appellant's expert, testified about the critical delays to the Project... The analysis about the critical delays was based on appellant's original schedule, the schedule updates, the daily reports, Project correspondence, and the contract documents. Mr. Maurer described his analysis as a step-by-step process, beginning with the original schedule and proceeding chronologically through the Project, updating the sequence at intervals to see what happens as the Project progressed [(tr. 262) ASBCA No. 34, 645, 90–3 BCA ¶ 12, 173 (1990)].*

The key point in this decision is that the analysis took into account all relevant project information as it became available to the project participants as the project progressed. In doing an analysis from this at-the-time perspective, the actual progress of all of the work is compared to the plan for the work as the project progresses, and considers all that the project participants knew at the time. In this manner, all delays that occur on the project are identified by the analysis instead of the delays being predetermined and tested by an analysis that either ignores other delays (beginning-of-the-project perspective) or ignores the plan (end-of-the-project perspective).

From this and other discussions in this book, it should become apparent that the only valid perspective for the analyst is the view of the project contemporaneous to the delay itself—not from the beginning of the project or the end of the project—but, at the time of a given delay. An analysis from this perspective is greatly aided by the project schedule.

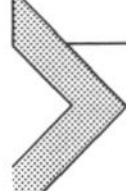

USE THE CONTEMPORANEOUS SCHEDULE TO MEASURE DELAY

A contemporaneous schedule is the project schedule, which typically consists of the baseline schedule and schedule updates that were used to manage the construction project.

These contemporaneous project schedules are essentially snapshots of the project's status at specific moments in time. As snapshots in time, the schedule updates identify what work has been done and the order in which it was completed. These contemporaneous project schedules also capture changes made to the construction plan in reaction to ever-evolving project conditions.

The contemporaneous project schedules are the preferred tool to measure project delay because they were used by the project participants to manage the project and, therefore, provide the most accurate model of the plan to complete the project. They are, also, the only management tool that forecasts when the project will finish based on the then-known project conditions. These attributes provide the analyst with an at-the-time perspective of the team's plan to complete the project and enable the analyst to identify, measure, and assign responsibility for project delay using the same information available to the project participants at the time the delay occurred. By using the contemporaneous schedules, and by tracking delays as they occur throughout the project, there is no need to attempt to inject information that is known at a later date. Information is incorporated into the analysis based on contemporaneous information throughout the analysis.

DO NOT CREATE SCHEDULES AFTER THE FACT TO MEASURE DELAYS

In the absence of contemporaneous schedules, an analyst may feel it would be acceptable to create a schedule after the fact that he or she believes better portrays the contractor's intended construction plan. Although the analyst may rely on project documentation and, perhaps, firsthand knowledge of the type of construction being performed, creating a schedule for the sole purpose of measuring and identifying project delay after the project is complete undermines the perceived objectivity of the analysis. Even though the analyst may do his or her best to remain objective, because the actual events of the project are known, whether intentional or not, this after-the-fact knowledge influences the creation of the after-the-fact schedule and ignores, or at least significantly diminishes, the contemporaneous knowledge and thinking of the project participants before and during the project.

The analyst may argue that creating an after-the-fact schedule will allow the analysis to be more precise, containing all the facts of the project. However, there is almost always more than one way to build a project, and the analyst may choose an approach different from the approach chosen by the original planner. Even seemingly small differences in a schedule could affect the results of an analysis.

Using a schedule created after the fact to measure and identify project delay has at least two basic weaknesses: The schedule does not depict the original construction plan and the schedule may include predetermined conclusions concerning delays. There are many ways a construction plan can be represented in a schedule. Preparing one after the fact merely shows the plan the analyst believes was intended. This does not make it correct.

When possible, it is always best to use the contemporaneous project schedules to measure project delay. While the analyst may make very minor modifications to the contemporaneous schedule to account for obvious errors, such changes must be made judiciously. To make this point, consider this judge's decision, which describes the use of the contemporaneous schedules as they existed at the time.

> *In the absence of compelling evidence of actual errors in the CPMs, we will let the parties "live or die" by the CPM applicable to the relevant time frames [Santa Fe, Inc. VABCA No. 2168, 87–3 BCA ¶ 20677].*

WHAT TO DO WHEN THERE IS NO SCHEDULE?

There are instances when contemporaneous project schedules cannot be used to measure the project delay. In these cases, the project schedules either did not exist or the analyst has determined that the contemporaneous schedules did not appropriately depict the plan to construct the project and, thus, would not be a reliable tool to identify and measure the project delay.

When a contemporaneous schedule is not available as a tool with which to identify and measure the critical project delays, the analyst should perform an as-built analysis to identify the critical project delay. An as-built analysis usually starts with the preparation of an as-built diagram. An as-built diagram is prepared using the project's contemporaneous documents. Such documents may include, but are not limited to, timesheets, inspector daily reports, meeting minutes, project photos, and so on. When complete, an as-built diagram should depict the order and durations of the project work activities. The as-built analysis is described in more detail in a later section of this book.

WHAT IS AS-BUILT INFORMATION?

Most, if not all, analysis methods are based in significant part on information that indicates how the project was built. As-built information consists of the reported actual start and finish dates of the Project work activities and the progress made each day on these activities. One of the best places to find as-built information is in the project schedule updates, because the periodic updates typically record the dates that specific activities start and finish. In addition to containing the activities' actual start and finish dates, schedule updates also record the remaining duration of activities that have started but not finished in each update period. Even if the updates contain the project's as-built information, it is always wise to verify information in the updates, using as many independent sources as possible. For example, the analyst might review the project daily reports to verify that specific activities started and finished on the dates indicated in the updates.

Note that the schedule updates do not usually provide information sufficient to determine how much work was performed on an activity each day. However, this can often be approximated by comparing the planned and remaining duration and reviewing other data, such as daily reports and meeting minutes.

If the updates do not provide the information required or if the updates simply do not exist, then the analyst has no alternative but to prepare an as-built diagram, using the contemporaneous project documents. The following documents should be reviewed as possible sources of as-built information:

- Project daily reports
- Project diaries
- Meeting minutes
- Pay applications or estimates
- Inspection reports by the designer, owner, lending institution, construction manager, or other parties making periodic inspections of the project
- Correspondence
- Memos to the file
- Dated project photos

When preparing an as-built diagram, the analyst should document every day that work is recorded for each activity. It is not enough to merely record the start and finish dates. While the start and finish dates are extremely important, the determination of whether work was performed continuously or interrupted will also be significant.

A CONCEPTUAL APPROACH TO ANALYZING DELAYS

While we present several analysis methods in detail later in this book, the following conceptual approach may be used to gain an initial understanding of how to properly analyze delays on a construction project. Also, because the specific steps in each analysis will vary depending on the nature of the available information, the concepts presented in the following figures may also be used as a guide to ensure that the method being used will result in a reliable answer.

The first step in any analysis is to determine the contractor's plan, generally depicted in the as-planned schedule. For purposes of this discussion, a simple, single bar chart network is used to demonstrate the analysis. Because the schedule consists of a single path, this path is the critical path. Fig. 5.1 is the contractor's as-planned schedule for a project.

To determine what occurred on the project, the analyst may create an as-built diagram or rely on the as-built dates from the last submission of the project's CPM schedule. For this example, Fig. 5.2 depicts the as-built or actual performance of the project work.

At this stage of an analysis, there may be a desire to simply compare the as-planned schedule with the as-built diagram, which is depicted in Fig. 5.3, and an attempt to reach conclusions concerning what was delayed.

When we look at Fig. 5.3, the project was planned to have finished on Day 35, but actually finished 30 days later on Day 65. With the knowledge of both the project's planned and actual completion dates, the goal of any delay analysis is to determine why the project finished 30 days late. Clearly, Activity E was added to the project and was the last work to finish. It might be tempting to simply conclude that this new activity and the added work it represented was responsible for the project finishing 30 days late. Fight this temptation.

When analyzing delays, start at the beginning of the project and move through the project chronologically. This will allow the identification of

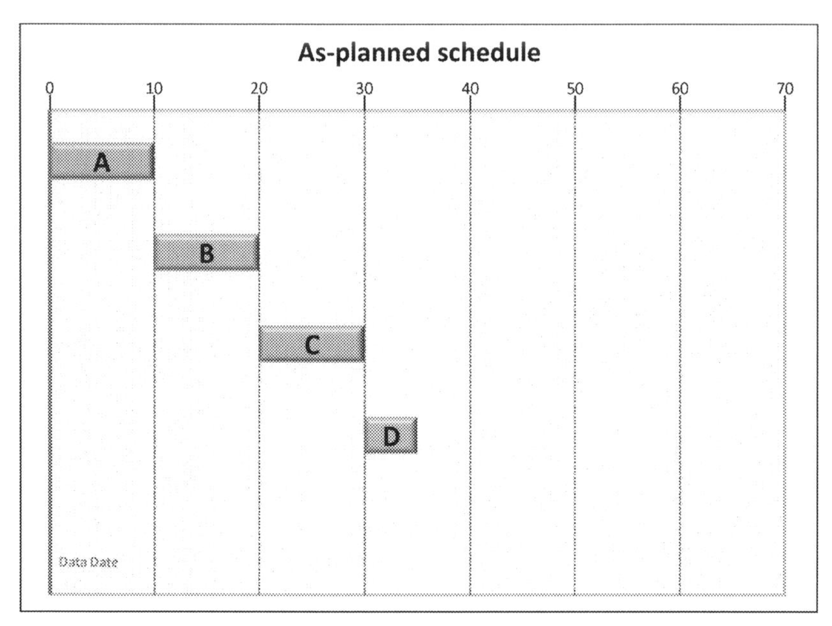

Figure 5.1 As-planned schedule.

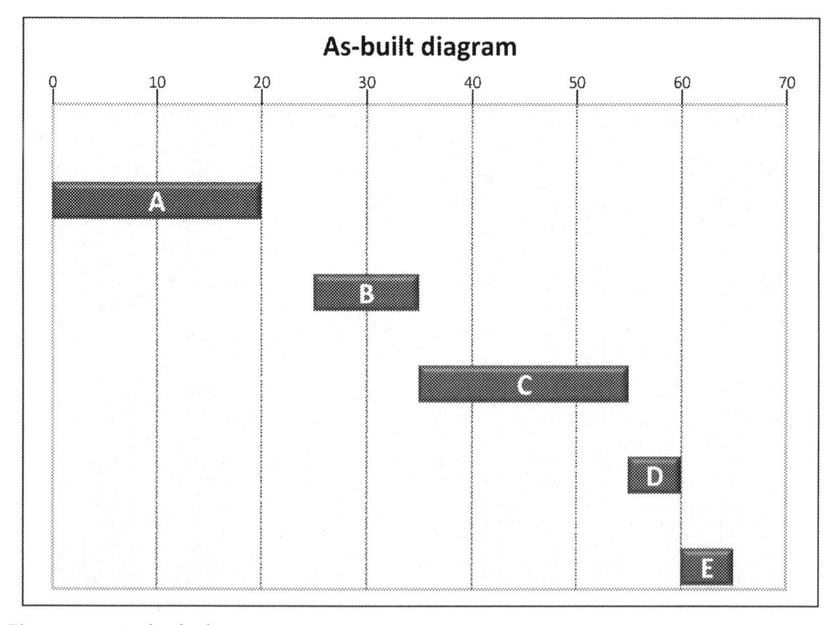

Figure 5.2 As-built diagram.

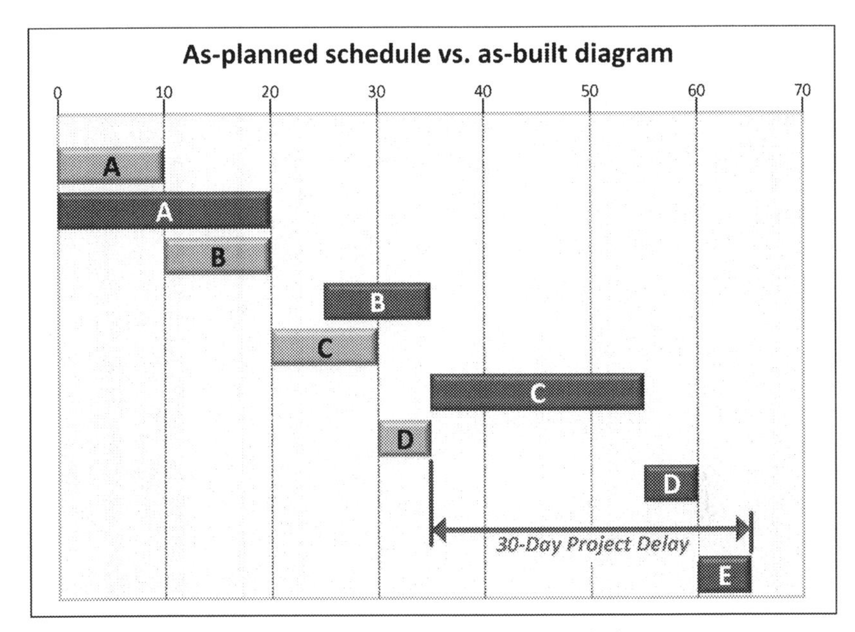

Figure 5.3 Comparison of as-planned schedule and as-built diagram.

each delay as it occurs. Therefore, the analysis will proceed starting with Fig. 5.4.

Looking at Fig. 5.4, start with Activity A, which is the project's first activity and the initial critical activity. The planned yellow (gray in print versions) bar shows that Activity A should have begun immediately and finish on Day 10. However, when comparing Activity A's planned and actual performance, we see that Activity A started on time, but took twice as long to finish than planned. As opposed to 10 days, Activity A took 20 days to complete. The proper conclusion from this comparison is that Activity A was delayed for 10 days as a result of its late finish.

Because an activity delay may or may not be equal to the project delay, it is important to update the schedule to determine the effect that the late finish of Activity A had on the remaining work. Fig. 5.5 depicts the updated schedule based on the actual performance of Activity A.

Note that moving the data date of the schedule from Day 0 to immediately after the completion of Activity A updates the schedule for the remaining work. As a result of the late finish of Activity A, the planned start of Activity B and remaining work have been pushed out and the project was delayed 10 days.

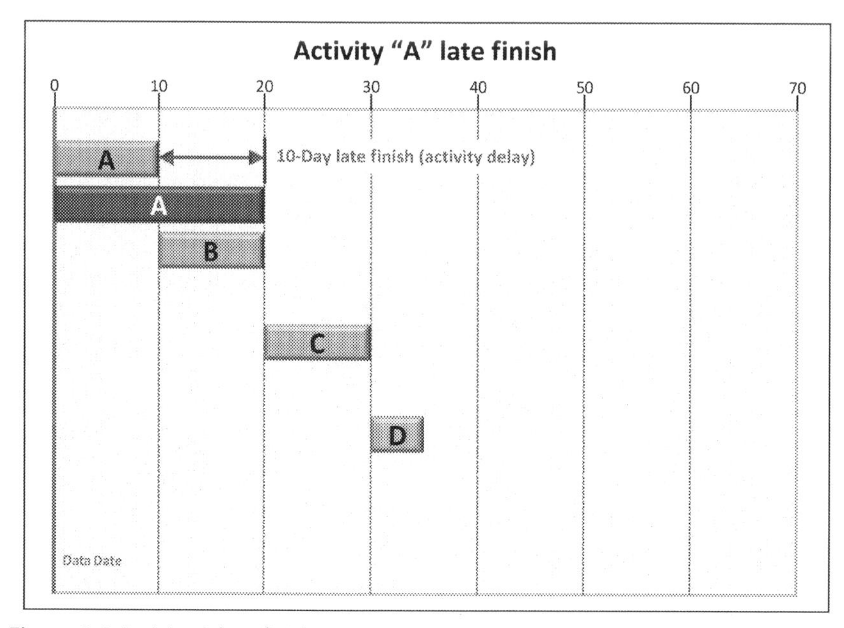

Figure 5.4 Activity A late finish.

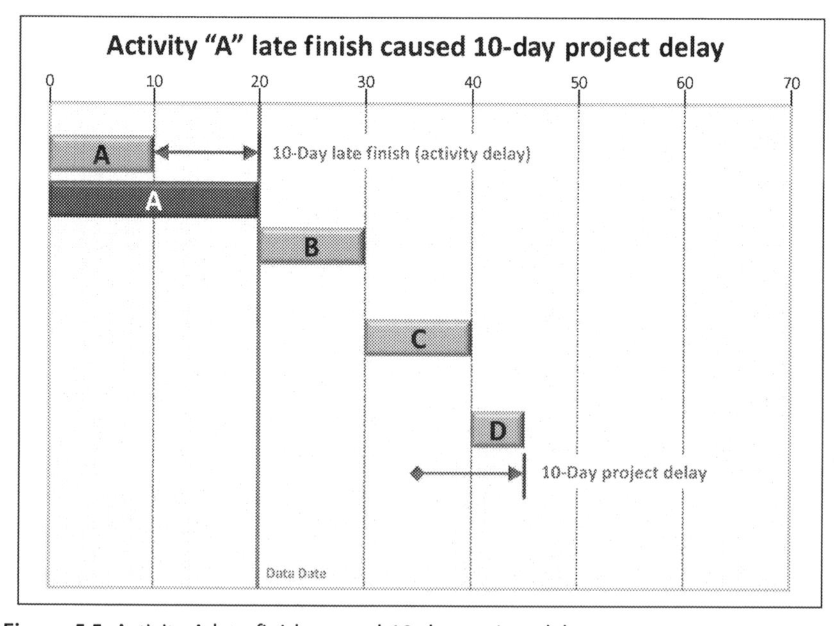

Figure 5.5 Activity A late finish caused 10-day project delay.

At this point, let us stop and consider the delay to Activity A. Activity A is critical and the delay to Activity A is a critical delay. The 10-day critical delay to Activity A happens to also cause a 10-day critical delay to the scheduled project completion date. Think of this critical delay the same way classical physicists thought of matter. It can be neither created nor destroyed. It can be mitigated if the contractor and owner can figure out a way to complete the remaining critical project work more quickly, but this initial 10-day critical delay will never go away.

Note, also, that this 10-day critical delay is based on the change in the forecast or scheduled project completion date. In this case, the schedule becomes the proper measuring tool for the delay. The schedule analyst is putting the same level of confidence in the schedule that the nuclear reactor operator puts into the gauge showing reactor temperature. If the gauge shows the reactor temperature going up, the operator assumes that the reactor is heating up, not that the gauge is broken. Similarly, the analyst assumes that the schedule is correct and that the project has been delayed, not that the schedule is incorrect and the project is likely to finish on time.

Finally, note that in this example, the activity delay and the project delay are equal. That obvious result is because we have simple finish-to-start logic on a calendar with no nonwork days. However, even most simple construction project schedules will have nonworkday weekend days and so it is common for the activity delay and the project delay to differ by a few days.

Fig. 5.5 shows that Activity B was expected to start immediately after the completion of Activity A and finish on Day 30. The next step is to evaluate whether Activity B was completed as expected.

Fig. 5.6 shows that Activity B did not start immediately after the completion of Activity A as expected. Activity B actually started 5 days later than its "adjusted" planned start date. This is the second delay. Activity B was delayed 5 days because it started late. However, this activity then completed within its planned duration.

To determine the effect that the late start of Activity B had on the project's completion, the schedule is updated to Day 35. Fig. 5.7 depicts this schedule update.

Fig. 5.7 shows that the late start of Activity B was responsible for pushing out Activities C and D and, ultimately, delaying the project 5 days. Fig. 5.7 also shows that Activity C was expected to start

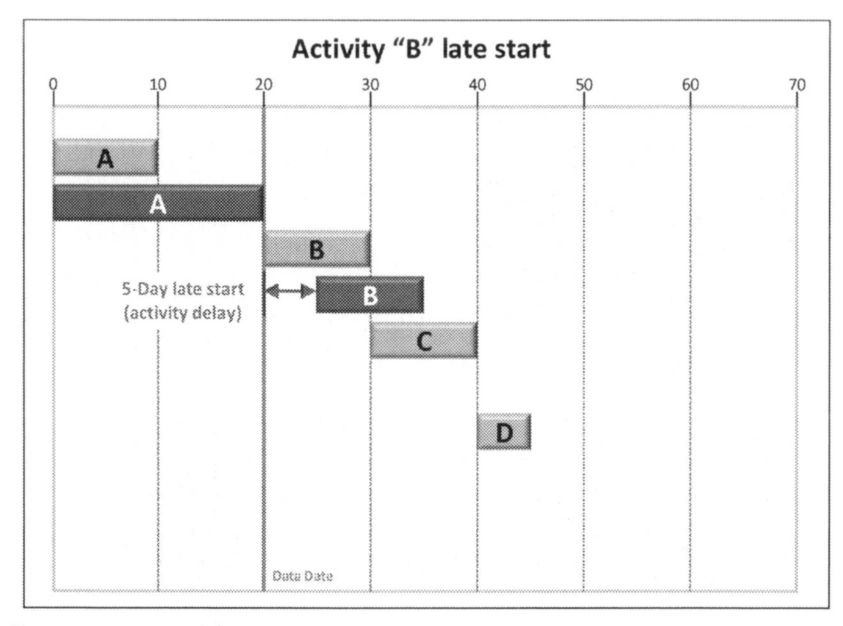

Figure 5.6 Activity B late start.

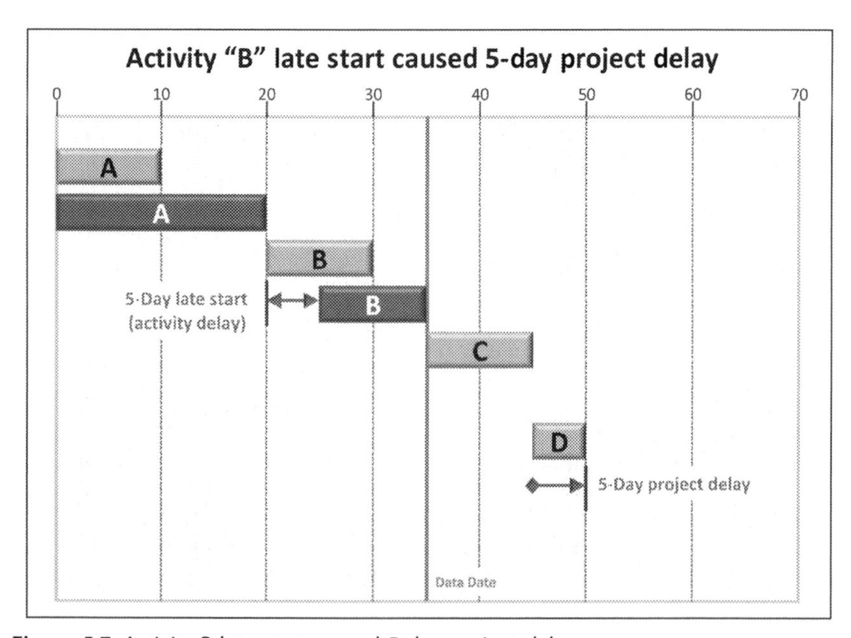

Figure 5.7 Activity B late start caused 5-day project delay.

immediately after the completion of Activity B and finish on Day 45. The next step is to evaluate the performance of Activity C. Fig. 5.8 shows that Activity C was not performed as expected.

Fig. 5.8 shows that Activity C started as expected, which was immediately after Activity B finished, but that it finished 10 days late. Perhaps more importantly with regard to the evaluation of Activity C's performance, Activity C was completed intermittently. Again, to measure the effect that the late finish of Activity C had on the project, the schedule is updated to Day 55. The result is depicted in Fig. 5.9.

Fig. 5.9 shows that the late finish of Activity C resulted in a 10-day project delay.

On Day 55, the owner added work to the project and the parties agreed to add an activity to the schedule to represent the added work. The result is depicted in Fig. 5.10.

Fig. 5.10 shows that after Activity E is inserted into the schedule, the project is now forecast to finish on Day 70, a 10-day delay.

The next step in the analysis is to compare the planned and actual completion of Activity D to determine whether its actual progress caused any project delay. Fig. 5.11 depicts the actual progress of Activity D.

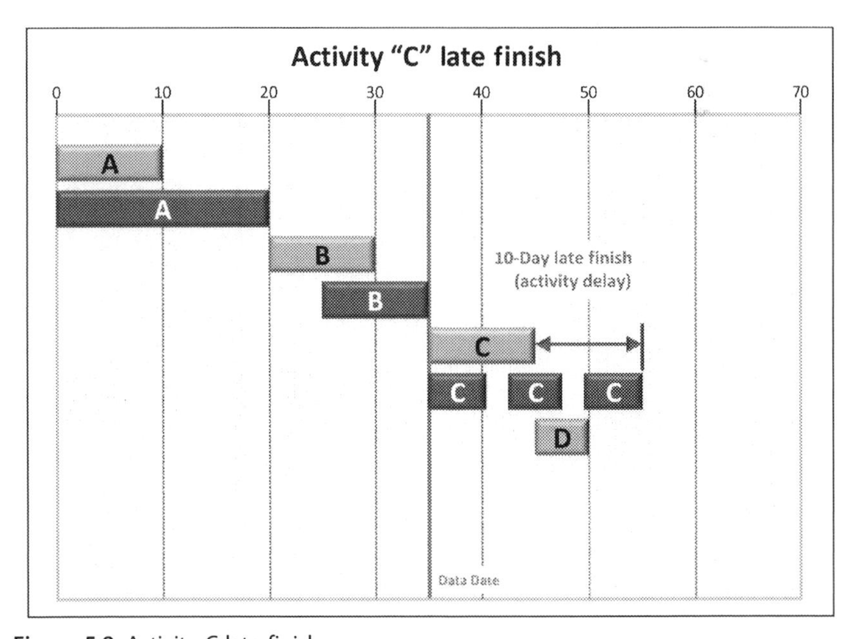

Figure 5.8 Activity C late finish.

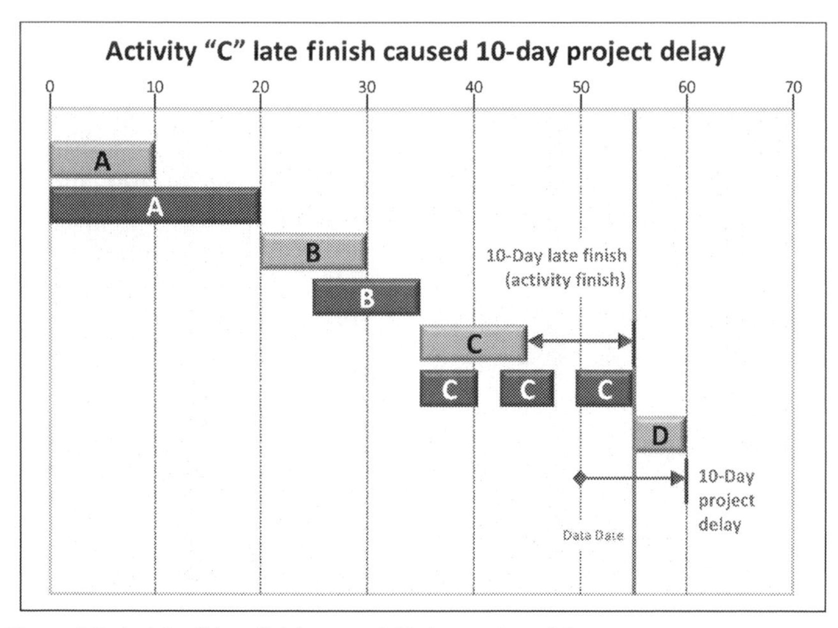

Figure 5.9 Activity C late finish caused 10-day project delay.

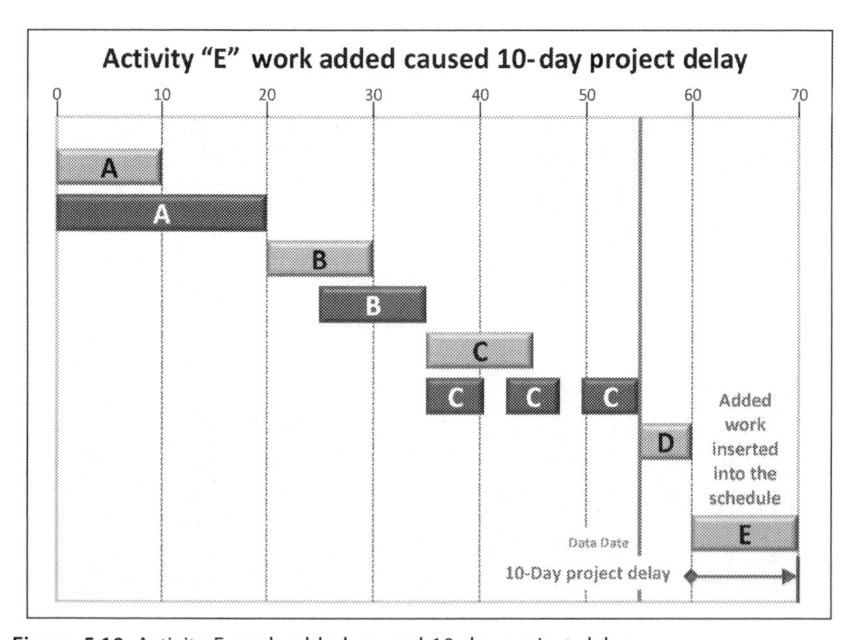

Figure 5.10 Activity E work added caused 10-day project delay.

Fig. 5.11 shows that Activity D started and finished as expected, resulting in no additional project delay. The next step is to compare the planned and actual performance of Activity E to determine whether it was responsible for any project delay. The result of this comparison is depicted in Fig. 5.12.

Fig. 5.12 shows that Activity E was actually completed in 5 days less than expected. As a result of Activity E finishing 5 days early, the project experienced a 5-day savings and the project was completed on Day 65. Lastly, Fig. 5.13 summarizes the assignment of the project delay to the responsible activities.

In one form or another, this stepwise conceptual approach starting at the beginning of the project should be used as a guide in almost all analyses of delays. Note that the precise charting of the as-built information is very helpful when the analyst moves to a determination of the cause of the delay or the liability for the delay. For instance, by knowing that the performance of Activity C was interrupted, as opposed to just taking longer, the analyst can focus on the available project documentation to identify the reasons why the work was performed in an intermittent fashion.

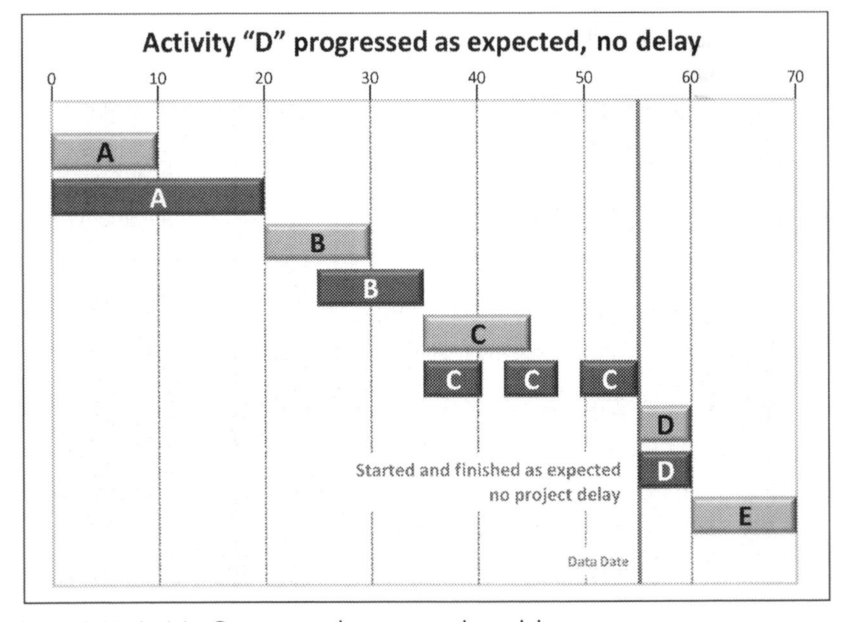

Figure 5.11 Activity D progressed as expected, no delay.

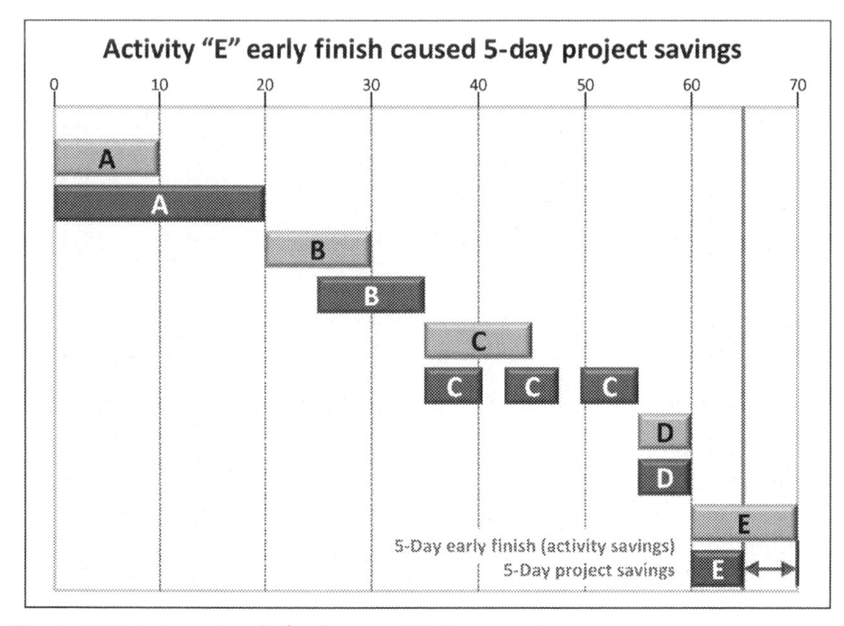

Figure 5.12 Activity E early finish caused 5-day project savings.

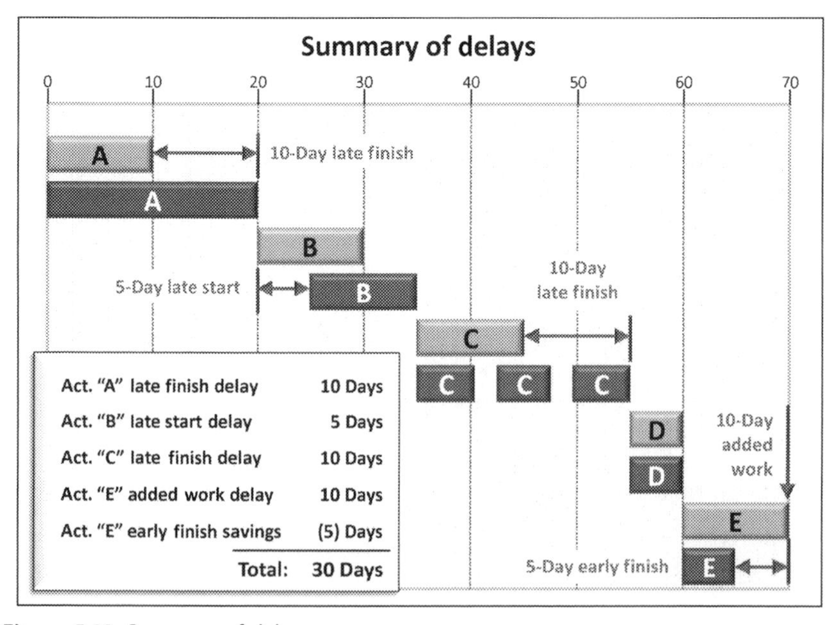

Figure 5.13 Summary of delays.

THE UNIQUE POSITION OF SUBCONTRACTORS

Because the duration of a project can only be extended by delays to activities on the project's critical path, a contractor's performance period can only be extended when the project experiences a critical delay. However, this is not necessarily the case for the performance period of a subcontractor.

While certain subcontractors may have work to perform during the entire project period, it is more typical for a trade subcontractor to have its work become available sometime after the project work has begun and be required to complete its subcontract work before all the project work can be or has been completed. As a result, the trade subcontractor's work may or may not show up on the project's critical path. Still, delays that extend the subcontractor's work will require the subcontractor to be on the job longer, thus, extending the subcontractor's performance period.

For example, a masonry subcontractor may not be able to begin its subcontract brick veneer work until the exterior sheathing has been installed on a building that is expected to take 18 months to construct. The as-planned schedule may show that after the exterior sheathing had been completed on one elevation, the masonry work could begin and would take 3 months to complete, followed by other exterior and interior finish work. Because the masonry work is planned to follow the expected pace of the exterior sheathing installation, the masonry may never show up on the critical path.

Presuming that the exterior sheathing work experiences delays, the exterior sheathing work is more likely to show up on the critical path than is the masonry work. Yet, the masonry work will be delayed because it will not be able to proceed at the planned pace. As a result, the masonry work takes 5 months to complete instead of the planned 3-month period. The mason claims that its performance period was extended by 2 months through no fault of its own. It requests additional compensation for extended overhead costs and other delay damages.

In this example, an analysis of delays along the critical path of the project may not support the mason's request. However, from the facts presented, it is evident that the mason's performance period was extended, and, depending on the provisions of its subcontract, the mason may be able to recover the delay costs caused by others.

If a critical path analysis of the project does not support the mason's claim, what type of analysis should be performed to determine if the mason's claim has merit? In the preceding simple example, the answer appears straightforward. For most subcontractors, however, their work is integrated with many aspects of the project work. Often, the relationships among the various work activities of the prime contractor and the various subcontractors are more complex than the preceding simple example. To complicate matters, when the subcontractor's work is not on the critical path of the project, unless constrained in some other way, it will have float. Therefore, any analysis of subcontractor delays will also involve an examination of the subcontractor's obligations with respect to activity float.

When investigating potential delays to a subcontractor's performance, begin by evaluating the performance requirements of the subcontract. The objective of this evaluation is to determine the period of performance for which the subcontractor is obligated under the terms of the subcontract. Often, a subcontractor is required to perform its work according to the project schedule. Typically, the prime contractor reserves its right to modify the schedule as necessary to complete the work in a timely fashion. The subcontract will then require the subcontractor to perform according to these modifications as well. Because these modifications are not known at the time of the subcontract agreement, there is a certain expectation that the parties have regarding the subcontractor's performance period. This expectation will usually be a product of the particular negotiation that led to the signing of the subcontract.

While the subcontractor typically takes on some risk regarding the prime contractor's right to modify the schedule, this risk is typically not without limits. The contract CPM schedule will identify early and late dates for all of the subcontractor's work activities. When obligated to perform according to the contract schedule, it is reasonable to conclude that the subcontractor is obligated to be on site from the projected early start date of its first activity to the late finish of its last work activity. This conclusion recognizes that the work activities do not need to be performed on the early dates for the project to complete on time. Thus, performance of the work within the early and late date ranges are foreseeable because such performance is, in fact, "according to" the schedule. This remains true, even if such performance affects the continuity of the trade subcontractor's work activities.

The prime contractor, however, may argue that the subcontract allows for modifications to the sequence and duration of the work. Here again, there may be a question as to the degree such modifications are foreseeable. It may be reasonable for the prime contractor to argue that, because it is responsible to the owner to complete the project on time, it must continually assess progress against its plan to complete the work. When the actual progress differs from that planned, it must modify the sequence and duration of future work to ensure an on-time completion. As a result, modifications to the schedule that change the sequence and duration of the subcontractor's work activities within the original project performance period may be foreseeable. Much of this argument, however, will depend on the nature and extent of these changes. Unlimited modifications to the sequence and duration of the work are generally not anticipated by the parties.

Through careful evaluation of the subcontract and the understandings and circumstances leading to the subcontract agreement, the subcontractor's planned performance period can be determined. Unlike the prime contractor's contract performance period, which is generally expressed in the contract, the parties may be unable to agree on the subcontract period of performance. In such cases, the parties will prepare their respective arguments based on the subcontract performance period they believe to be correct.

Once the subcontract period of performance has been established, a comparison to the subcontractor's actual performance period provides a measure of the total delay experienced by the subcontractor. But this is only the beginning of the story. Next, it is necessary to determine the causal link between the actions of the parties and the delays incurred in order to determine if the subcontractor's delays were caused by others.

In order to determine the cause of delays to the subcontract period of performance, it is necessary to determine the critical path of the subcontractor's work. The critical path of the subcontractor's work is the longest path of activities leading from the first work activity to be performed by the subcontractor to the last. This path may consist of some or all of the subcontractor's work activities, as well as work activities performed by others. Because these activities are integrated within the entire schedule network, the analyst cannot simply isolate the subcontractor's work activities and evaluate the paths among these in a vacuum. Many of the subcontractor's work activities will be driven by activities being performed by others, and all of these relationships must be considered in the analysis.

As a result of these complexities, it may not be possible to determine the longest path between the subcontractor's start and end points through an analysis of the electronic schedules. As an alternative, it may be necessary to determine the subcontractor's critical path through a detailed evaluation of the subcontractor's daily work progress. This evaluation is similar to the As-Built Delay Analysis discussed later in this book.

This process begins with the preparation of a detailed as-built diagram that tracks the subcontractor's actual performance. This performance is then compared to the available planned performance information. To begin with, the analyst determines if the subcontractor was able to meet planned durations for its work and, if not, why not. Did the subcontractor provide sufficient resources to accomplish the work? Was the subcontractor given access to the work as anticipated or was it required to perform its work under conditions that differed from those it expected to encounter? Was the subcontractor in control of the pace of the work, or was something else controlling the pace?

It is also necessary to look at the sequence of the subcontractor's work to see if it differed from that planned and, if so, why. As the work progresses, it is also necessary to consider all of the subcontract work remaining and the precedent requirements of that work. For example, if the subcontractor was delayed in one area of the project, was there other available work for it to perform?

By evaluating the subcontractor's as-built work performance moving forward through the project and considering the work that remains, the critical path of the subcontractor's work can be determined in addition to the factors that extended the work along this path.

When a project is managed by a well-thought-out and periodically updated CPM schedule, the analyst has many tools at his or her disposal to help determine the delays to the prime contractor's performance period. To begin with, the contract will usually state the contract performance period, and the longest path through the Project can easily be determined by the scheduling software. However, in the case of delays to the subcontractor's performance period, these tools are less effective. As a result, the analyst must apply a more in-depth knowledge of both the subcontracting process and of the project management process in order to determine the most appropriate way to resolve disputes related to subcontractor delays.

CHAPTER SIX

Delay Analysis Using Bar Chart Schedules

Later in this book, we will explain how to perform a delay analysis when a detailed Critical Path Method (CPM) schedule was created as the original as-planned schedule for the project. However, many projects are scheduled using a bar chart schedule (bar chart). For projects with many interrelated activities, a bar chart is not as desirable as a CPM schedule, because a CPM schedule records and preserves the relationships among the activities. However, a meaningful and accurate delay analysis can still be performed using a bar chart. With any schedule, as the level of detail and the quality of information decrease, the delay analysis becomes more subjective. Therefore, a delay analysis that is based on a bar chart requires the analyst to guard against assumptions that favor one outcome or another and to work to be as objective as possible. This chapter describes how a delay analysis is performed when the project schedule is a bar chart.

There is nothing inherently wrong with scheduling a project with a bar chart. Bar charts were in use long before the Critical Path Method was ever created. As some professionals are quick to point out, the Empire State Building was scheduled with a bar chart and not a CPM. In fact, a detailed bar chart can provide almost as much information as a CPM schedule.

Fig. 6.1 is a simple bar chart for the construction of a bridge. Though it does not contain a significant number of activities, it does show the general sequence of work for the construction of the bridge. Using this simple bar chart as a starting point, the project manager could easily define each activity in more detail.

Fig. 6.2 is a more detailed bar chart of the project depicted in Fig. 6.1. This more detailed bar chart more clearly defines the contractor's proposed work plan. In this bar chart, each major activity is broken down into the work on the respective piers and spans, providing the contractor and owner with a more detailed illustration of the plan for construction.

With very little effort, the project manager or project scheduler can modify the bar chart in Fig. 6.2 to show the interrelationships among the activities, as shown in Fig. 6.3.

Construction Delays.
DOI: http://dx.doi.org/10.1016/B978-0-12-811244-1.00006-9

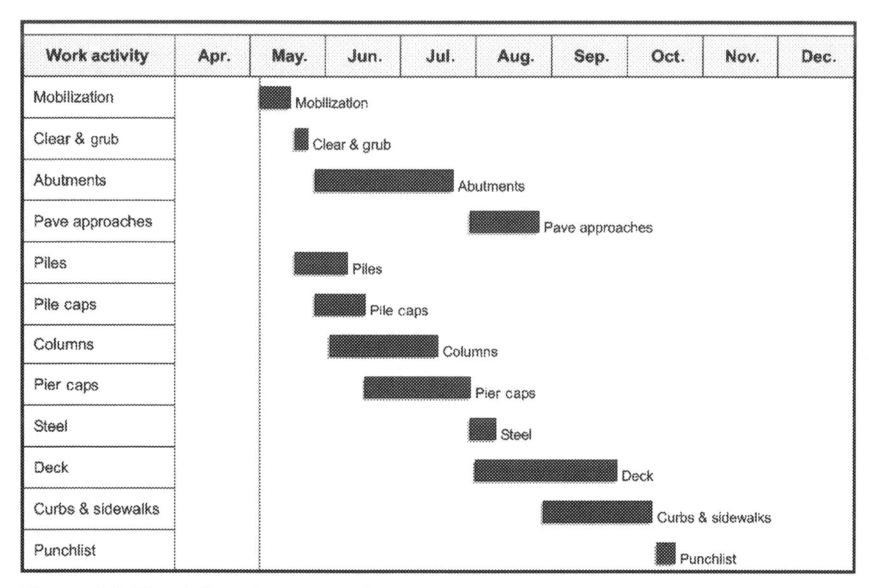

Figure 6.1 Simple bar chart example.

Using this schedule as a foundation, a CPM schedule for the project could be produced with little effort.

Unfortunately, most bar charts for projects do not contain as much detail as that in Fig. 6.3, and often not even as much as the bar chart in Fig. 6.2. In general, most bar charts suffer from the following major shortcomings that diminish their usefulness as a management tool and their effectiveness in measuring delays:

- Lack of detail—too few activities for the amount and complexity of the work
- No indication of the interrelationships among the activities
- No definition of the critical path of the project

Obviously, these weaknesses hamper the ability of the analyst to perform a delay analysis, but they do not make it impossible. If nothing else, the bar chart is helpful in that it defines the plan for constructing the project, and it can be used as the basis for an analysis of delays.

DEFINING THE CRITICAL PATH

The first step in analyzing a bar chart is to define the critical path. Every project has a critical path, including a project that was scheduled with a bar chart. The following definitions illustrate this point.

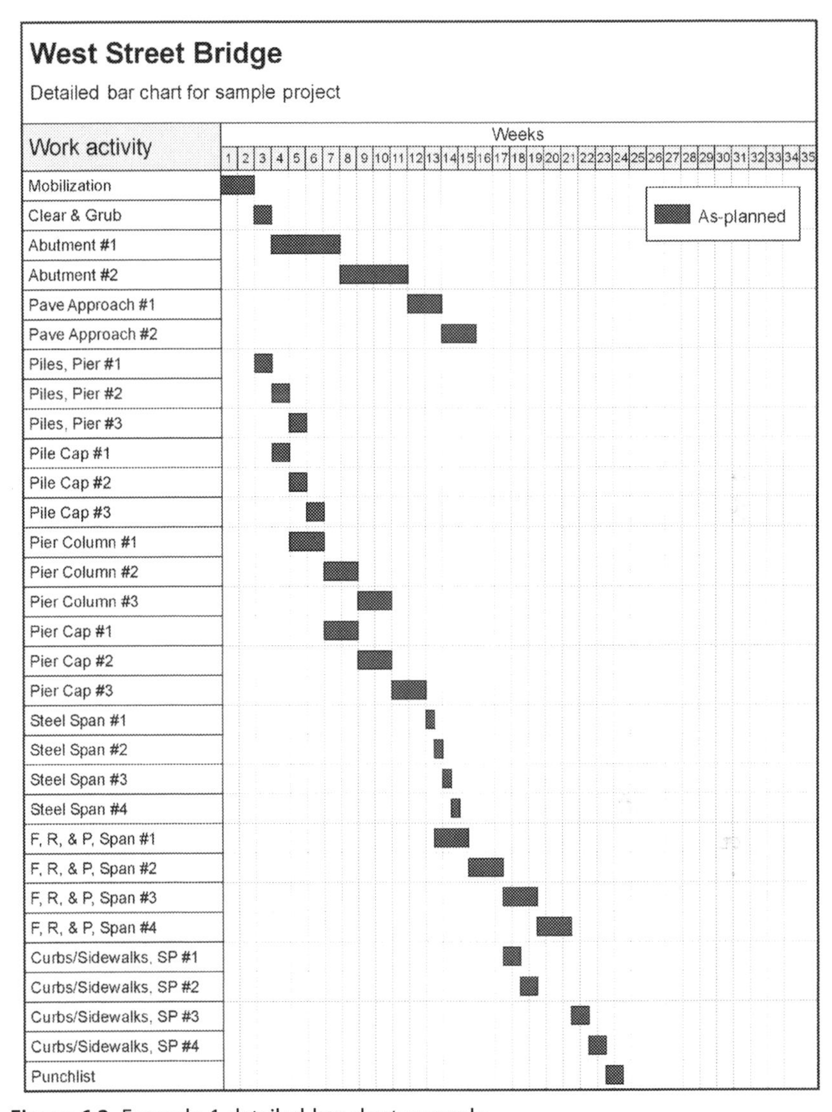

Figure 6.2 Example 1 detailed bar chart example.

BASIC CRITICAL PATH METHOD

In CPM scheduling, the drafter of the schedule prepares a logic or network diagram. As presented in Chapter 2, Float and the Critical Path, once durations are assigned to the activities in the network logic diagram, the critical path can be calculated. It is a purely arithmetic process.

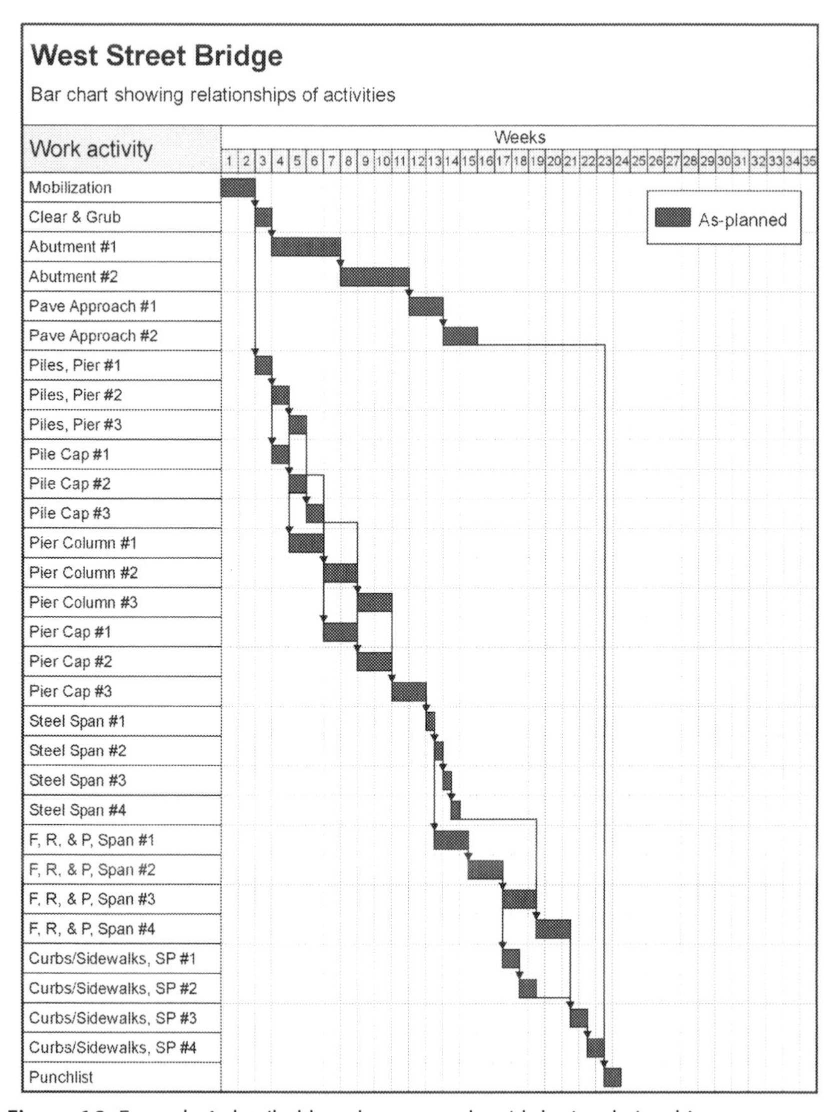

Figure 6.3 Example 1 detailed bar chart example with logic relationships.

The definition of the critical path is *the longest path of work activities through the network diagram that forecasts the date when the project will finish.* The project cannot finish until every path of work has been completed. Whether the critical path is defined in a CPM schedule or a bar chart, every project has a series of interrelated activities that will control the project completion date.

Also, the only way to delay the project is to delay an activity on the critical path of the project. In understanding this concept, it is essential to recognize that the critical path of a project is dynamic. In this manner, delays to noncritical activities that persist will cause the critical path to shift to the path containing that activity and, thus, will then delay the project.

IDENTIFYING THE CRITICAL PATH ON A BAR CHART

Because we know that a critical path exists in a bar chart schedule, the delay analyst should first identify this critical path. The analyst must review the bar chart in detail for obvious conclusions about the sequence of work. These conclusions may be based on project documentation that might clarify the thought process that went into creating the bar chart or defining the planned work sequence. Documentation that can be helpful includes the contract, which may dictate staging or phasing, the pre-bid or preconstruction meeting minutes, internal contractor or subcontractor documentation, project correspondence exchanged before the bar chart was prepared, and any other documentation that might shed light on the how the project team approached the planning and scheduling of the project.

Practical knowledge of the type of project and the physical construction requirements is also necessary to reach a reasonable conclusion regarding the project's critical path. For example, to analyze a bar chart of a high-rise structure, the analyst may need to know that interior finishes usually are not planned to start until the building or a portion of the building is "dried-in" or "watertight," that a common sequence of the progression of trades is from the bottom up, and that it is common for trades to follow behind one another as the building progresses upward, instead of waiting until the preceding trade has finished all of its work in the tower.

Given the variations in possible work sequences, analysts should resist the temptation to interpret the bar chart schedule based solely on their own experience. Just because a contractor has performed work in a particular sequence in the past does not mean that the contractor on the project being analyzed has planned to perform the work the same way. Unless the bar chart is extremely brief, analysts should be able to glean the best indication of the overall plan and sequence of activities to

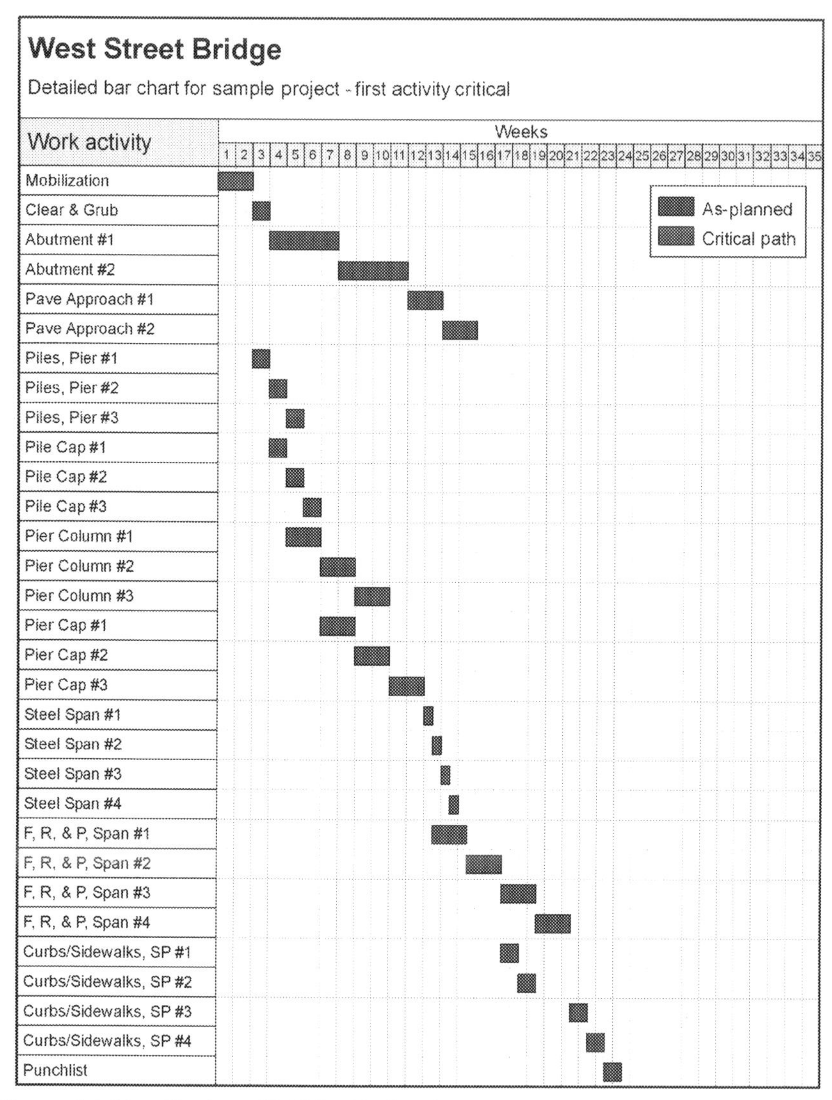

Figure 6.4 Identification of initial critical path activity.

determine the critical path using the bar chart as the primary indicator of the project team's plan for execution of the project, rather than analysts proposing their own version of a plan.

Referring to Fig. 6.3, we can define the critical path for the sample bridge project. The critical path starts with the mobilization activity, with a duration of 2 weeks. This is obvious, since no other activity is scheduled to occur during this period. We show this first critical activity in Fig. 6.4.

The next two activities on the schedule are the clear and grub activity and the piles at Pier #1. In reviewing the sequence of activities, the clear and grub activity is related to the abutment and approach work. The abutment and approach work is scheduled to finish well before the end of the project and does not appear to be related to the schedule of activities for bridge construction. The analyst can determine from the contract that the approaches are to be constructed using asphalt paving, but the bridge deck is to be paved with concrete. Consequently, there is no physical reason to coordinate the concrete placement for the bridge with the approach construction. The only possible relationship might be the need to move the concrete placing equipment onto the bridge superstructure. However, because the schedule reflects that the deck work is to start before workers complete either of the approaches, the analyst concludes that the equipment can be moved onto the bridge independently of the approach work. Therefore, it appears that the abutment/approach path is not on the critical path for the project.

Therefore, the critical path must be through the piles and piers. When viewing this bar chart, the analyst sees that the work is "stair-stepped" through the specific activities for each pier. Thus, after the piles at Pier #1 are completed, the piles at Pier #2 can start. While the piles at Pier #2 are being driven, the pile cap at Pier #1 is concurrently constructed. Based on the graphic representation, the critical path appears to follow these activities (shown in Fig. 6.5):

- Piles, Pier #1
- Pile Cap #1
- Pier Column #1
- Pier Cap #1

At this point, the analyst recognizes that the pier columns and the pier caps each have 2-week durations, and the next activity—steel erection—does not begin until all pier caps are completed. Therefore, all pier column and pier cap activities are most likely on the critical path, not just the first piles, piers, columns, and caps. Using similar reasoning, the steel for Span #1 is critical, and then the path continues through deck placement for all spans. Next are the curbs and sidewalks for Spans #3 and #4 and, finally, the punch list work.

Thus, the overall critical path from the bar chart (Fig. 6.6) is:

- Mobilization
- Piles, Pier #1
- Pile Cap #1

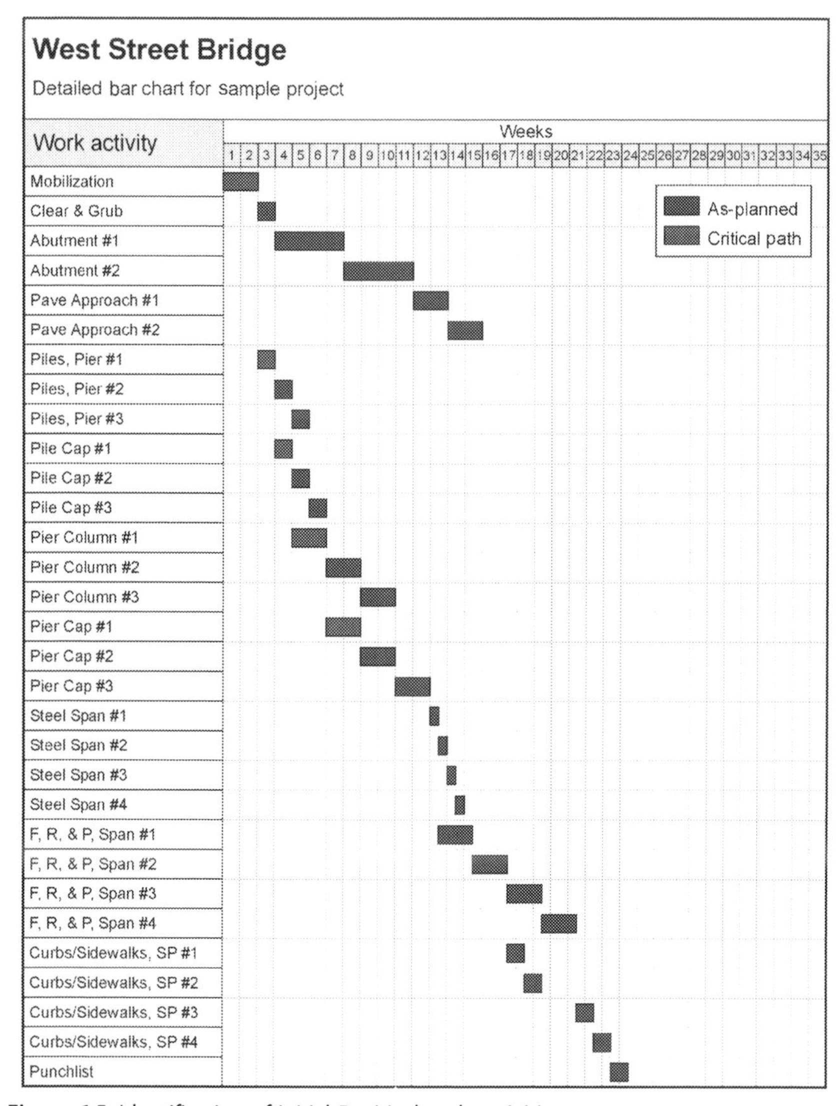

Figure 6.5 Identification of initial 5 critical path activities.

- Pier Column #1
- Pier Column #2
- Pier Cap #1
- Pier Column #3
- Pier Cap #2
- Pier Cap #3

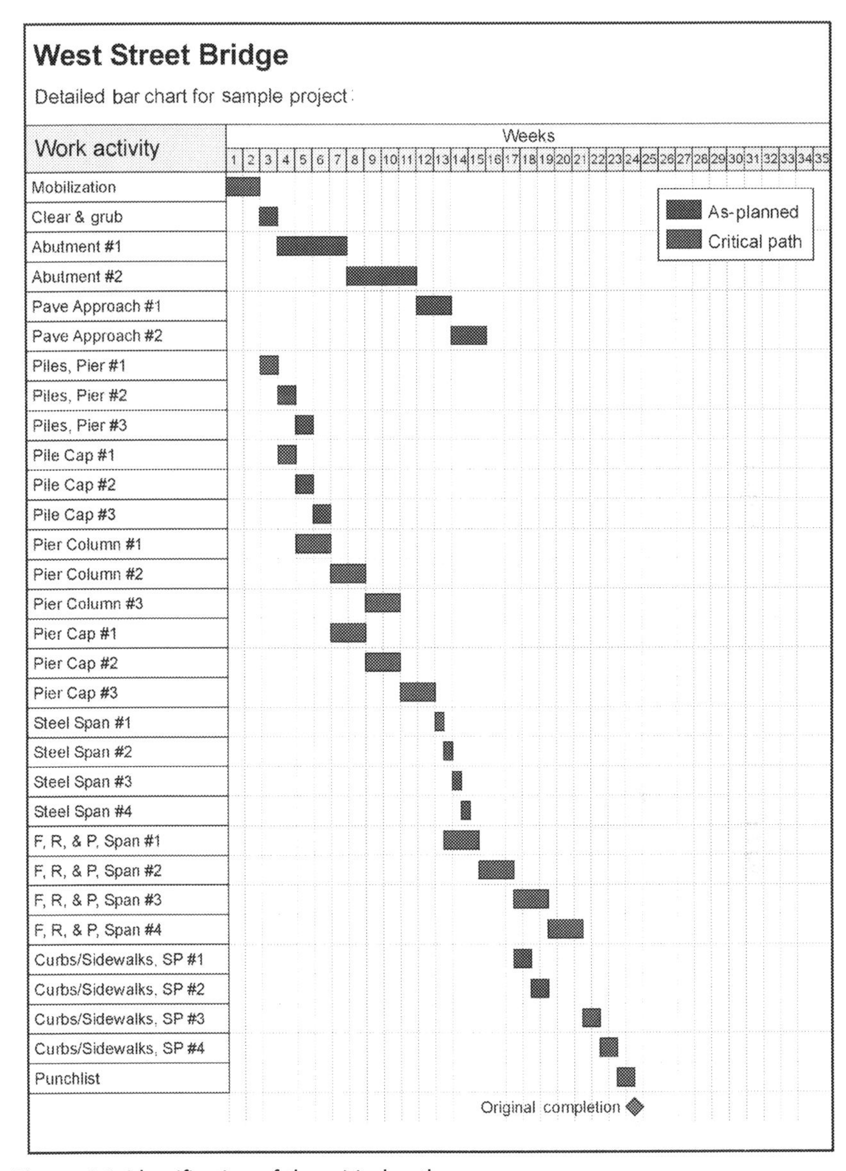

Figure 6.6 Identification of the critical path.

- Steel Span #1
- F, R, & P, Span #1
- F, R, & P, Span #2
- F, R, & P, Span #3
- F, R, & P, Span #4

- Curbs & Sidewalks, Span #3
- Curbs & Sidewalks, Span #4
- Punch list

The analyst could reach a similar conclusion working with the less detailed bar chart alone (see Fig. 6.1). This would, however, require that the analyst make more assumptions about the work on the separate piers. As was noted in the preceding discussion, contemporaneous documentation can help the analyst define the contractor's planned sequence in more detail. The less detailed the bar chart, the more assumptions are required by the analyst to determine the project's critical path.

QUANTIFYING DELAYS USING BAR CHART SCHEDULES

The process of quantifying delays using a bar chart is similar to the process that is described later in this book when a CPM schedule is available. To start the process, the analyst must prepare a detailed as-built diagram that shows as specifically as possible when the project work was actually performed. Fig. 6.7 is the as-built diagram for the West Street Bridge project. Once the as-built has been prepared, the analysis can proceed. As the as-built diagram (Fig. 6.7) shows, the mobilization activity started on schedule (the first day of Week 1) and finished on schedule (by the end of Week 2). The remaining activities, however, did not proceed in the same manner as the as-planned schedule had predicted.

As the as-built diagram (Fig. 6.7) shows, the pile driving at Piers #1, #2, and #3; the pile caps at Piers #1 and #2; and Pier Column #1 were accomplished as-planned in 3 weeks immediately following the mobilization activity. However, the clear and grub activity did not proceed as planned, but started 2 weeks late and finished in the 1-week planned duration. If the previous conclusions concerning the critical path were correct, the delay to the start of clearing and grubbing should not have resulted in a delay to the project. To check this conclusion, the analyst can "update" the bar chart as of the end of Week 5, as shown in Fig. 6.8.

As can be seen in Fig. 6.8, the project is still on schedule, but the abutment and approach work has been delayed, or "pushed out" in time, because of the delay to the clear and grub activity. As expected, there is no delay to the critical path. The adjusted schedule (Fig. 6.8) shows the as-built condition for the first 5 weeks of the project and the adjusted as-planned activities for the remainder of the work.

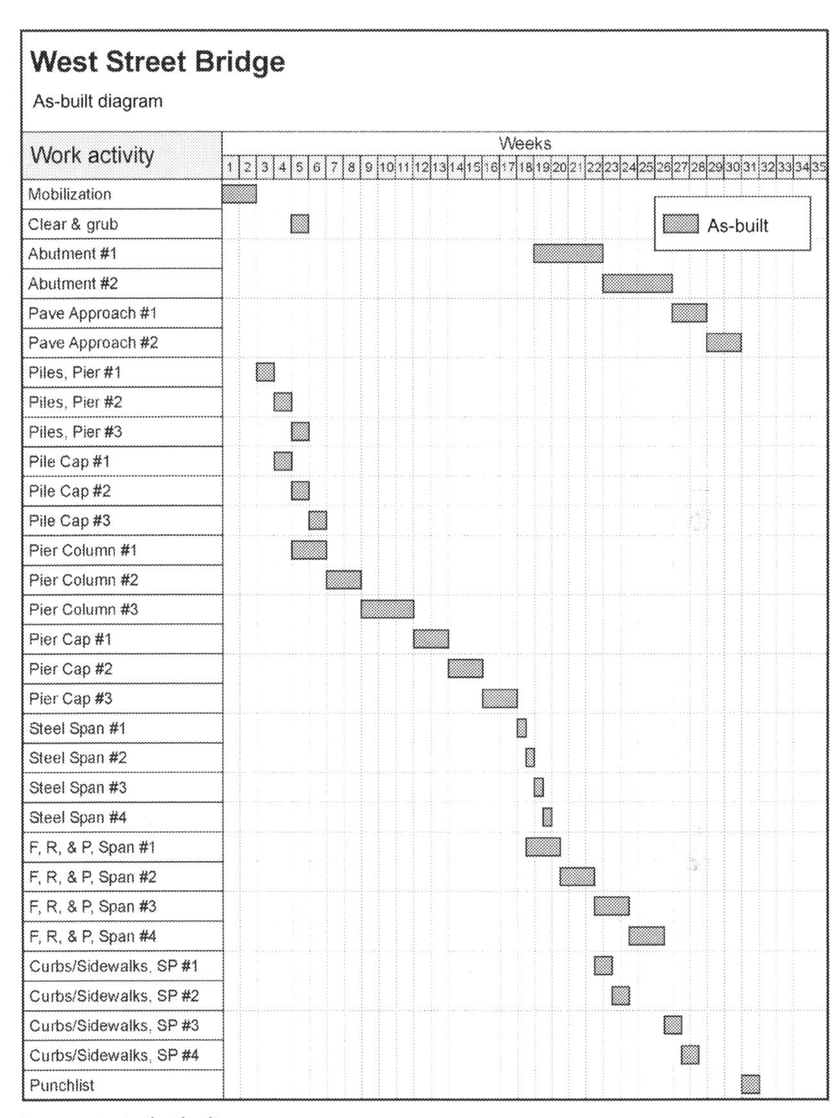

Figure 6.7 As-built diagram.

Based on the as-built information, the analyst decides to update the schedule as of the end of Week 11. The as-built diagram (Fig. 6.7) shows that the abutment and approach work has not yet begun and that the pier cap work also has not yet begun. Pier Columns #1 and #2 were completed on schedule. Pier Column #3, however, took 1 week longer to complete than planned. The adjusted schedule for Week 11 is shown in Fig. 6.9.

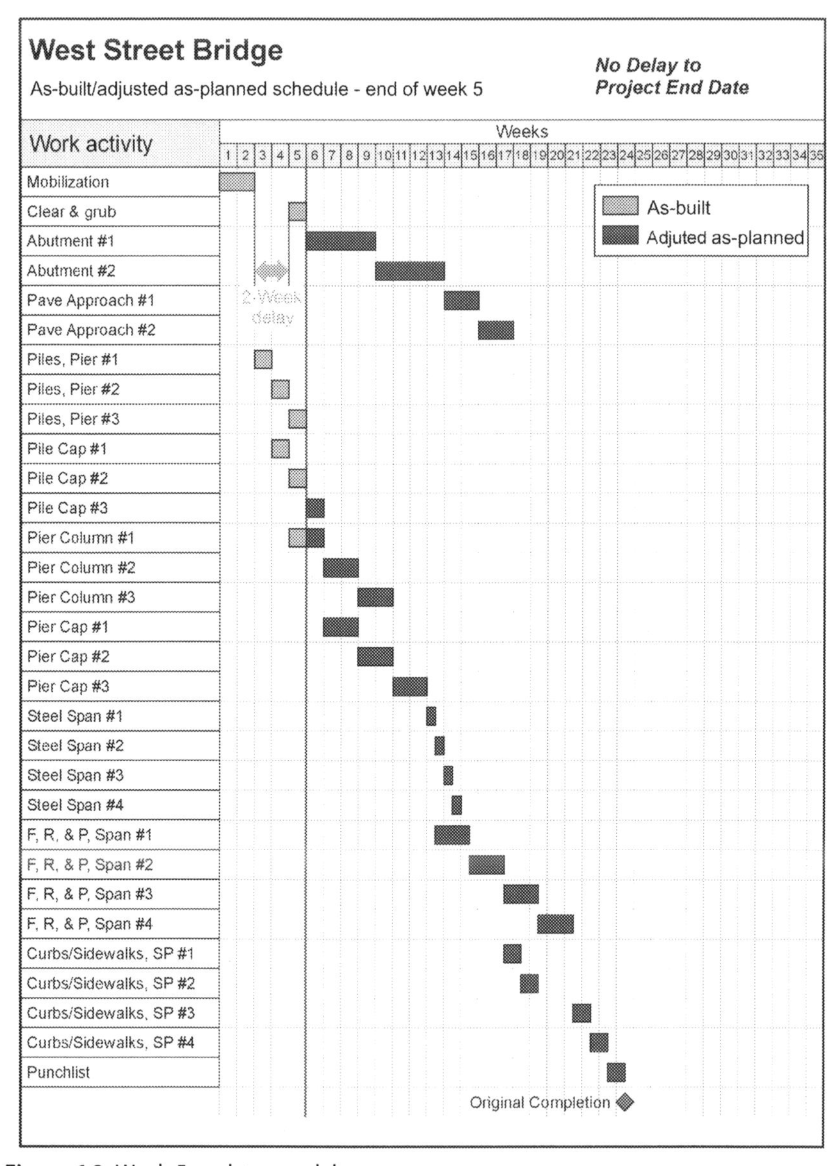

Figure 6.8 Week 5 update, no delay.

Based on the updated and adjusted schedule presented in Fig. 6.9, the analyst concludes that the project is now 5 weeks behind schedule. The delay was caused by the late start of Pier Cap #1 work, which was planned to start at the beginning of Week 7, but actually started at the beginning of Week 12. Although Pier Column #3 was late in finishing and

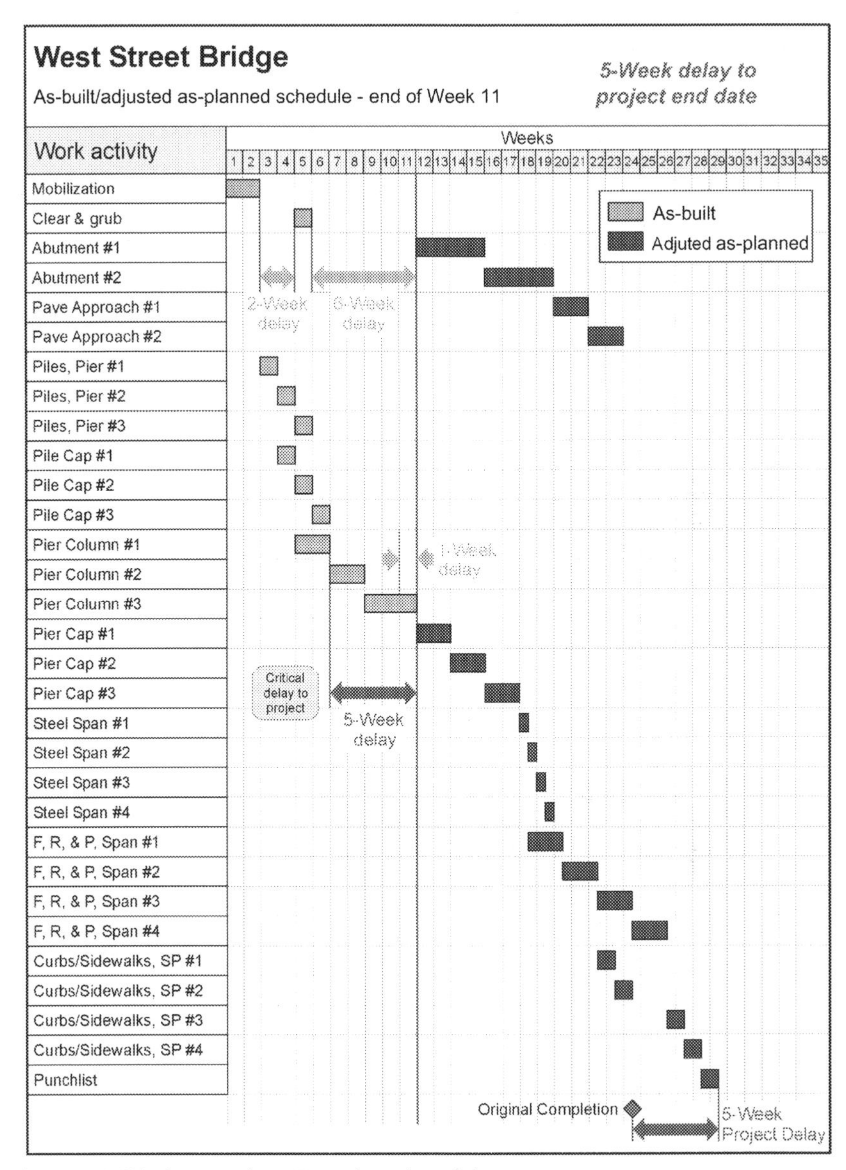

Figure 6.9 Week 11 update, 5-week project delay.

was on the original critical path, once the Pier Cap #1 activity did not start on time, the critical path shifted solely to the pier cap work. Pier Column #3 activity was effectively given float by this shift in the critical path.

Next, the analyst decides to update the schedule at the end of Week 15. This point is chosen because the activities along the bridge pier and deck path continued in accordance with the adjusted schedule, but the abutment and approach work did not. By tracking the two activities that are precedent to punchlist, specifically the curbs and sidewalks of Span #4, which are currently driving the punchlist activity, and the paving of Approach #2, it appears that the critical path shifts in the middle of Week 15. The schedule updated for the end of Week 15 is shown in Fig. 6.10.

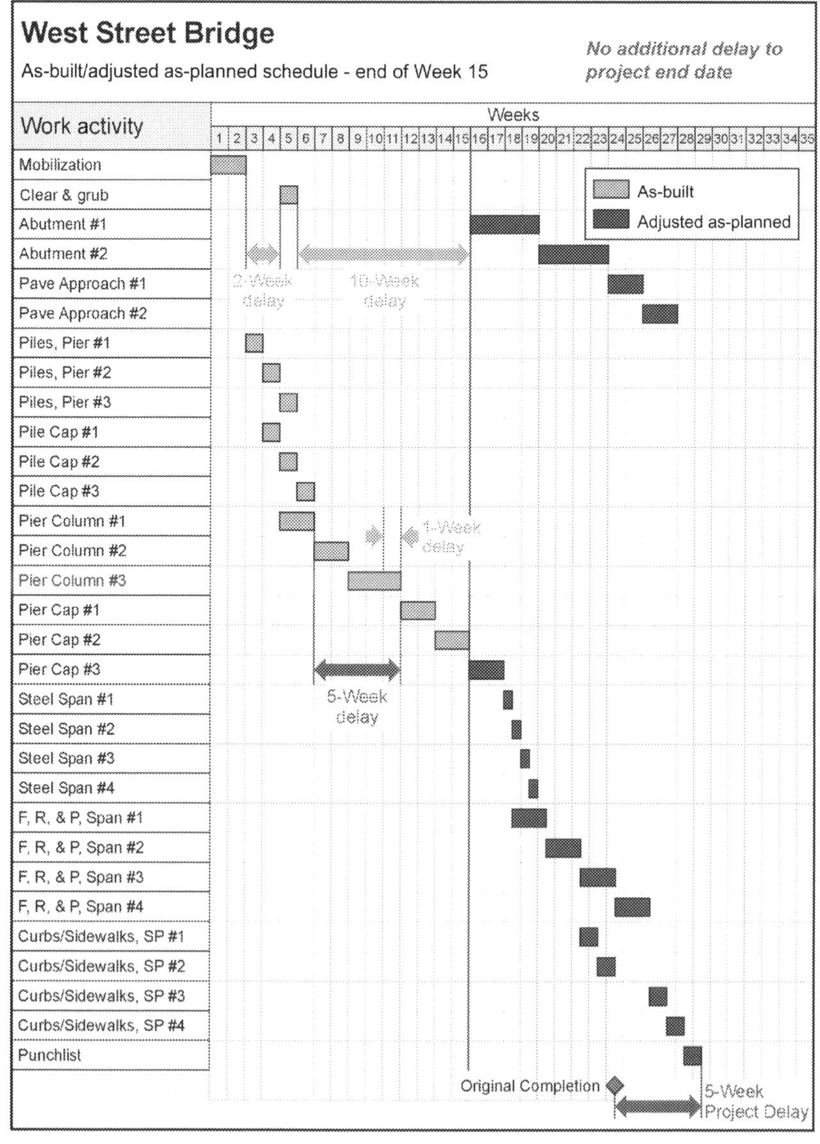

Figure 6.10 Week 15 update, no additional project delay.

As shown in Fig. 6.10, from the end of Week 11 to the end of Week 15, there was no additional delay, despite the fact that the abutment and approach work continued to be delayed.

To test the previous observation that the critical path shifts in the middle of Week 15, the analyst updates the schedule at the end of Week 16. The schedule updated for the end of Week 16 is shown in Fig. 6.11.

As can be determined from Fig. 6.11, for the first half of Week 16, the bridge pier and deck path continued to be the driving activity of the punchlist work, while the abutment and approach path continued to consume its last half-week of float. Because the pier cap work was progressing as planned, there was no additional delay for the first half of the week.

Then, the abutment and approach work path became longer than the bridge pier and deck path and, as the longest path, the abutment and approach work became critical. Because this path of work continued to be delayed for the remainder of Week 16, by the end of the week, the project had been delayed an additional half-week. As a result, by the end of Week 16, the project had been delayed a total of 5.5 weeks.

Next, the analyst decides to update the schedule at the end of Week 18. This point is chosen because the as-built schedule indicates that the abutment and approach work actually started at this time. The schedule updated for the end of Week 18 is shown in Fig. 6.12.

Because the project had been delayed 5.5 weeks as of the last update, the additional delay since that update is 2 weeks. The activities on the bridge pier and deck path were not delayed further since the last update. Instead, the additional 2-week delay was the result of the continued lack of progress on the abutment and approach path.

Next, the analyst decides to update the schedule at the end of Week 31, which is the time that the as-built schedule shows that the project actually completed. The schedule updated for the end of Week 31 is shown in Fig. 6.13.

As shown in Fig. 6.13, no additional delay was experienced during the completion of the project. At the completion of the analysis, all project delays have been identified. As a final check, the analyst ensures that the total net delay identified during the analysis equals the number of days that the project was completed late.

This example follows the conceptual approach to analyzing delays outlined in Chapter 5, Measuring Delays—The Basics. The exact method used to perform an analysis and the accuracy of the results depend on the level of detail of the as-planned schedule and of the available as-built information.

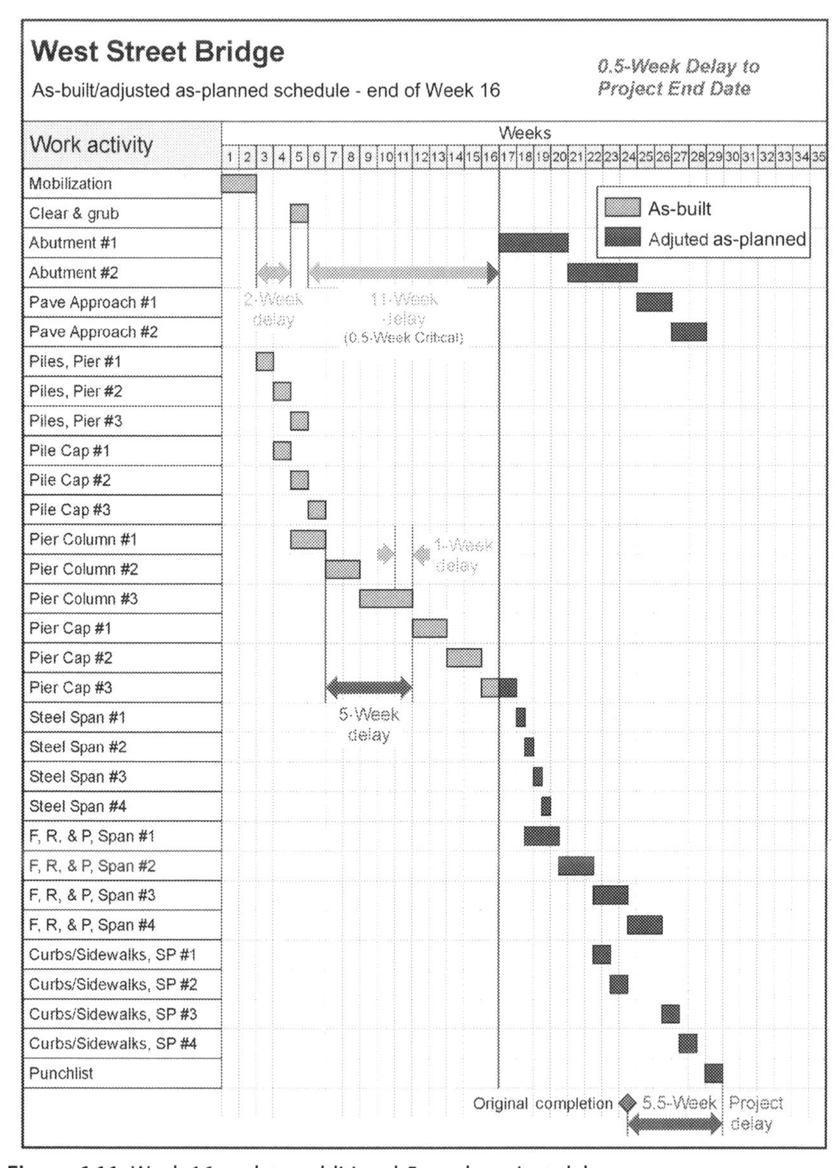

Figure 6.11 Week 16 update, additional 5-week project delay.

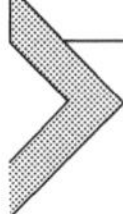

EXAMPLE DELAY ANALYSIS OF POTENTIAL CHANGES WITH BAR CHARTS

To further illustrate the process of determining delays with a bar chart, the following example of the construction of a simple, three-story

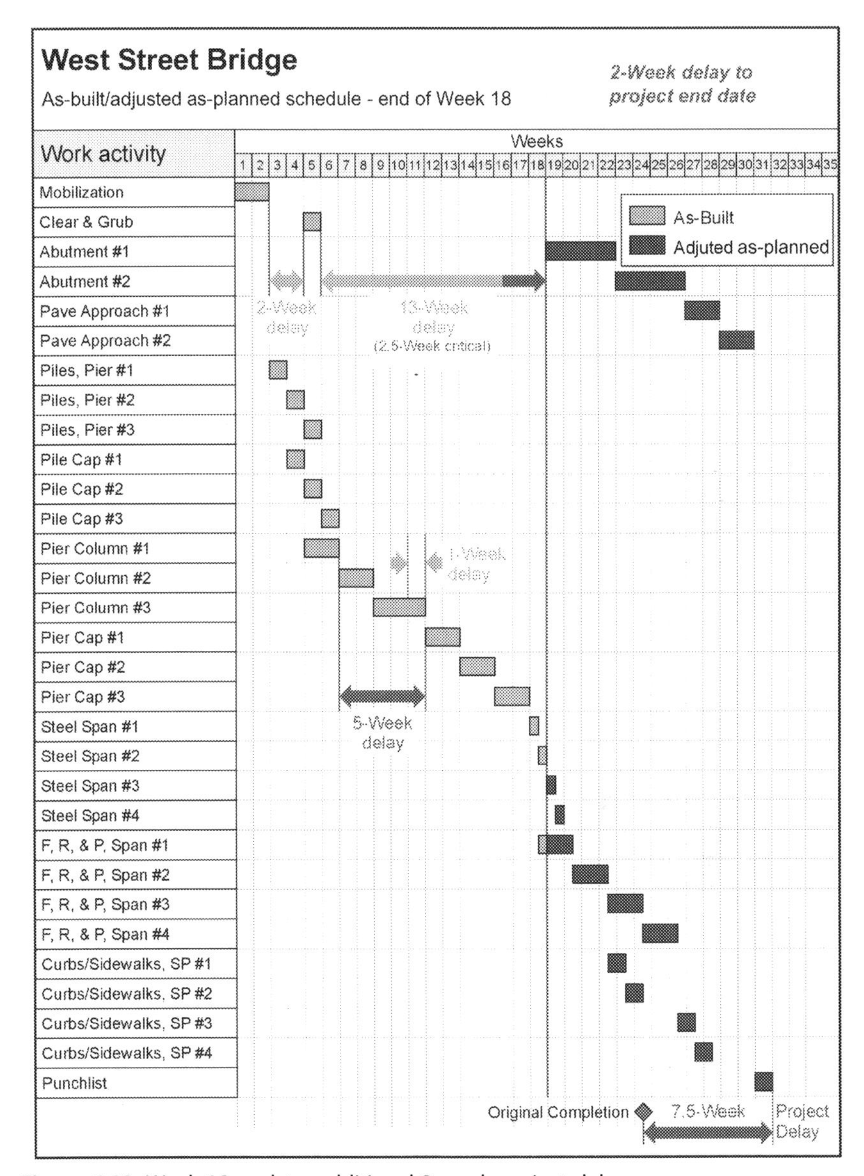

Figure 6.12 Week 18 update, additional 2-week project delay.

building is used. Fig. 6.14 is a bar chart schedule for the project. The critical path is indicated by the red (dark gray in print versions) bars.

During the project, there were four changes that occurred. The analyst has been asked to analyze each of these and determine what delay, if any, each caused to the project completion date. To perform the analysis,

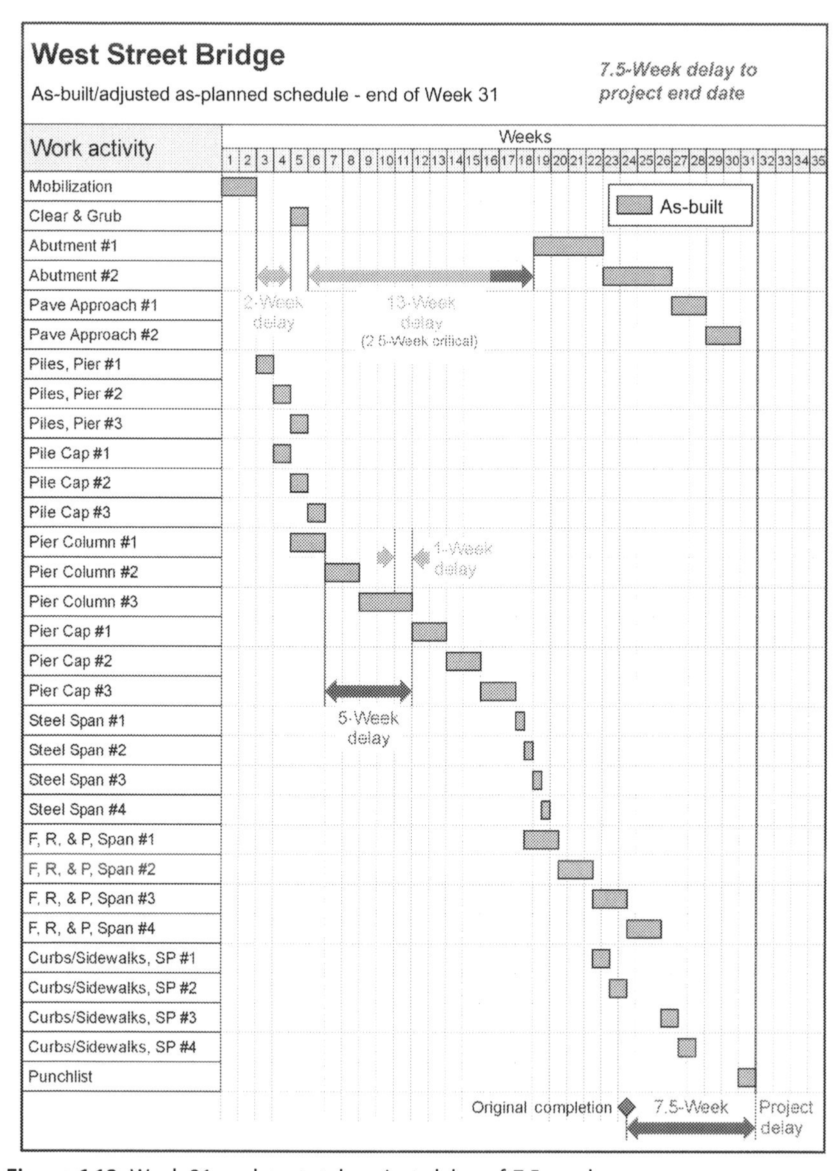

Figure 6.13 Week 31 update, total project delay of 7.5 weeks.

the analyst prepared an as-built diagram based on the project daily reports and other contemporaneous information. The as-built diagram, with the changed work highlighted in yellow (white in print versions), is shown in Fig. 6.15.

The analyst, having read this book, understands that the delay cannot be determined simply by comparing the as-planned schedule and

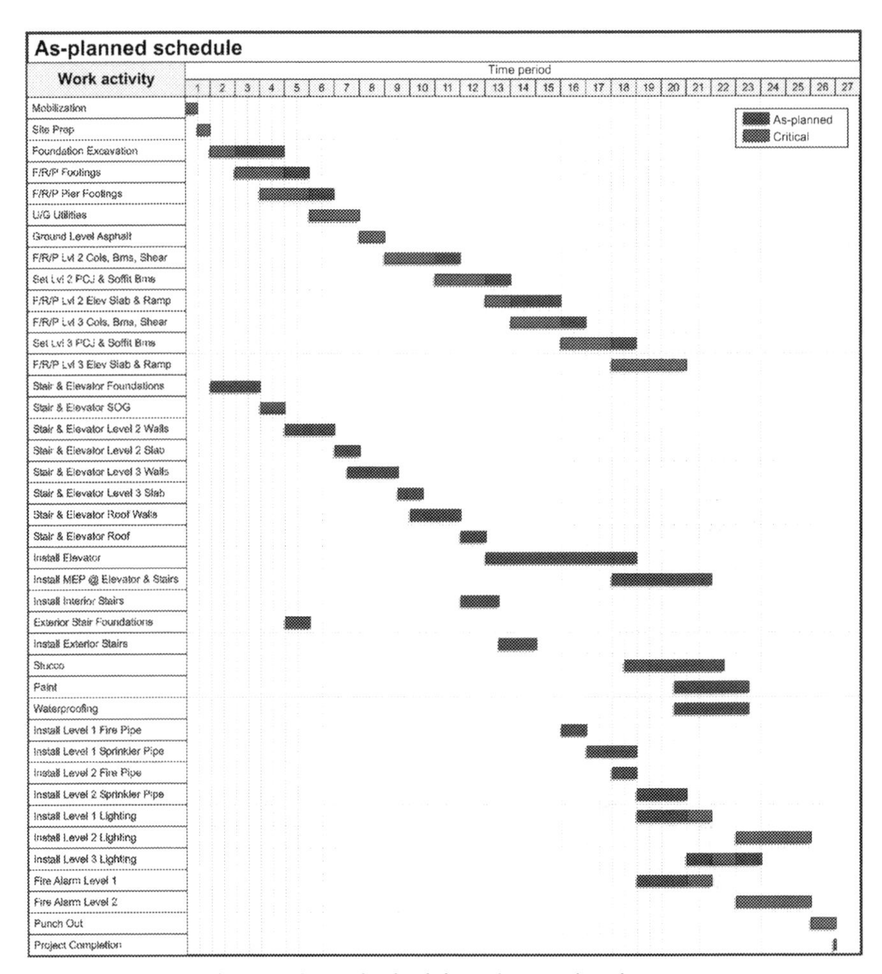

Figure 6.14 Example 2 as-planned schedule with critical path.

the as-built diagram. Instead, to measure the potential delays caused by each change, the analyst must apply the changes as they occur. Alternatively, the analyst could be performing the analysis contemporaneously as the project progresses. In this case, the as-built diagram would be prepared up to the date of the change and any delays determined at that time.

Fig. 6.16 shows the project through time period 4.5. The actual progress is plotted in green (light gray in print versions), the change in yellow (white in print versions), and the remaining work in red (dark gray in print versions) and blue (black in print versions) bars, similar to the as-planned schedule.

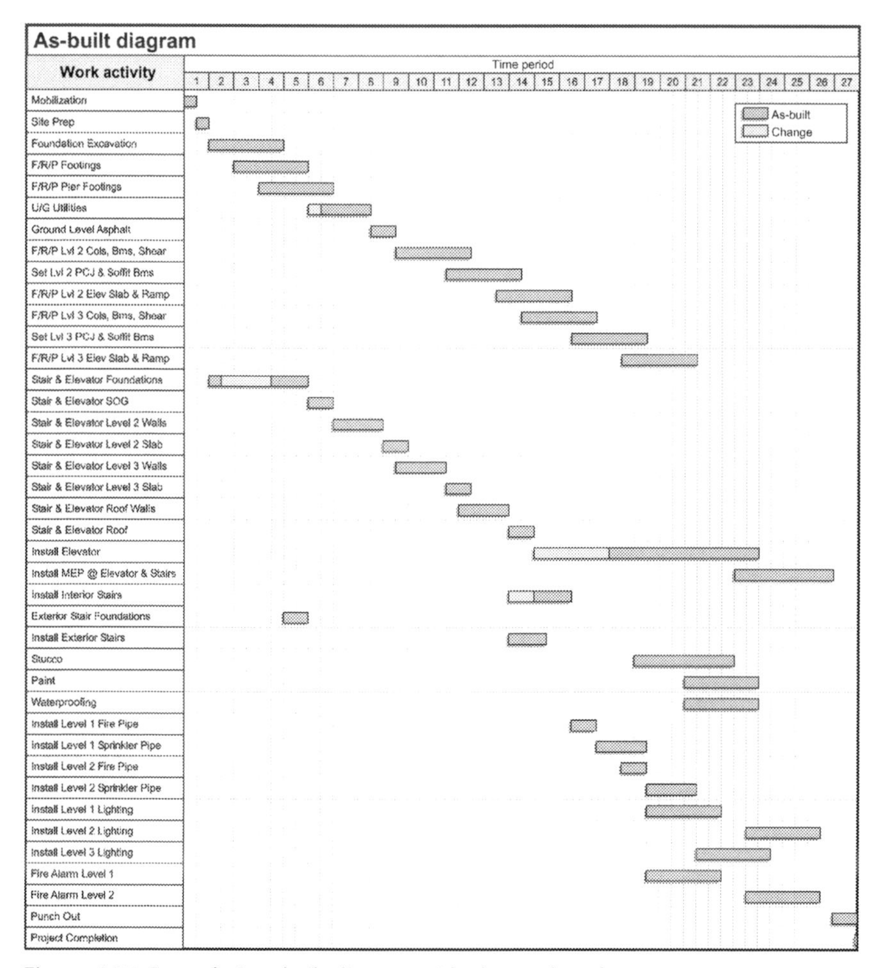

Figure 6.15 Example 2 as-built diagram with changed work.

As can be seen from a comparison of the as-planned schedule, Fig. 6.14, and the updated schedule with the changed work, Fig. 6.16, the project is still scheduled to finish at the end of time period 26. Therefore, no delay was caused by Change #1.

The next change to the project occurs during time period 6. The analyst has updated the bar chart through time period 6.5, as shown in Fig. 6.17, with the actual progress in green (light gray in print versions), the changed work in yellow (white in print versions), and the future work in red (dark gray in print versions) and blue (black in print versions) bars. As can be seen from a comparison of the first update of

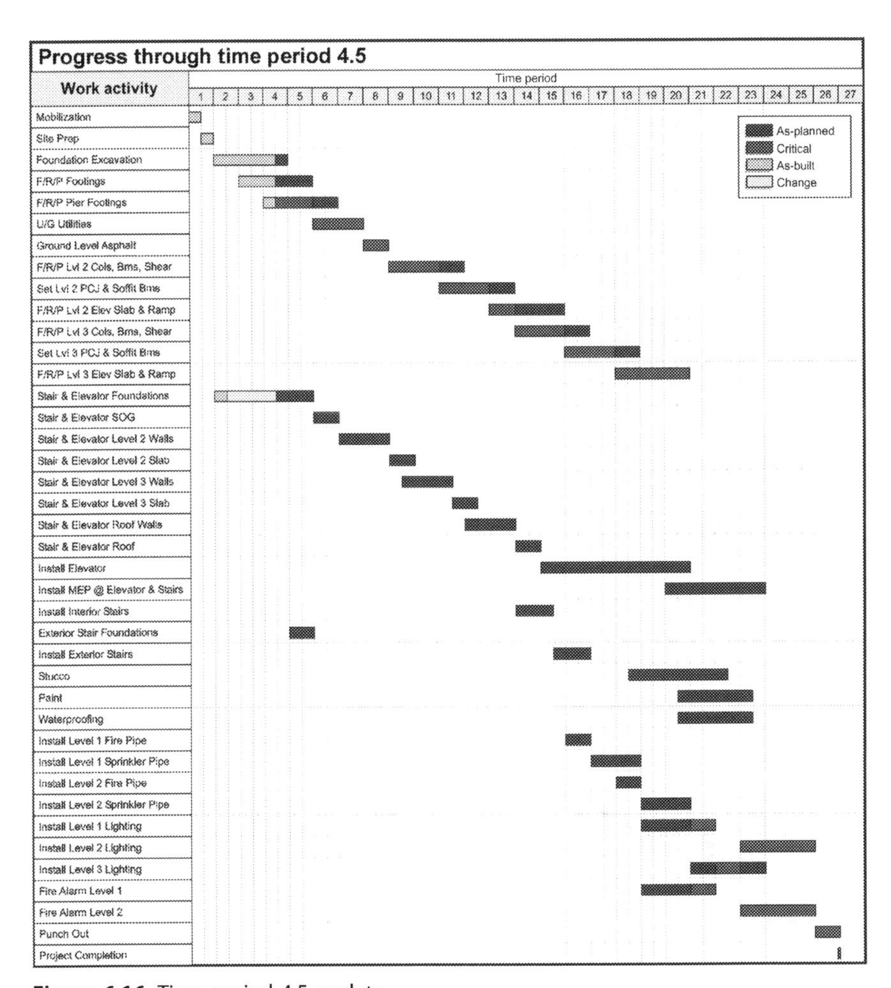

Figure 6.16 Time period 4.5 update.

the schedule, Fig. 6.16, and the present update of the schedule, Fig. 6.17, the project has been delayed one-half time period and will now finish in the middle of time period 27. It is also noted that Change #2 affected the critical work of the underground utilities. As a result of these observations, the analyst can conclude that a delay has occurred because of Change #2 and that the delay is one-half time period in duration.

The next change to the project, Change #3, occurred during time period 14. The analyst has updated the bar chart through time period 14 to include the changed work. This is shown in Fig. 6.18, with the actual

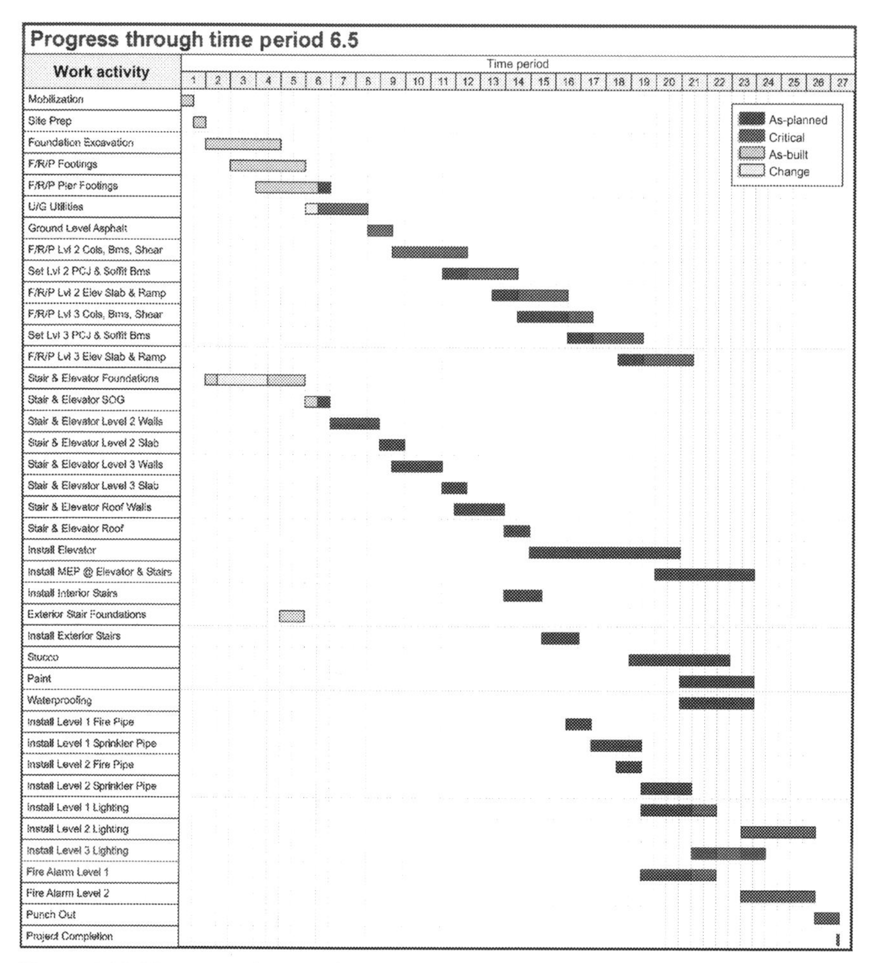

Figure 6.17 Time period 6.5 update.

progress in green (light gray in print versions), the changed work in yellow (white in print versions), and the future work in red (dark gray in print versions) and blue (black in print versions) bars. As can be seen from a comparison of the preceding update and this update, the end date of the project has not changed; therefore, Change #3 did not affect the critical path, and no project delay resulted.

The next change, Change #4, occurred during time periods 15 through 17. The analyst has updated the bar chart schedule through time period 17 to include the changed work. This is shown in Fig. 6.19, with the actual progress in green (light gray in print

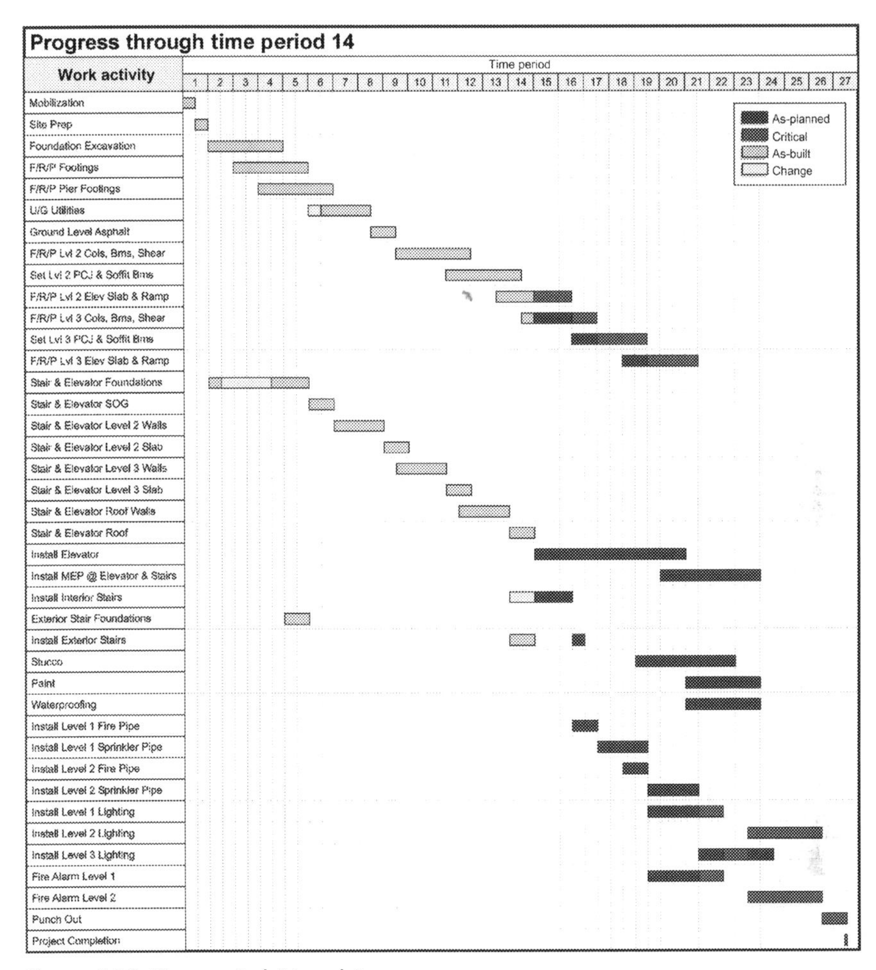

Figure 6.18 Time period 14 update.

versions), the changed work in yellow (white in print versions), and the future work in red (dark gray in print versions) and blue (black in print versions) bars. Comparing this update with the preceding update shows that the project end date has moved to a later date. The project will now complete at the end of time period 27 and has been delayed an additional one-half time period. Note that the critical path has changed. Because of the duration of the changed work, a path that previously had float is now the critical path and caused the delay that was measured.

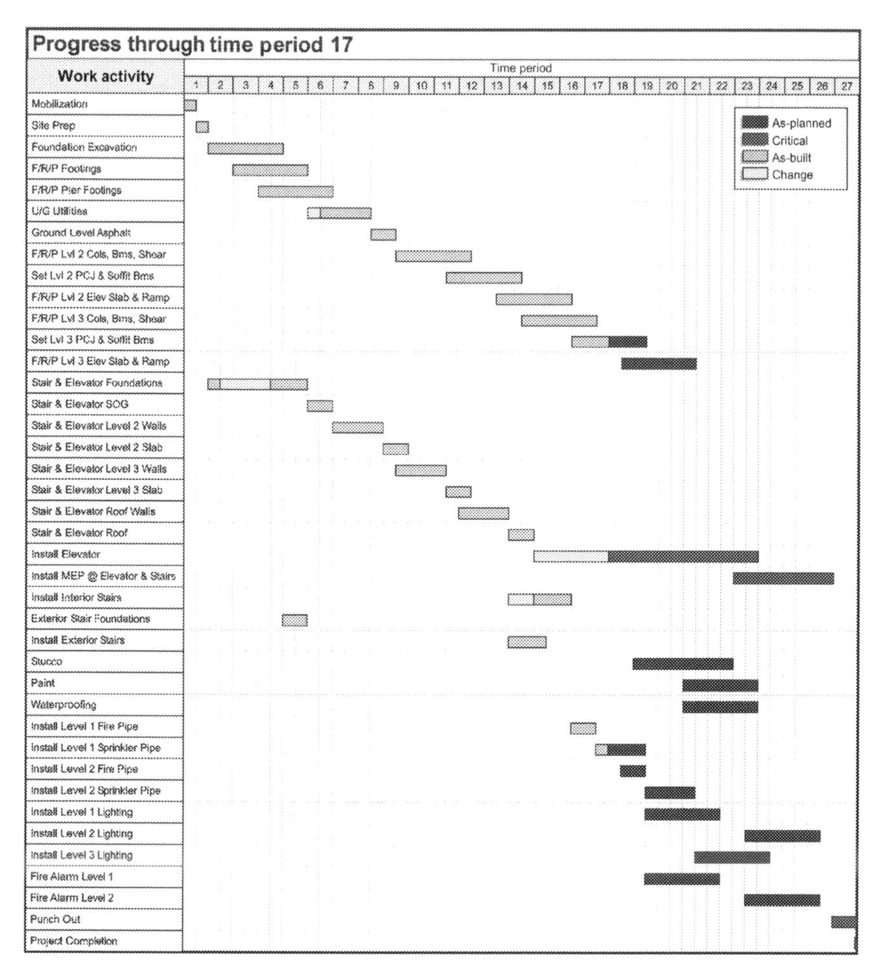

Figure 6.19 Time period 17 update.

This simple example demonstrates how a bar chart can be updated contemporaneously to determine the effect of changes while the project progresses or after the project work is complete.

CHAPTER SEVEN

Delay Analysis Using Critical Path Method Schedules

USING CRITICAL PATH METHOD SCHEDULES TO MEASURE DELAYS

In this chapter, we explore the proper way to perform a delay analysis using a Critical Path Method (CPM) schedule. While a detailed explanation of every nuance of delay analysis using a CPM schedule is beyond the scope of this book, this chapter covers the basic principles in sufficient detail to allow most analysts to measure most delays on most projects.

The theory behind CPM scheduling is that the logic network of activities is designed to model the way the project will be constructed. If the network thoroughly models the project's plan, the predictions calculated from the schedule will be reliable. Therefore, the better the model is, the better the predictions of the schedule updates will be.

CPM software developers have worked to improve the modeling capability of CPM scheduling software. Chapter 2, Float and the Critical Path, identified some of their innovations, including the ability to assign activities to different work calendars, the ability to constrain the performance of work activities, and the ability to link activities with more than one type of logic relationship. While improving the ability of the CPM schedule to model the project work plan, each of these tools have complicated the task of the schedule analyst measuring delay.

Another aspect of CPM scheduling that complicates the analyst's job is the fact that the CPM schedule is a dynamic planning tool that evolves throughout the duration of the project in response to changing project conditions, changes to the project's scope of work, and the contractor's performance, among many other variables. The critical path is equally dynamic. This means that the delay analyst often cannot rely on a single schedule to evaluate all project delays. Rather, the analyst must track the critical path as the project makes progress, using the schedule updates to

Construction Delays.
DOI: http://dx.doi.org/10.1016/B978-0-12-811244-1.00007-0

identify both the actual progress of the work and any changes made to the plan to complete the remaining work.

Schedule logic revisions are made for many reasons. For example, they may be necessary to reflect a change in the contractor's plan. The contractor may change its plan to take advantage of an alternative that avails itself or to mitigate previous delays. Other revisions to the logic may be necessary even though the plan has not changed. This may be because, as the actual progress is entered and the schedule is recalculated, certain logic may need to be refined to improve the model. For example, the previous update may have forecast a start date for work in an environmentally sensitive area within the time period allowed in the permit. However, in the current update, the schedule is now forecasting the work to start during a restricted period, highlighting the need to constrain the start to occur after the restricted period has passed or to assign the activity to a different calendar. This does not necessarily mean that the original schedule was flawed. Rather, such revisions may reflect refinements made to the plan as the project work progresses.

Because the critical path of the project is dynamic, it is possible for it to change from day to day. While such frequency would be unusual, critical path shifts between updates are quite common. This results from the fact that, as the project progresses, the lengths of the work paths relative to one another change. For example, as the steel erection work on the longest path makes progress, the remaining duration of that path becomes less. Conversely, as the masonry work on a shorter work path fails to make progress, the remaining duration of that path remains the same. If this condition continues, the day will come when the remaining duration of the masonry work path will equal the remaining duration of the steel erection path. On that day, both steel erection and masonry are concurrently critical. On the following day, the lack of progress on the masonry work causes the critical path to shift solely to the masonry work and its continued slow progress will begin to delay the project.

Understanding how to identify shifts in the critical path is essential to properly allocating critical project delays. In the preceding example, the lack of progress on the masonry work does not delay the project until its path of work becomes the longest path or critical path. A more detailed discussion of why critical path shifts occur is presented later in this chapter. The analysis techniques employed by the analyst should be such that the critical path of the project is known for every day of the project.

Use of scheduling software and other software tools to quantify delays

Advances in computer technology and software have improved the capabilities of construction scheduling software over the years. Today's scheduling software runs faster, is more powerful, and contains numerous features that allow the project manager and scheduler to organize their specific plan for resource allocation, cost forecasts, and work sequences to complete the project.

With the multiple needs of project managers and the variance in capabilities and cost, software companies have diversified their products to provide viable and cost-effective software for each type of project. Some of the more popular construction software applications on the market today are produced by Oracle and Microsoft. Software from other companies is also available, but most projects these days use software products made by one of these two companies.

Regardless of the power of the software, the capabilities of the user are key to using scheduling software as an effective management tool. As a result, no matter what software is chosen, the project manager must be aware of the different scheduling capabilities and options of the software they are using. This is because selecting or unselecting certain software options can mean a world of difference in how the software mathematically forecasts the plan to complete the project.

As with creating and updating a schedule, the analyst must be familiar with scheduling terminology and be able to accurately interpret the data and results predicted by the schedule. It is also important for the analyst to be familiar with the specific software used to create and update the schedules, given the different scheduling options available in each software package. However, no matter what software was used to manage the project schedule, the basic principles of analyzing a project for delays remains the same.

Once analysts have familiarized themselves with the software used to create, update, and manage the project schedule, the analyst should gather all of the contractor's schedules throughout the duration of the project—the as-planned or baseline schedule and all subsequent schedule updates. If possible, the analyst should obtain the electronic computer files in native format for each of the schedules used on the project. These electronic files allow the analyst to access all of the activity and project data contained within the schedules, whereas "hard copies" or paper copies only allow the analyst to view the information that is available on the printout. Hard-copy printouts can be easily manipulated to show only the information the hard-copy provider wants the analyst to see, and they often lack information, such as logic ties, relationship lags, scheduling option selections, which is vital to analyzing the schedule for delay.

For the remainder of this chapter, the discussion assumes that the analyst has obtained the native electronic files of all of the project schedules used on the project. Because Oracle's Primavera P6 Project Management software is the most widely used scheduling software in the construction industry, terminology from Oracle's Primavera P6 Project Management scheduling software is used in this chapter and throughout this book.

Identifying and quantifying critical delays using the Critical Path Method schedule

The project's CPM schedule is the best tool to use to identify and measure critical project delays. This is because the project CPM schedule:

- Shows the contractor's plan to complete the project
- Captures the alterations to the contractor's plan
- Forecasts when the project will finish

Most construction contracts recognize this fact and require the contractor to perform a schedule analysis using the project schedule to measure the project delay when requesting a time extension.

Though this book is not a legal treatise and the analysis of delays is not governed chiefly by the law, judges sometime use both colorful and wise words when describing basic concepts. For example, a Veterans Administration Board of Contract Appeals Judge once explained the importance of using the actual project schedules to identify and quantify project delays this way:

> *... in the absence of compelling evidence of actual errors in the CPMs we will let the parties 'live or die' by the CPM applicable to the relevant time frames.*
>
> *VABCA Nos. 1943, 1944, 1945, and 1946, 84–2 BCA. ¶ 17,341 at 86,411*

In other words, unless the schedules were obviously and seriously flawed, delays would be measured using the schedules developed and used by the parties to manage the project.

Measuring delays based on perspective

Delays can be categorized many ways. One important categorization relates to when the delay occurs. Some delays can be predicted. Identifying these delays before they occur is an important skill for the project manager. An example of a delay that can be predicted is the delay that might occur if the owner decides to change the windows from one manufacturer to another after the submittal for the windows has already been approved. This change will likely require the contractor to identify a new window supplier. It may also necessitate the resubmission of the window submittal. It may also mean that the windows will arrive on site later than the contractor had originally planned. It is helpful to both the contractor and the owner to be able to predict the delay that might result if the owner chooses to change the window manufacturer. Because such delays have not yet occurred, they are known as prospective delays. Such delays require a forward-looking or prospective analysis.

Some delays are predictable, but others are not. Examples of delays that are harder to quantify in advance include delays in obtaining owner-furnished permits, unanticipated inclement weather, or a subcontractor's failure to mobilize. Because the duration of these are difficult to predict or quantify in advance, they are typically identified and measured after they occur, or retrospectively.

In Chapter 5, Measuring Delays—The Basics, we introduced two basic methods for identifying and quantifying delay—prospective and retrospective methods. As a reminder, a prospective delay analysis is performed before the changed work is performed or before the delay has occurred, whereas a retrospective delay analysis is performed after the changed work is completed or after the delay has occurred.

Prospective measurement of delays

As described in Chapter 5, a "prospective" schedule delay analysis estimates or forecasts the project delay resulting from added or changed work before that work is performed. A prospective delay analysis is the time equivalent of the contractor's cost estimate prepared before the work is added or changed. This estimate becomes the basis for the owner's and contractor's negotiation and agreement with regard to the cost of work before the work is actually performed. Project management best practices recommend that the contractor and owner agree on the cost of added or changed work before the contractor begins the work. It is also a best practice to agree on the time needed to complete the added or changed work before it is performed.

There is sometimes resistance to the idea that delays should be evaluated before they occur and that time extensions be granted based on schedule forecasts. Some believe that the owner should wait to the end of the project before granting a time extension because the "actual" project delay is not known until the project is completed. Such an approach is not a project management best practice. Just as it is often better to get agreement as to price before executing a change, it is best to get agreement on time, as well. The reasoning is the same. Coming to agreement before the work is performed and memorializing this agreement in a change order, modification, or supplemental agreement that both parties sign is the best way to ensure that the issue is resolved. Such bilateral agreements also prevent disputes and claims.

The difference between agreeing to a price in advance and agreeing to a time extension in advance is that the owner's representative will probably never know if they overpaid or underpaid for a change. With regard to time, however, if the owner's representative is willing to do a little analysis, they will know how much delay a particular change actually caused. This means that the owner's representative will know if they gave too much or too little time for the change. But do not let this fact persuade you that you should not address both cost and time in every change order. Just as with the cost of a change, agreeing on the time of a change provides both parties with certainty. In that regard, it is good practice and will prevent problems later on.

As also discussed in Chapter 5, Measuring Delays—The Basics, there is nearly universal agreement that the Prospective Time Impact Analysis (TIA) is the best method to forecast the delay resulting from added or changed work before the added or changed work is performed. It is important to note that the term "Time Impact Analysis" is a term of art in the realm of schedule delay analysis. As described in Chapter 5, Measuring Delays—The Basics, a Prospective TIA describes a specific type of analysis that consists of modeling the added or changed work using a "fragnet," which is the term used for a "fragmentary network." A fragmentary network is a model of the changed or added work represented by an activity or collection of activities linked to one another and designed to be inserted into the project schedule. By inserting the fragnet into the version of the project schedule that is in effect at the time the owner and the contractor are contemplating adding the work to the contract and calculating it, the modified schedule will predict the effect of the change.

A significant consideration when using the Prospective TIA is that it must be performed before the added or changed work is started. This is because it estimates the effect caused by the added or changed work based on planned logic and estimated durations for all remaining work activities. The comparison of the planned logic and the estimated durations of the fragnet activities to the planned logic and estimated durations of the original, uncompleted activities ensures an apples-to-apples comparison.

This is in contrast to a Retrospective TIA, which is performed in a similar manner to a Prospective TIA in which it consists of the insertion of a fragnet representing the changed or added work into the version of the project schedule in effect when the changed or added work occurred. The difference between the Prospective and Retrospective TIAs is that a Retrospective TIA compares the actual logic and durations for the fragnet activities to the planned logic and durations of the other schedule activities that had not yet been completed as of the data date of the schedule into which the fragnet is inserted. In other words, the model considers the actual progress of the fragnet in a schedule environment that pretends that all the other work proceeded as planned. The result is that the comparison upon which the Retrospective TIA is based is actual logic to planned logic and actual durations to planned durations rather than planned-to-planned. The comparison is no longer apples-to-apples. Significantly, even though the work on the "unchanged" paths of work is also complete, the actual progress of that work is not considered in the Retrospective TIA. That can be a significant problem and is one of several reasons why Retrospective TIAs should be used with great caution or not used at all. The analyst can severely overestimate the effect of a change and severely underestimate the effect of other project delays because the Retrospective TIA ignores these other delays. Note, also, that the Retrospective TIA cannot be used to evaluate concurrent delays.

If the parties can agree on the time extension to be granted for a change before the changed work is performed, then the contractor has an incentive to complete the added or changed work as quickly and efficiently as possible. This has a benefit to both the contractor and the owner. In contrast, if the contractor's time extension is to be evaluated after the added or changed work is completed by inserting a fragnet representing the actual logic and duration of the changed work into the project schedule, the contractor has less incentive to complete the added or changed work as quickly as possible. This is identical to the problem created when the owner and contractor cannot agree on the cost of the added or changed work before the contractor performs it. In such circumstances, the contractor may be paid for the change on a time and materials, cost-plus, or force account basis. Owners are often reluctant to proceed on this basis because they are concerned that the contractor may not have an incentive to control the cost of the added or changed work. Reaching agreement before the work is performed avoids these concerns.

Prospective time impact analysis

As stated above, the Prospective TIA should be used to estimate the project delay that would result from added or changed work. The following procedure will guide the proper performance of a Prospective TIA:

1. The first step is to identify the effective date that the added or changed work will be made part of the contract. This date is usually identified as the date that the owner will execute the change order, the date the owner gives the contractor a directive to perform the work, or the date that the contractor begins to perform the added or changed work.
2. The second step is to identify the project schedule in effect at the time the changed work is being contemplated, goes into effect, or begins to be performed. For example, using the earlier example of an owner who has decided to use a different window manufacturer than the manufacturer submitted and previously approved by the owner, if the owner decides that it wants to investigate the consequences of using a new manufacturer on March 15, then the contractor should select the project schedule in effect on March 15. Typically, this schedule will be the most recently issued schedule update, optimally an update with a data date in early to mid-March.
3. The reason for using the schedule in effect when the change is being contemplated is to ensure that the change is evaluated against the current plan for completion of the project.
4. The third step consists of the contractor's development and the owner's review and approval of the fragnet. As noted earlier in this chapter, the fragnet consists of an activity or multiple activities that represent the added or changed work. Note that it is not uncommon for the activities, logic, and durations of the fragnet to be negotiated, just as the parties negotiate the price of the change.
5. The fourth step consists of making a copy of the selected schedule and inserting the agreed-upon fragnet into the copied schedule. The insertion of the fragnet also includes agreement between the parties with regard to how the fragnet activities are linked or connected to the existing activities in the schedule.

6. Finally, the schedule with the fragnet inserted is rerun or recalculated and its forecast completion date is compared to the forecast completion date of the original version of the same schedule without the fragnet. If the schedule with the fragnet inserted has a forecast completion date later than that of the schedule without the fragnet added, then the difference is the project delay resulting from the added or changed work.

One problem that analysts often face when performing a Prospective TIA is the significant amount of time that can exist between the data date of the schedule in effect on the project and the date the work addition or change occurs. When too much time has passed between the data date of the schedule and the changed work, we recommend that the project schedule be updated or statused to the day that the changed work is affected. This updated or statused schedule should then be the unimpacted schedule that will be used as the basis of the analysis. The need for such updating is a judgment call based on the type of work being performed, the logic of the schedule, and the effect of the progress that has occurred during the period. The fragnet representing the added or changed work should be inserted into the updated schedule to measure the critical project delay resulting from the change order.

As an example of a Prospective TIA, if a contractor was directed to install an additional wall in an office, the fragnet might include the following activities:

- Install metal studs
- Install electrical rough-ins
- Install and finish drywall
- Paint

These work activities would then be logically tied to each other in series, and then tied logically into the existing CPM. For example, the installation of metal studs might be tied to the existing metal stud installation activity for the building. In addition to identifying added work, fragnets are also prepared to measure the effect of distinct features of work within a complex project, such as added requirements for the construction of a clean room at a new pharmaceutical development and manufacturing installation.

Many private and public owners require contractors to use fragnets to express, in a CPM format, the activities associated with change orders and to use these fragnets as a basis for requesting time extensions. The US Army Corps of Engineers, the US Department of Veterans Affairs, and Florida Department of Transportation are just a few of the public owners that require contractors, when appropriate, to use fragnets as part of their requests for time extensions. Typically, the effect of changes on the project schedule is measured by developing a fragnet for the change and inserting this fragnet into the schedule. The measure of the delay caused by the change is the difference between the scheduled project completion date before the fragnet is inserted into the schedule and the completion date after the fragnet is inserted.

Returning to the previous example of installing a new wall in an office, if the original drywall installation was critical, and the new drywall activity required two workdays and was inserted in series with the existing drywall work, the additional time required is easy to estimate. Prior to inserting the fragnet, the predicted project completion date was September 19, 2015. After the fragnet is inserted into the CPM, the predicted project completion date is September 21, 2015. Thus, the added drywall work caused a critical delay to the project of 2 calendar days. This delay was quantified by inserting a fragnet and measuring the difference between the predicted completion dates before and after the fragnet was inserted.

- Predicted completion date prior to inserting fragnet: September 19, 2015
- Predicted completion date after inserting fragnet: September 21, 2015
- September 21, 2015−September 19, 2015 = 2 calendar days

Using fragnets to measure delays has advantages in that both parties will have agreed to the activities and logic of the fragnet. Typically, the fragnet is required to be submitted as part of a contractor's change order proposal. The fragnet is negotiated along with the estimated costs of the change. Ideally, the parties will have discussed the labor and equipment and the time required to complete the work, and how the fragnet activities are logically tied into the CPM. This negotiation process allows the parties to assure themselves that they fully understand the logic of the fragnet and are in agreement as to the most efficient and effective way to perform the changed work.

While it is advantageous for both parties to understand the fragnet prior to its insertion, there are challenges. The two biggest challenges are the time it takes to develop and negotiate a fragnet and, if necessary, the time it takes to identify and update the project schedule into which the fragnet will be inserted. Many contracts allow the contractor 30 calendar days to provide its change order proposal. Once the proposal has been received and reviewed, it may take an owner several days or even weeks before it is prepared to negotiate the costs and the time. The negotiations themselves may take several weeks depending on the amount in question and the support provided by the contractor. Therefore, it could be weeks or months after a change is contemplated before the parties can agree upon the fragnet and the associated time extension. Sometimes this means that the owner will have to direct the contractor to proceed before an appropriate time extension can be analyzed and agreed to.

Another issue that may arise relates to identifying the project schedule into which the fragnet should be inserted. The fragnet should be inserted into the CPM update that is in effect at the time the change is being contemplated, or at the time both parties understood that there was a change, or the date the contractor began its work related to the change, whichever was the earliest. In the case where the owner issues a directive to change the work, the schedule in affect when the directive is issued usually becomes the schedule into which the fragnet is inserted. However, lacking a directive, it is not necessarily clear which CPM update should be used. Many times the CPM update to be used must be negotiated along with the fragnet itself.

The fragnet approach is advantageous in that the analysis is focused only on the portion of the work that was changed. One major disadvantage to the fragnet approach is the time it takes to reach agreement on the logic of the fragnet and how it is to be inserted into the overall schedule. Both parties will benefit when they work together to keep this time to a minimum.

Prospective time impact analysis example

D-Tunneling Company (D-Tunneling), a tunneling Subcontractor, was performing the tunneling and drainage piping installation for a large terminal expansion at an airport in New Mexico. In its contract with the airport authority, D-Tunneling received its notice-to-proceed on February 1, 2017, and was required to complete the project by June 5, 2017. Before starting construction, D-Tunneling created a baseline schedule reflecting its original plan for the work. After construction began, D-Tunneling updated its schedule on a monthly basis.

Baseline schedule

D-Tunneling's baseline schedule, shown in Fig. 7.1, identifies a notice-to-proceed on February 1, 2017, and a forecast project completion of June 5, 2017, the same date as the contract completion date.

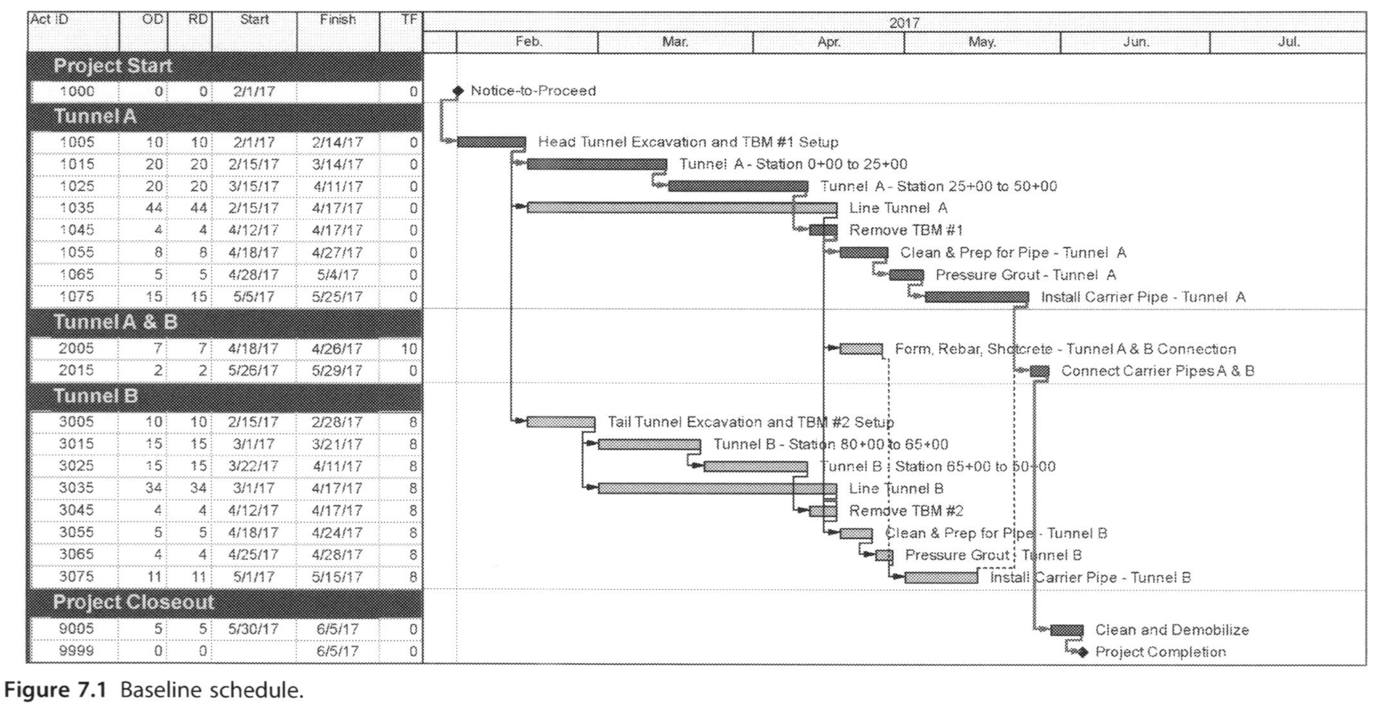

Act ID	OD	RD	Start	Finish	TF
Project Start					
1000	0	0	2/1/17		0
Tunnel A					
1005	10	10	2/1/17	2/14/17	0
1015	20	20	2/15/17	3/14/17	0
1025	20	20	3/15/17	4/11/17	0
1035	44	44	2/15/17	4/17/17	0
1045	4	4	4/12/17	4/17/17	0
1055	8	8	4/18/17	4/27/17	0
1065	5	5	4/28/17	5/4/17	0
1075	15	15	5/5/17	5/25/17	0
Tunnel A & B					
2005	7	7	4/18/17	4/26/17	10
2015	2	2	5/26/17	5/29/17	0
Tunnel B					
3005	10	10	2/15/17	2/28/17	8
3015	15	15	3/1/17	3/21/17	8
3025	15	15	3/22/17	4/11/17	8
3035	34	34	3/1/17	4/17/17	8
3045	4	4	4/12/17	4/17/17	8
3055	5	5	4/18/17	4/24/17	8
3065	4	4	4/25/17	4/28/17	8
3075	11	11	5/1/17	5/15/17	8
Project Closeout					
9005	5	5	5/30/17	6/5/17	0
9999	0	0		6/5/17	0

Figure 7.1 Baseline schedule.

D-Tunneling's baseline schedule identifies that it is planning to perform tunneling work on two separate work paths, Tunnels A and B, at the same time. The plans show that Tunnels A and B will combine to form one, continuous, straight, drainage tunnel when finished. D-Tunneling has decided to use two tunnel boring machines (TBMs) to complete its work. The TBMs would be set up at opposite ends of the drainage tunnel and work toward each other, meeting at Station 50 + 00. TBM #1 will be boring Tunnel A from Station 0 + 00 to 50 + 00 and TBM #2 will be boring Tunnel B from Station 80 + 00 to 50 + 00.

D-Tunneling's schedule update on the morning of March 6, 2017, identified that between February 1, 2017 and March 6, 2017, the project had progressed as expected. D-Tunneling's March 6, 2017 Update did not contain any schedule revisions. Fig. 7.2 depicts the status of D-Tunneling's work when it submitted the schedule update to the airport authority on the morning of March 6, 2017.

On the afternoon of March 6, TBM #2 encountered rock at Station 75 + 00 that was harder than the geotechnical report indicated in the contract documents. In addition, the rock was considered "mixed face" and contained a mixture of rocks with varying compositions and hardnesses. As a result, D-Tunneling stopped work on Activity 3015 and was not able to progress work in Tunnel B until it resolved the issue.

D-Tunneling's contract with the airport authority stated that time extensions must be substantiated by a fragnet analysis, comparing the project completion date before the fragnet was inserted into the schedule to the project completion date after the fragnet was inserted into the schedule.

On March 14, D-Tunneling sent a letter to the airport authority identifying its new plan to complete the changed work, along with a fragnet schedule to substantiate D-Tunneling's requested time extension for PCO #1, TBM #2 Shutdown. D-Tunneling's March 14th letter stated:

1. The airport authority/engineer was responsible for all soil borings and site testing in the Contract.
2. D-Tunneling had encountered "mixed face" rock at Station 75 + 00 that was harder than the tolerance limits identified by the engineers' soil boring tests in the contract documents. As a result, D-Tunneling had to shut down its TBM #2 operations in Tunnel B. D-Tunneling has and will continue work in Tunnel A, as expected.
3. The engineer has completed additional testing of the area and determined that the out-of-tolerance rock exists between Station 75 + 00 and Station 69 + 00.

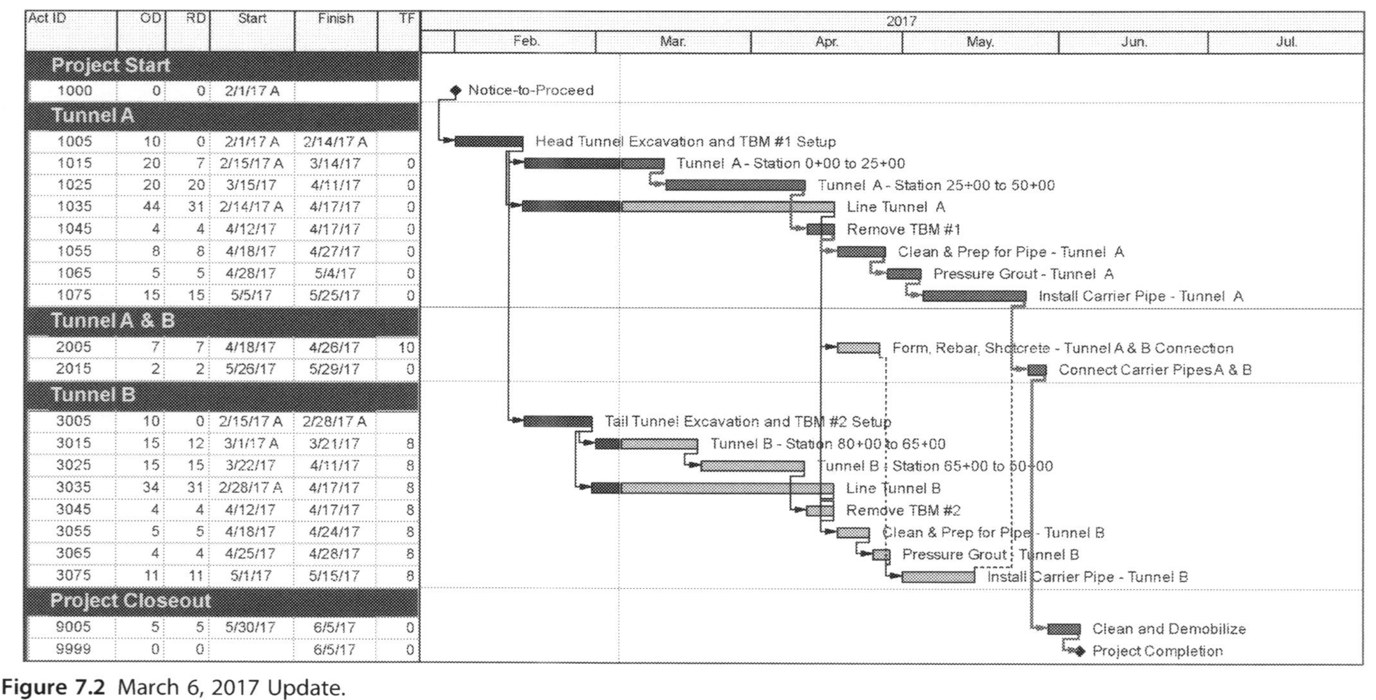

Figure 7.2 March 6, 2017 Update.

4. D-Tunneling has purchased a new cutterhead that is expected to be delivered on April 3, 2017. The cutterhead will take 4 workdays to assemble. D-Tunneling will then return to Station 75 + 00 to resume tunneling. In addition, it will take 13 workdays to complete tunneling from Station 75 + 00 to 65 + 00, due to the hardness of the rock.
5. Using the above timeline, D-Tunneling has inserted a fragnet into its March 6, 2017 Update, containing the following changes:
   a. Activity 3015 received an actual finish of March 5, 2017, and a changed work scope to only cover the tunneling work that had already been completed between Station 80 + 00 and Station 75 + 00.
   b. Activity 3015A, Order/Ship/Rec. New Cutterhead for TBM #2, was added to the schedule to denote the time it will take to receive the new cutterhead. Activity 3015A was given a 21-workday duration to take its finish date through April 3, 2017, the date that D-Tunneling expects to receive the new cutterhead.
   c. Activity 3015A1, Assemb. Cutterhead and Return to Station 75 + 00, was added to the schedule to represent the time needed to assemble the new cutterhead. Also, Activity 3015A1 was given a duration of 4 workdays.
   d. Activity 3015B, Complete Tunnel B—Station 75 + 00 to 65 + 00, was added to the schedule to denote the remaining tunneling work after the new cutterhead is assembled. Also, Activity 3015B was given a duration of 13 workdays.
   e. Logic was revised to reflect finish-to-start relationships of the new activities in the following sequence: Activity 3015 followed by 3015A followed by 3015A1 followed by 3015B followed by 3025.

Fig. 7.3 depicts D-Tunneling's new March 6, 2017 Update, including the addition of the new fragnet.

As a result of the insertion of the fragnet into the March 6, 2017 schedule, which is depicted in Fig. 7.3, the critical path of the project shifted to Tunnel B, and extended the project completion date from June 5, 2017 to June 29, 2017. D-Tunneling requested a time extension of 24 days (June 5, 2017 to June 29, 2017 = 24 calendar days) for the differing site condition.

The airport authority requested that D-Tunneling reorganize its work, if possible, to mitigate some of the delays to the new critical path of the Project. D-Tunneling revised its schedule to perform a portion of Activity 3025, Tunnel B—Station 60 + 00 to 50 + 00 (Fig. 7.4) with TBM #1 instead of TBM #2. This mitigated schedule is depicted in Fig. 7.4.

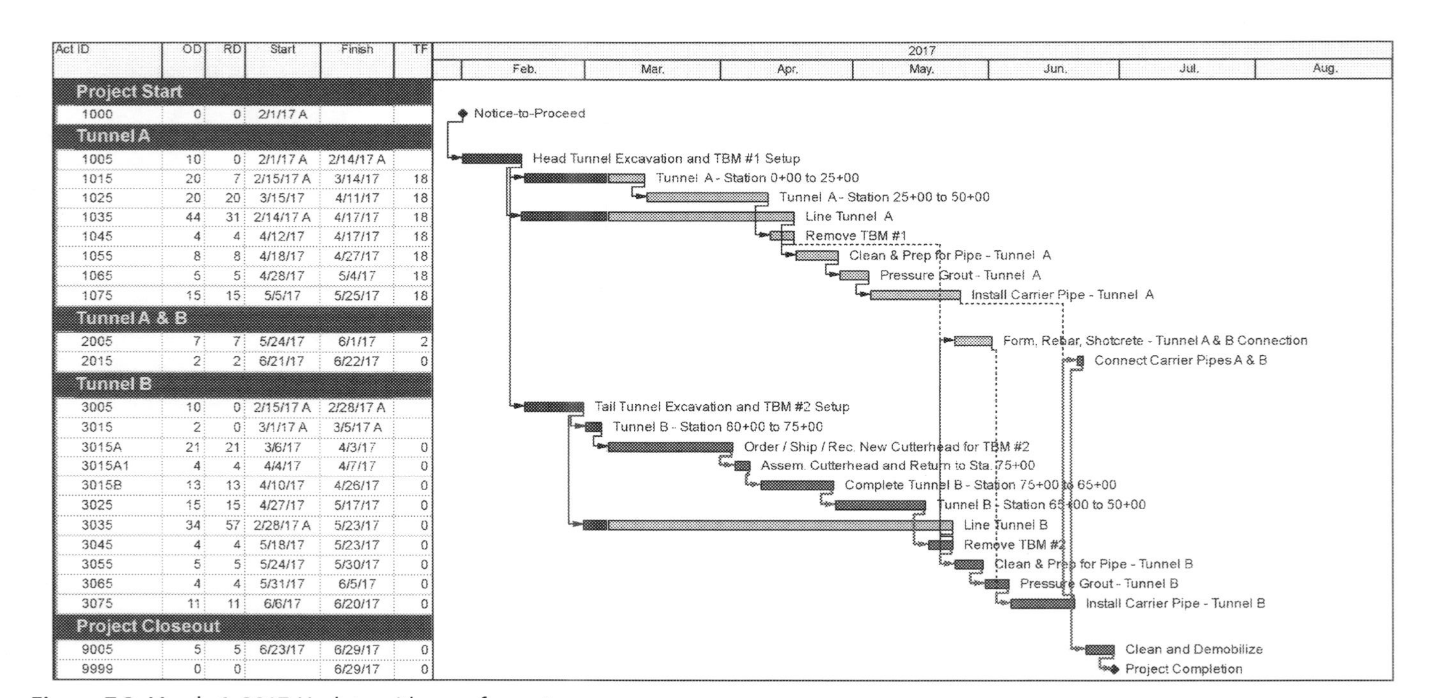

Figure 7.3 March 6, 2017 Update with new fragnet.

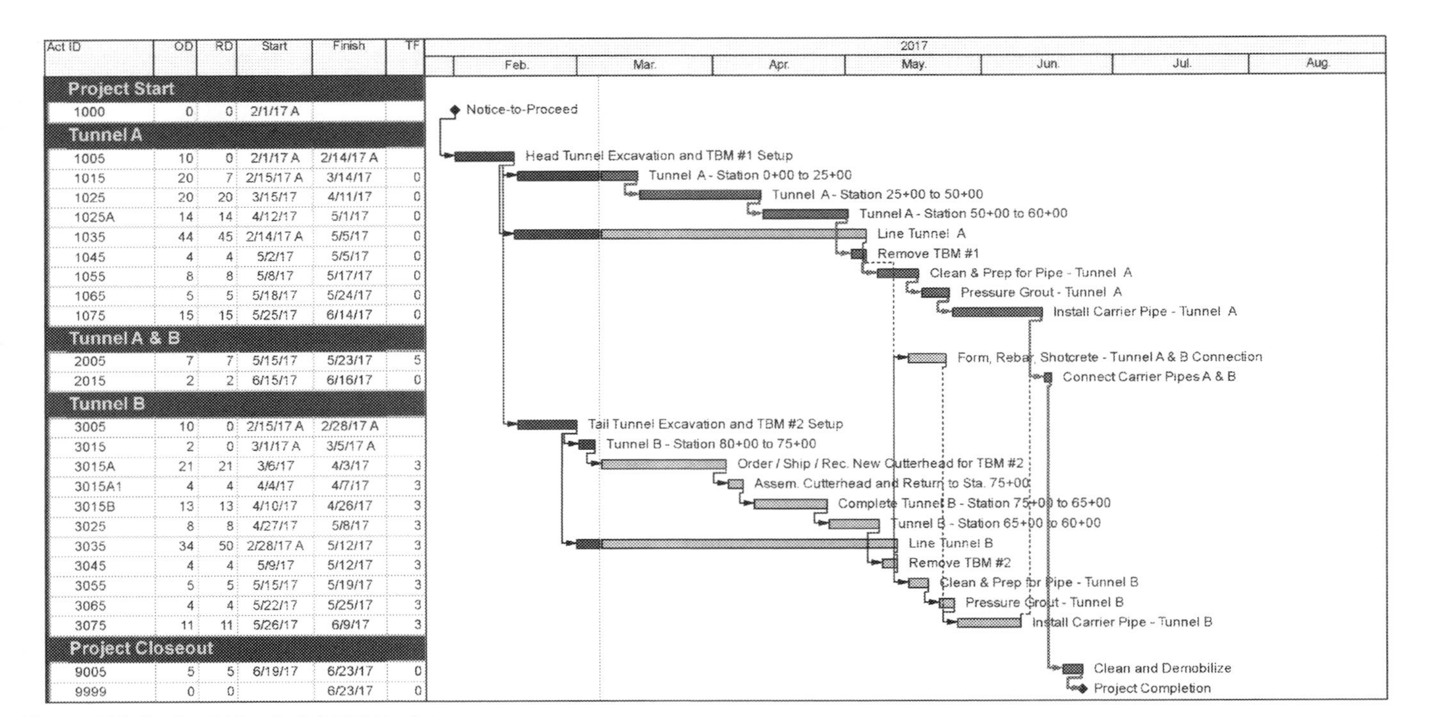

Act ID	OD	RD	Start	Finish	TF
Project Start					
1000	0	0	2/1/17 A		
Tunnel A					
1005	10	0	2/1/17 A	2/14/17 A	
1015	20	7	2/15/17 A	3/14/17	0
1025	20	20	3/15/17	4/11/17	0
1025A	14	14	4/12/17	5/1/17	0
1035	44	45	2/14/17 A	5/5/17	0
1045	4	4	5/2/17	5/5/17	0
1055	8	8	5/8/17	5/17/17	0
1065	5	5	5/18/17	5/24/17	0
1075	15	15	5/25/17	6/14/17	0
Tunnel A & B					
2005	7	7	5/15/17	5/23/17	5
2015	2	2	6/15/17	6/16/17	0
Tunnel B					
3005	10	0	2/15/17 A	2/28/17 A	
3015	2	0	3/1/17 A	3/5/17 A	
3015A	21	21	3/6/17	4/3/17	3
3015A1	4	4	4/4/17	4/7/17	3
3015B	13	13	4/10/17	4/26/17	3
3025	8	8	4/27/17	5/8/17	3
3035	34	50	2/28/17 A	5/12/17	3
3045	4	4	5/9/17	5/12/17	3
3055	5	5	5/15/17	5/19/17	3
3065	4	4	5/22/17	5/25/17	3
3075	11	11	5/26/17	6/9/17	3
Project Closeout					
9005	5	5	6/19/17	6/23/17	0
9999	0	0		6/23/17	0

Figure 7.4 Revised March 6, 2017 Update.

The portion of the tunnel operation from Station 50 + 00 to 60 + 00 was moved from Tunnel B to Tunnel A. To depict this change, Activity 1025A, Tunnel A—Station 50 + 00 to 60 + 00, was added to the Tunnel A portion of the schedule between Activity 1025 and Activity 1035, and the description of Activity 3025 was changed from Station 65 + 00 to 50 + 00 to Station 65 + 00 to 60 + 00. As a result of this change, Tunnel A became the critical path of the project. However, Fig. 7.4 shows that reorganizing the tunneling work from Station 60 + 00 to 50 + 00 result in a 6-calendar-day improvement to the project completion date (June 29, 2017–June 23, 2017 = 6-CD improvement).

In addition to the resequencing of the tunneling work from Station 60 + 00 to 50 + 00, which resulted in a 6-calendar-day improvement, the resequencing also created 3 days of total float for Tunnel B. By creating 3 days of total float in Tunnel B, D-Tunneling is ensuring that if the delivery of the new cutterhead is a few days late, it will use up some of its total float, and not negatively affect the forecast project completion date.

As a result of the newly revised schedule submitted in the March 6, 2017 Update, D-Tunneling again resubmitted its time extension request due to the differing site condition in Tunnel B, but this time for 18 calendar days. D-Tunneling also reserved its right to request additional days of delay related to the differing site condition in Tunnel B should the hard digging prove to be more difficult than is currently anticipated. The airport authority granted D-Tunneling its requested 18-calendar day time extension from the fragnet added to the March 6, 2017 Update.

D-Tunneling submitted its next schedule update on the morning of April 1, 2017. All activities made as-expected progress between March 6, 2017 and April 1, 2017, and no schedule revisions were made. Fig. 7.5 shows D-Tunneling's April 1, 2017 Update.

As of the April 1, 2017 Update, the project completion date remained June 23, 2017, which is depicted in Fig. 7.5. Thus, D-Tunneling made expected progress during March resulting in no delay.

D-Tunneling submitted its next schedule update on the morning of May 1, 2017. All activities made as-expected progress between April 1, 2017 and May 1, 2017, with no schedule revisions. Fig. 7.6 shows D-Tunneling's May 1, 2017 Update.

D-Tunneling's May 1, 2017 Update still identified a project completion date of June 23, 2017, and showed that all fragnet activities related to the TBM#2 shutdown were completed.

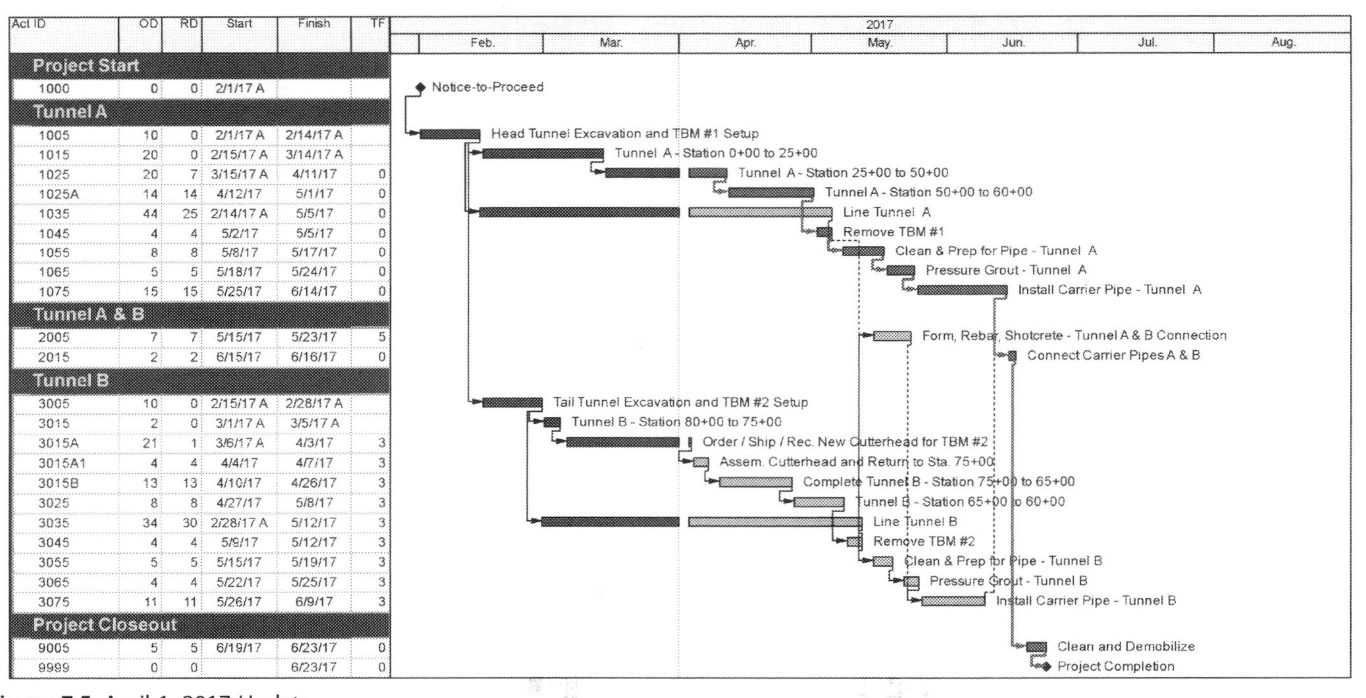

Figure 7.5 April 1, 2017 Update.

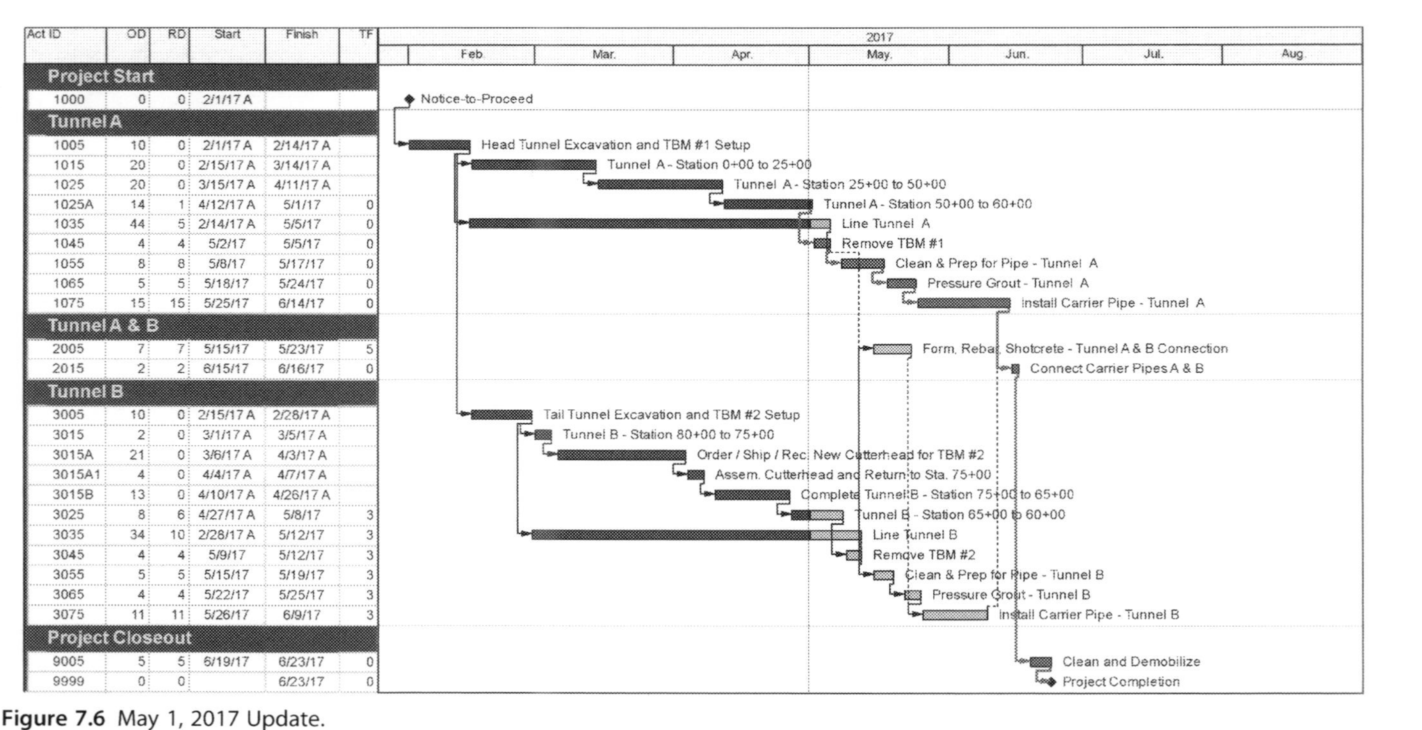

Figure 7.6 May 1, 2017 Update.

This example represents the proper way to address project changes in a proactive, forward-looking manner using fragnets. D-Tunneling encountered a change in the project, adjusted its schedule to reflect the project change, and then timely submitted a complete request for a time extension. In addition, the airport authority also benefited from swift resolution of project changes. The airport authority was presented with accurate project status information and, therefore, was better prepared to adjust its future budget and planning concerns, alert future tenants of project changes, and consider acceleration options to mitigate the effects of the delay. It is usually in the best interests of all parties to resolve project changes in a timely manner, as they occur.

Retrospective measurement of delays

Owners should expect, and contractors should want to maintain an updated CPM schedule at intervals throughout the project. When CPM updates are available, the analyst can readily perform the delay analysis for the entire project. In doing so, the analyst should utilize all of the project schedules that were used to manage the project to identify and quantify project delays.

When identifying and measuring project delays using the schedule updates, the analyst should perform the analysis in two separate and distinct steps: first, determining delays and improvements due to the actual progress of the work, and second, determining delays and improvements due to revisions to the project schedule. Often, these two sources of delays and improvements are not segregated, leading to an inaccurate determination of what is delaying or improving the project's completion date. However, work progress delays and improvements and schedule revision delays and improvements are easy to separate and a proper analysis will do this inherently. The schedule analysis method that independently analyzes these sources of delay is called a Contemporaneous Schedule Analysis, also sometimes referred to as the Contemporaneous Period Analysis. This schedule analysis method is described as Method Implementation Protocol 3.4, Observational/Dynamic/Contemporaneous Split (MIP 3.4) in the AACE International's Recommended Practice No. 29R-03, Forensic Schedule Analysis.

The proper performance of this schedule delay analysis method involves using the contemporaneous project schedule submissions to identify and measure the project delay, starting at the beginning of the project and moving forward in time from the baseline schedule to the first update and then to successive updates, comparing the actual performance of the project work to the contractor's evolving plan.

Work progress delays and improvements

The first step in identifying and measuring the critical project delays using the Contemporaneous Schedule Analysis method is to evaluate the effect that the actual progress of work had on the project. The following procedure describes how to identify and measure the project delays or improvements caused by the actual work progress that occurs between two schedules:

1. Begin with the plan (activities, durations, logic relationships, resources, etc.) depicted in the schedule with the earlier data date (Schedule 1).
2. Define the critical path and near-critical paths in Schedule 1. Near-critical paths are work paths where the total duration of the path is not as long as the critical path, but could become the critical path if progress is not maintained on its activities during the update period.
3. Progress Schedule 1 on a daily basis using the actual start dates, actual finish dates, and remaining durations from the next schedule with the later data date (Schedule 2).
4. Assess how the progress, or lack of progress, made to the Schedule 1 plan affected the critical path of the project on a daily basis between Schedule 1 and Schedule 2. Remember that the critical path is dynamic and can change between Schedule 1 and Schedule 2, based on the progress or lack of progress of activities.
5. Determine how the progress or lack of progress on the critical path between Schedule 1 and Schedule 2 changed the forecast completion date of the project between the data dates of Schedule 1 and Schedule 2.
6. Assign the calculated project delay or improvement to the critical activities that were responsible for changes to the forecast completion date between the data dates of Schedule 1 and Schedule 2.
7. Continue with the analysis until Schedule 1 has been fully progressed with all of the actual dates and remaining durations from Schedule 2 through to the data date of Schedule 2.

Schedule revision delays and improvements

Following the analysis of the progress-caused delays and improvements described in the preceding section, the projected dates in Schedule 1 and

Schedule 2 are compared. If they are the same, then there are likely no schedule revisions made in Schedule 2 that affected the critical path. We say "likely" because having the same projected dates does not mean that no schedule revisions were made. It simply means that, if there were revisions made, they either did not affect the critical path or they combined to have no net effect on the critical path.

However, if the projected dates between the two schedules differ, this is an indication that there were schedule revisions made in Schedule 2 that did affect the critical path causing delays or improvements to the schedule. In this case, the analyst should determine how the critical path was delayed, improved, or shifted, due to schedule revisions. Schedule revisions are changes made to the schedule logic, such as added and deleted logic relationships; changed logic relationships; increased or decreased durations; changed activity descriptions; added and deleted activities; changes in the work calendar; and changed, added, or deleted constraints. At a minimum, these schedule revisions should be analyzed to determine how they affect the critical path of the project. Through careful evaluations, the revisions that were responsible for causing delays or improvements to the forecast completion date can be identified and quantified.

To simplify the identification of schedule revisions, there are software packages on the market that compare two schedules, identify all differences between the schedules, and then provide a detailed report of the differences. For example, Claim Digger, a software application that is now imbedded in Oracle's P6 Project Management software, makes identifying schedule revisions much easier.

Contemporaneous schedule analysis example

The bridge example introduced earlier is used here to demonstrate the Contemporaneous Schedule Analysis method using CPM updates to identify and measure project delay.

The project is a simple four-span bridge with two reinforced concrete abutments and three piers. The piers have pile foundations, concrete pile caps, concrete pier columns, and concrete pier caps. The bridge has a steel superstructure, stay-in-place metal deck forms, and reinforced concrete decks, curbs, and sidewalks.

Fig. 7.7 is bar chart printout of the project's baseline schedule that groups or categorizes the activities by location and, then, sorts the

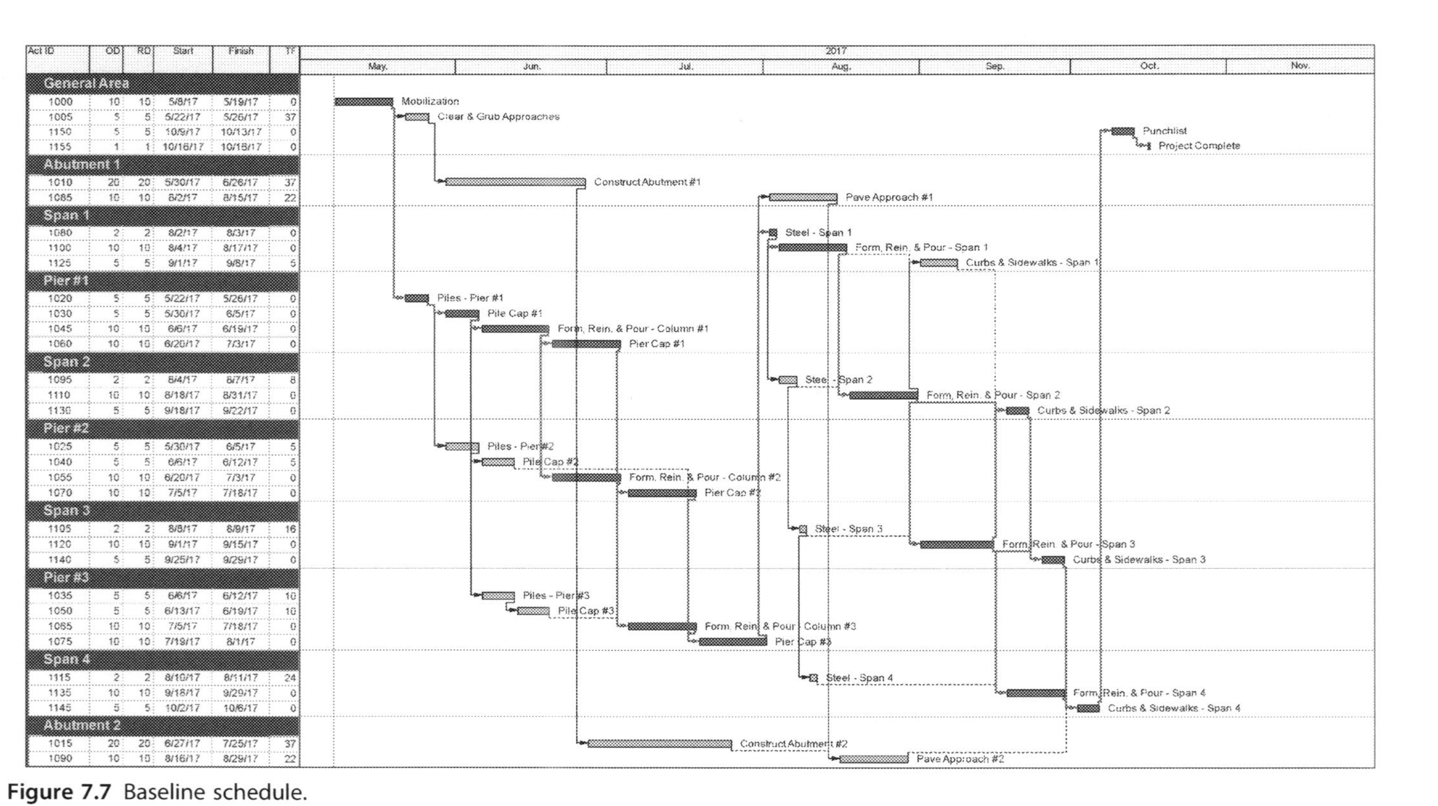

Figure 7.7 Baseline schedule.

activities within each category by their finish dates in descending order, as it was submitted by the contractor.

Fig. 7.7 depicts and conveys a considerable amount of information. First, the six columns on the left-hand side identify each of the activities' Activity ID, Original Duration, Remaining Duration, Start date, Finish date, and Total Float value. Next, the bar chart portion includes the Activity Name and depicts when each activity is expected to be performed according to their early dates, which are determined by their location in the network, and the logic relationships among the activity bars. Additionally, Fig. 7.7 also includes different colors for different activities, the *red* (dark gray in print versions) bars identify the critical path activities and the *green* (light gray in print versions) activities are noncritical activities.

The submitted schedule has a data date of May 8, 2017, one calendar, and no constraints. As a result, we can rely on float as an indication of the critical path; however, as explained previously, this may not always be the case. From a review of the schedule, it is evident that the contractor did not include activities for procurement or shop drawings. Other than this oversight, the logic and durations for the remaining activities appear reasonable based on the information available to the analyst. Also note that the project is forecast to complete on October 16, 2017.

Update No. 1

The first update of the CPM schedule for the example project has a data date of June 1, 2017, which is a month after the baseline schedule, and is shown in Fig. 7.8.

Fig. 7.8 depicts the first update in the same bar chart representation as the baseline schedule in Fig. 7.7. The first thing we notice is the status of the project's forecast completion date in the first update, which is October 23, 2017. A quick comparison of the project completion dates between the baseline schedule and Update No. 1 shows that the project was delayed 7 calendar days during the month of May 2017 from October 16, 2017 to October 23, 2017.

Earlier, float was defined as the difference between when an activity could start or finish and when the activity must start or finish so as not to delay the project. Note that not only do the critical paths of both the baseline schedule and Update No. 1 consist of the same work activities, but they also have total float values of zero. So, despite the fact that the

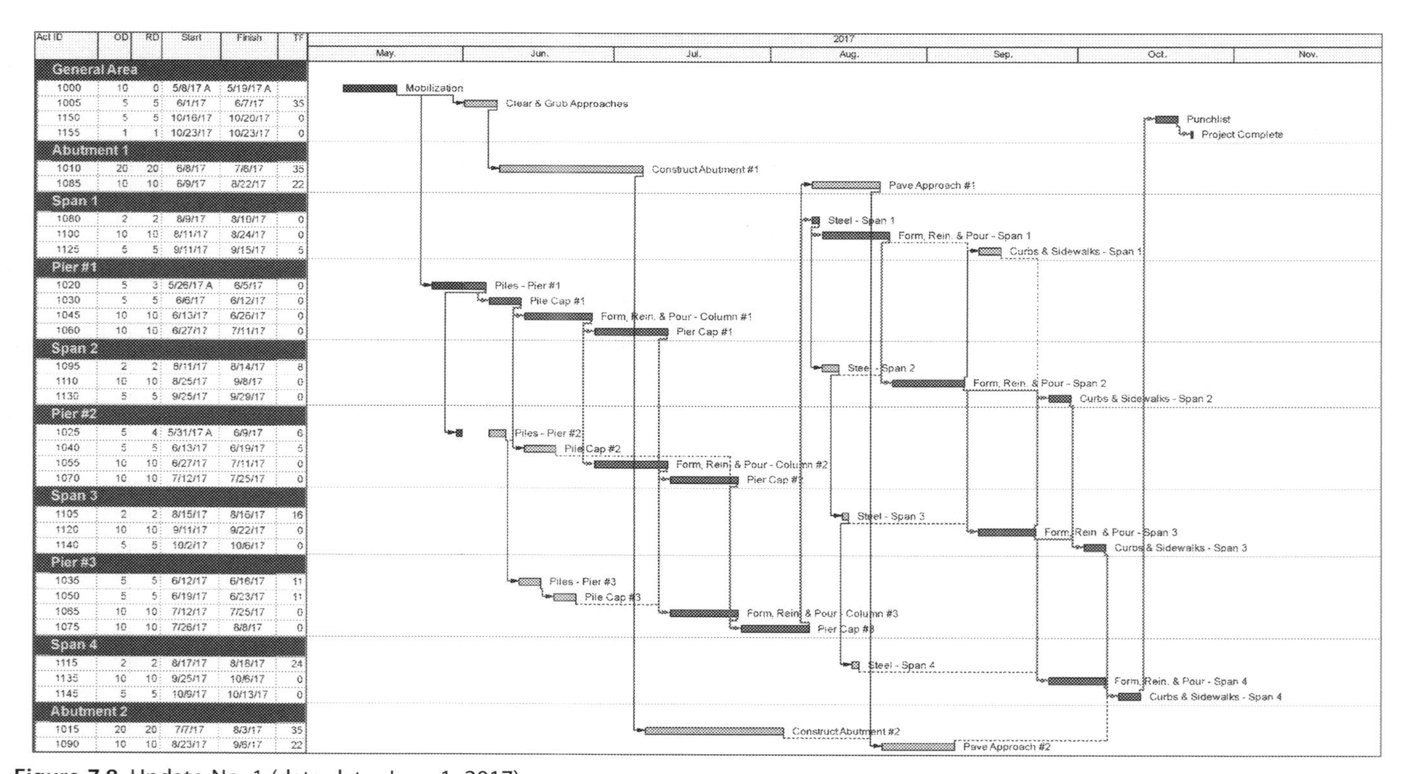

Figure 7.8 Update No. 1 (data date: June 1, 2017).

schedules forecast different project completion dates, the total float values of their critical paths are zero. This means that a finish-on-or-before constraint, also called a late-finish constraint, was not assigned to the schedule's completion activity. Still, in both the baseline schedule and Update No. 1, the critical path is identified as the schedule's longest path, not necessarily by the activities' total float values.

Despite both schedules not having their project completion dates constrained to a particular date, we are still able to measure the project delay of 7 calendar days. Now the analyst must determine why the project was delayed 7 calendar days in May. To do that, the analyst first needs to identify the critical path in the baseline schedule, which is identified using the *red* (dark gray in print versions) bars in Fig. 7.7. Next, the analyst needs to identify the critical path of Update No. 1, which we have already recognized is the same work path as the baseline schedule critical path. Because only delays to the critical path can delay the project, the next step is to compare the planned performance of the critical activities from the baseline schedule to how these activities actually progressed between May 8 and June 1. This comparison is depicted in Fig. 7.9.

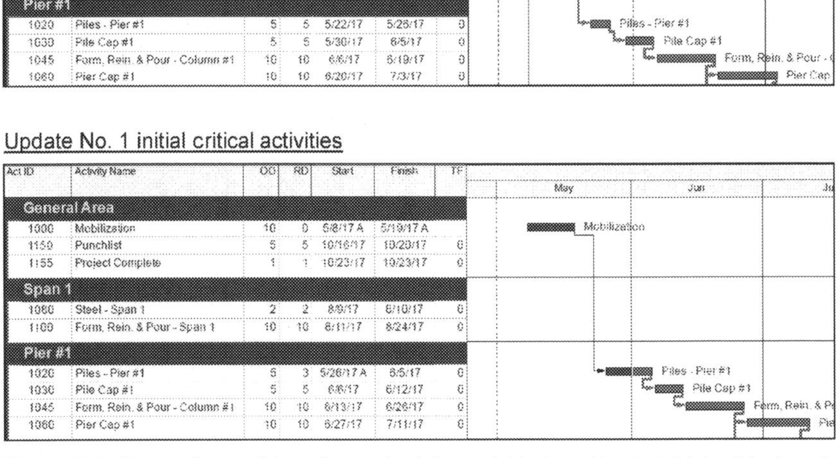
Baseline schedule initial critical activities
Act ID | Activity Name | OD | RD | Start | Finish | TF
General Area
1000 Mobilization 10 10 5/8/17 5/19/17 0
1150 Punchlist 5 5 10/9/17 10/13/17 0
1155 Project Complete 1 1 10/16/17 10/16/17 0
Span 1
1080 Steel - Span 1 2 2 8/2/17 8/3/17 0
1100 Form, Rein. & Pour - Span 1 10 10 8/4/17 8/17/17 0
Pier #1
1020 Piles - Pier #1 5 5 5/22/17 5/26/17 0
1030 Pile Cap #1 5 5 5/30/17 6/5/17 0
1045 Form, Rein. & Pour - Column #1 10 10 6/6/17 6/19/17 0
1060 Pier Cap #1 10 10 6/20/17 7/3/17 0
May
Jun
Mobilization
Piles - Pier #1
Pile Cap #1
Update No. 1 initial critical activities
Act ID | Activity Name | OD | RD | Start | Finish | TF
General Area
1000 Mobilization 10 0 5/8/17 A 5/19/17 A
1150 Punchlist 5 5 10/16/17 10/20/17 0
1155 Project Complete 1 1 10/23/17 10/23/17 0
Span 1
1080 Steel - Span 1 2 2 8/9/17 8/10/17 0
1100 Form, Rein. & Pour - Span 1 10 10 8/11/17 8/24/17 0
Pier #1
1020 Piles - Pier #1 5 3 5/26/17 A 6/5/17 0
1030 Pile Cap #1 5 5 6/6/17 6/12/17 0
1045 Form, Rein. & Pour - Column #1 10 10 6/13/17 6/26/17 0
1060 Pier Cap #1 10 10 6/27/17 7/11/17 0
May
Jun
Mobilization
Piles - Pier #1
Pile Cap #1

Figure 7.9 Comparison of baseline schedule and Update No. 1 initial critical path activities.

Fig. 7.9 shows that in the baseline schedule the contractor planned to work on three critical activities in May 2017, Activities 1000, *Mobilization*; *1020, Piles—Pier #1*; and 1030, *Pile Cap #1*. Table 7.1 compares the planned and actual performance dates of these activities between the baseline schedule and Update No. 1.

Table 7.1 shows that the mobilization activity started on May 8 and finished in May 19 as planned and, as a result, it did not cause any project delay. However, Act. 1020, *Piles—Pier #1*, did not start or finish as planned.

Despite the mobilization activity finishing as planned, the pile driving for Pier #1 did not start on May 22, 2017 as planned, it actually started 4 workdays later than planned on May 26, 2017. To calculate the project delay caused by the late start of Act. 1020, *Piles—Pier #1*, we add 4 workdays to the original completion date of Monday, October 16, 2017. The result is a 4-calendar-day delay to Friday, October 20, 2017.

After Act. 1020, *Piles—Pier #1*, started, it did not make expected progress. For example, when Act. 1020, *Piles—Pier #1*, started on Friday, May 26, 2017, based on its 5-workday original duration, it should have finished on Friday, June 2, 2017, as depicted in Table 7.2.

Table 7.1 Comparison of baseline schedule and Update No. 1 initial critical path activity dates.

Critical activities	**Baseline schedule**		**Update No. 1**	
	Start	**Finish**	**Start**	**Finish**
1000, *Mobilization*	5/8/17	5/19/17	5/8/17 A	5/19/17 A
1020, *Piles—Pier #1*	5/22/17	5/26/17	5/26/17 A	6/5/17
1030, *Pile Cap #1*	5/30/17	6/5/17	6/6/17	6/12/17

Table 7.2 Activity 1020 actual start date and expected progress.

Act. 1020, *Piles—Pier #1*, actual start date and expected finish date

Date	**Day of week**	**Workday**	**Comment**
5/26/17	Friday	X	Actual Start Date
5/27/17	Saturday		Weekend day, no work
5/28/17	Sunday		Weekend day, no work
5/29/17	Monday		Holiday, no work
5/30/17	Tuesday	X	
5/31/17	Wednesday	X	
6/1/17	Thursday	X	
6/2/17	Friday	X	Expected Finish Date

However, Fig. 7.9 shows that Act. 1020, *Piles—Pier #1*, had a remaining duration of 3 workdays and a planned finish date of June 5, 2017, which is 1 workday later than expected. This means that Act. 1020, *Piles—Pier #1*, did not make expected progress and caused a 1-workday delay. This 1-workday delay resulted in a 3-calendar-day delay to the project's completion date, delaying the project completion from Friday, October 20, 2017 to Monday, October 23, 2017.

In summary, 7 calendar days of project delay between the baseline schedule and Update No. 1 was caused by a combination of Act. 1020s, *Piles—Pier #1*, late start and its slow progress. Four calendar days of project delay was caused by the activity's late start and 3 calendar days of project delay was caused by the activity's slow progress. Note that it is necessary to separately attribute the project delay to both causes, because different parties may be responsible for each delay.

Update No. 2

The second update for the project has a data date of July 1, 2017 and appears in Fig. 7.10.

Again, the first thing to do is to determine whether the project experienced any delay during the month of June. A quick comparison of the project completion dates between Update No. 1 and Update No. 2 shows that the project completion date was delayed 4 calendar days in June from October 23, 2017 to October 27, 2017. Also, we note that the critical paths in Update Nos. 1 and 2 are the same.

The next step is to compare the expected progress of the Update No. 1 critical path to actual progress of the same activities in Update No. 2. This comparison is depicted in Fig. 7.11.

This comparison of the initial critical activities from Update Nos. 1 and 2 show that both Acts. 1020, *Piles—Pier #1*, and 1030, *Pile Cap #1*, started and finished as expected and that Act. 1045, *Form, Rein. & Pour—Column #1*, started as expected. However, despite starting as expected, Act. 1045, *Form, Rein. & Pour—Column #1*, finished 4 workdays late, with its completion being delayed from June 26, 2017 to June 30, 2017.

The 4-workday late finish of Act. 1045, *Form, Rein. & Pour—Column #1*, was responsible for the 4-calendar-day delay to the project's completion date, delaying the project completion from Monday, October 23, 2017 to Friday, October 27, 2017.

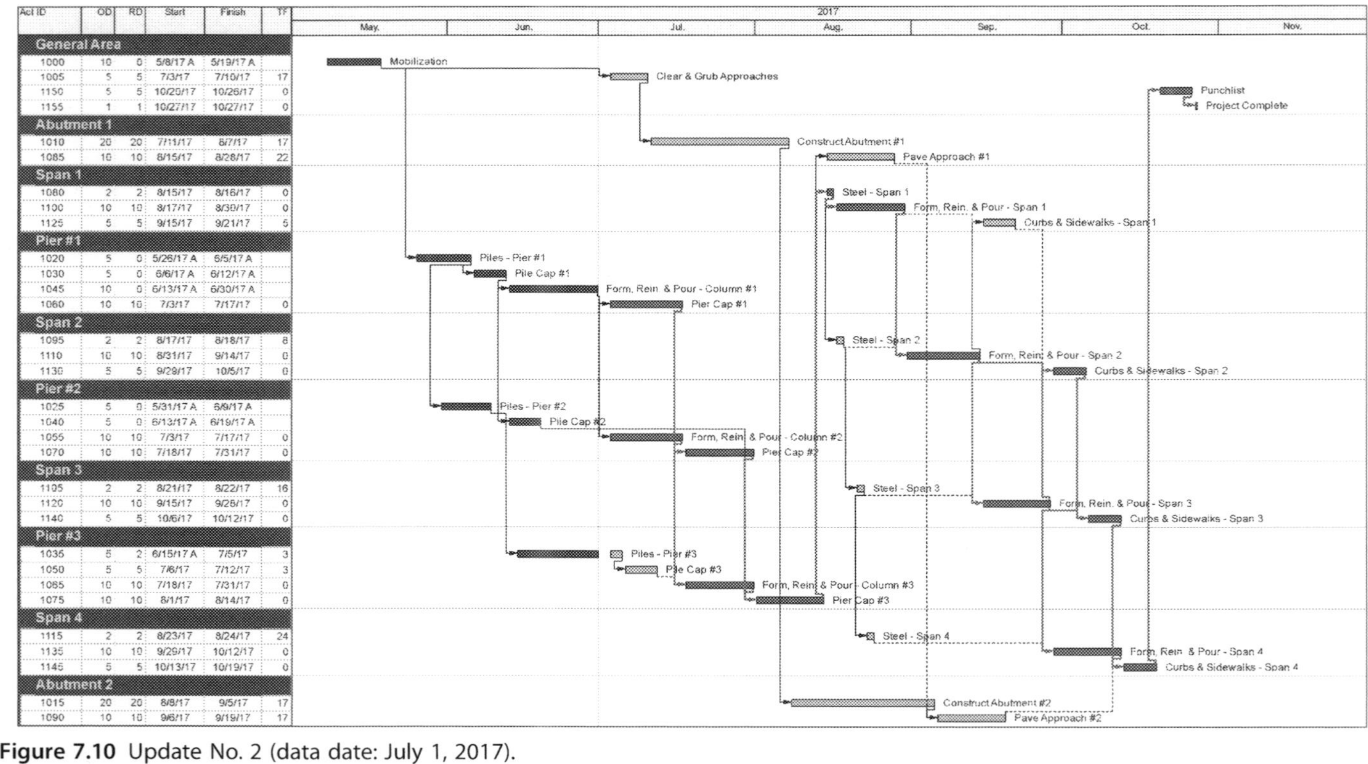

Act ID	OD	RD	Start	Finish	TF
General Area					
1000	10	0	5/8/17 A	5/19/17 A	
1005	5	5	7/3/17	7/10/17	17
1150	5	5	10/20/17	10/26/17	0
1155	1	1	10/27/17	10/27/17	0
Abutment 1					
1010	20	20	7/11/17	8/7/17	17
1085	10	10	8/15/17	8/28/17	22
Span 1					
1080	2	2	8/15/17	8/16/17	0
1100	10	10	8/17/17	8/30/17	0
1125	5	5	9/15/17	9/21/17	5
Pier #1					
1020	5	0	5/26/17 A	6/5/17 A	
1030	5	0	6/6/17 A	6/12/17 A	
1045	10	0	6/13/17 A	6/30/17 A	
1060	10	10	7/3/17	7/17/17	0
Span 2					
1095	2	2	8/17/17	8/18/17	8
1110	10	10	8/31/17	9/14/17	0
1130	5	5	9/29/17	10/5/17	0
Pier #2					
1025	5	0	5/31/17 A	6/9/17 A	
1040	5	0	6/13/17 A	6/19/17 A	
1055	10	10	7/3/17	7/17/17	0
1070	10	10	7/18/17	7/31/17	0
Span 3					
1105	2	2	8/21/17	8/22/17	16
1120	10	10	9/15/17	9/28/17	0
1140	5	5	10/6/17	10/12/17	0
Pier #3					
1035	5	2	6/15/17 A	7/5/17	3
1050	5	5	7/6/17	7/12/17	3
1065	10	10	7/18/17	7/31/17	0
1075	10	10	8/1/17	8/14/17	0
Span 4					
1115	2	2	8/23/17	8/24/17	24
1135	10	10	9/29/17	10/12/17	0
1145	5	5	10/13/17	10/19/17	0
Abutment 2					
1015	20	20	8/8/17	9/5/17	17
1090	10	10	9/6/17	9/19/17	17

Figure 7.10 Update No. 2 (data date: July 1, 2017).

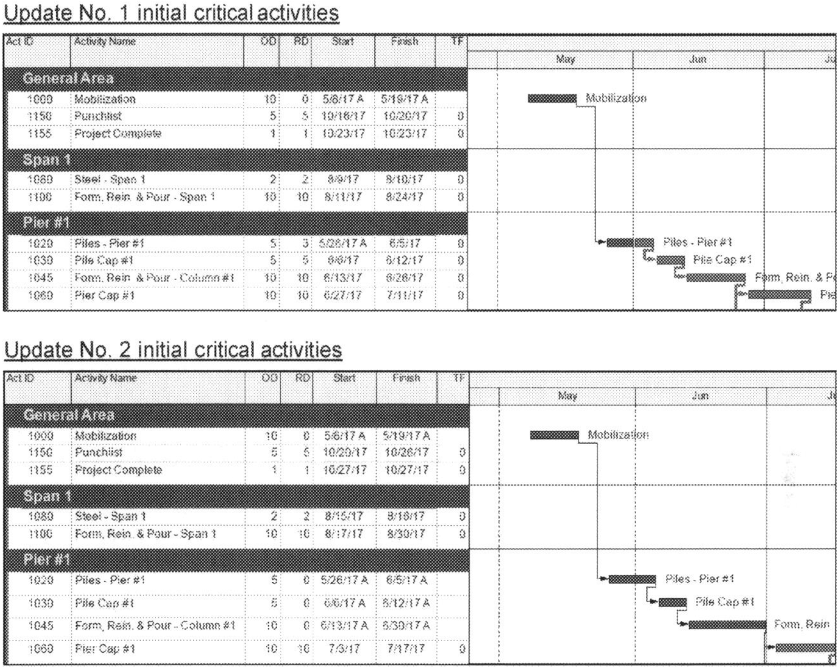
Update No. 1 initial critical activities
Act ID
Activity Name
OD
RD
Start
Finish
TF
May
Jun
General Area
1000 Mobilization 10 0 5/8/17 A 5/19/17 A
1150 Punchlist 5 5 10/16/17 10/20/17 0
1155 Project Complete 1 1 10/23/17 10/23/17 0
Span 1
1080 Steel - Span 1 2 2 8/9/17 8/10/17 0
1100 Form, Rein. & Pour - Span 1 10 10 8/11/17 8/24/17 0
Pier #1
1020 Piles - Pier #1 5 3 5/26/17 A 6/5/17 0
1030 Pile Cap #1 5 5 6/6/17 6/12/17 0
1045 Form, Rein. & Pour - Column #1 10 10 6/13/17 6/26/17 0
1060 Pier Cap #1 10 10 6/27/17 7/11/17 0
Mobilization
Piles - Pier #1
Pile Cap #1
Update No. 2 initial critical activities
Act ID
Activity Name
OD
RD
Start
Finish
TF
May
Jun
General Area
1000 Mobilization 10 0 5/8/17 A 5/19/17 A
1150 Punchlist 5 5 10/20/17 10/26/17 0
1155 Project Complete 1 1 10/27/17 10/27/17 0
Span 1
1080 Steel - Span 1 2 2 8/15/17 8/16/17 0
1100 Form, Rein. & Pour - Span 1 10 10 8/17/17 8/30/17 0
Pier #1
1020 Piles - Pier #1 5 0 5/26/17 A 6/5/17 A
1030 Pile Cap #1 5 0 6/6/17 A 6/12/17 A
1045 Form, Rein. & Pour - Column #1 10 0 6/13/17 A 6/30/17 A
1060 Pier Cap #1 10 10 7/3/17 7/17/17 0
Mobilization
Piles - Pier #1
Pile Cap #1
Form, Rein

Figure 7.11 Comparison of Update No. 1 and Update No. 2 initial critical path activities.

Update No. 3

The third update for the project had a data date of August 1, 2017, and is presented in Fig. 7.12.

A quick comparison of the second and third update shows that the project completion date in both schedule updates is October 27, 2017, indicating that the project did not experience any additional project during the month of July.

However, appearances are not always what they seem to be. Although the completion date of the third update suggests the project was not delayed in July, a more thorough evaluation should always be performed and, in this case, will indicate otherwise. Fig. 7.13 compares the initial critical path activities from both the second and third updates.

Fig. 7.13 shows that there are two sets of concurrently critical activities that were expected to progress during July. In the second update, Acts. 1060, *Pier Cap #1,* and 1055, *Form, Rein. & Pour—Column #2,* were expected to start and finish on July 3, 2017, and July 17, 2017, respectively. Then, Acts. 1070, *Pier Cap #2,* and 1075, *Form, Rein. & Pour—Column #3,* were expected to start and finish on July 18, 2017 and July 31, 2017, respectively.

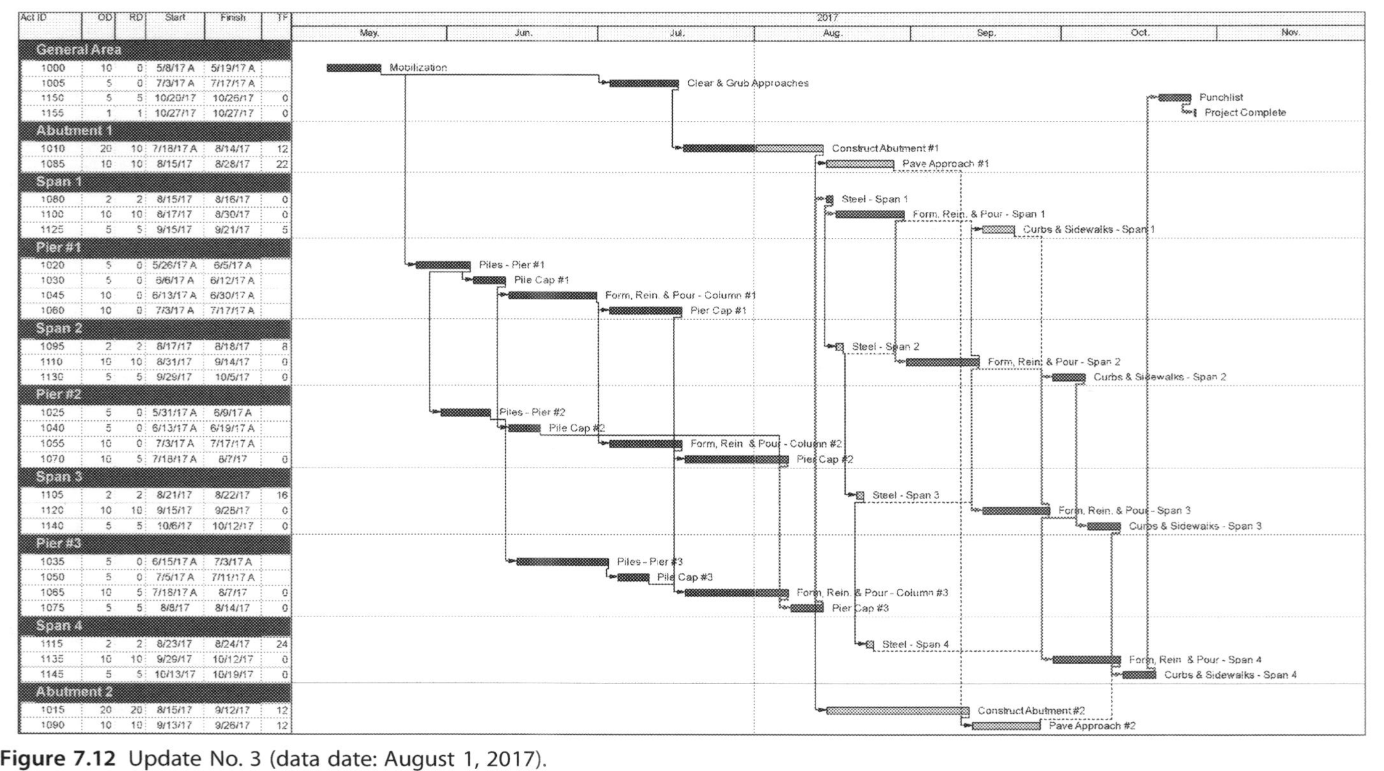

Act ID	OD	RD	Start	Finish	TF
General Area					
1000	10	0	5/8/17 A	5/19/17 A	
1005	5	0	7/3/17 A	7/17/17 A	
1150	5	5	10/20/17	10/26/17	0
1155	1	1	10/27/17	10/27/17	0
Abutment 1					
1010	20	10	7/18/17 A	8/14/17	12
1085	10	10	8/15/17	8/28/17	22
Span 1					
1080	2	2	8/15/17	8/16/17	0
1100	10	10	8/17/17	8/30/17	0
1125	5	5	9/15/17	9/21/17	5
Pier #1					
1020	5	0	5/26/17 A	6/5/17 A	
1030	5	0	6/6/17 A	6/12/17 A	
1045	10	0	6/13/17 A	6/30/17 A	
1060	10	0	7/3/17 A	7/17/17 A	
Span 2					
1095	2	2	8/17/17	8/18/17	8
1110	10	10	8/31/17	9/14/17	0
1130	5	5	9/29/17	10/5/17	0
Pier #2					
1025	5	0	5/31/17 A	6/9/17 A	
1040	5	0	6/13/17 A	6/19/17 A	
1055	10	0	7/3/17 A	7/17/17 A	
1070	10	5	7/18/17 A	8/7/17	0
Span 3					
1105	2	2	8/21/17	8/22/17	16
1120	10	10	9/15/17	9/28/17	0
1140	5	5	10/6/17	10/12/17	0
Pier #3					
1035	5	0	6/15/17 A	7/3/17 A	
1050	5	0	7/5/17 A	7/11/17 A	
1065	10	5	7/18/17 A	8/7/17	0
1075	5	5	8/8/17	8/14/17	0
Span 4					
1115	2	2	8/23/17	8/24/17	24
1135	10	10	9/29/17	10/12/17	0
1145	5	5	10/13/17	10/19/17	0
Abutment 2					
1015	20	20	8/15/17	9/12/17	12
1090	10	10	9/13/17	9/26/17	12

Figure 7.12 Update No. 3 (data date: August 1, 2017).

Update No. 2 initial critical activities

Act ID	Activity Name	OD	RD	Start	Finish	TF
General Area						
1150	Punchlist	5	5	10/20/17	10/26/17	0
1155	Project Complete	1	1	10/27/17	10/27/17	0
Span 1						
1080	Steel - Span 1	2	2	8/15/17	8/16/17	0
1100	Form, Rein. & Pour - Span 1	10	10	8/17/17	8/30/17	0
Pier #1						
1045	Form, Rein. & Pour - Column #1	10	0	6/13/17 A	6/30/17 A	
1060	Pier Cap #1	10	10	7/3/17	7/17/17	0
Span 2						
1110	Form, Rein. & Pour - Span 2	10	10	8/31/17	9/14/17	0
1130	Curbs & Sidewalks - Span 2	5	5	9/29/17	10/5/17	0
Pier #2						
1055	Form, Rein. & Pour - Column #2	10	10	7/3/17	7/17/17	0
1070	Pier Cap #2	10	10	7/18/17	7/31/17	0
Span 3						
1120	Form, Rein. & Pour - Span 3	10	10	9/15/17	9/28/17	0
1140	Curbs & Sidewalks - Span 3	5	5	10/6/17	10/12/17	0
Pier #3						
1065	Form, Rein. & Pour - Column #3	10	10	7/18/17	7/31/17	0
1075	Pier Cap #3	10	10	8/1/17	8/14/17	0

Update No. 3 initial critical activities

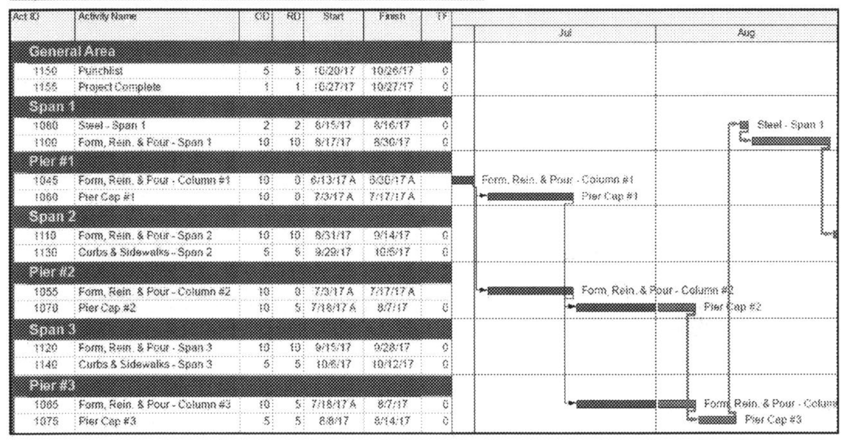

Act ID	Activity Name	OD	RD	Start	Finish	TF
General Area						
1150	Punchlist	5	5	10/20/17	10/26/17	0
1155	Project Complete	1	1	10/27/17	10/27/17	0
Span 1						
1080	Steel - Span 1	2	2	8/15/17	8/16/17	0
1100	Form, Rein. & Pour - Span 1	10	10	8/17/17	8/30/17	0
Pier #1						
1045	Form, Rein. & Pour - Column #1	10	0	6/13/17 A	6/30/17 A	
1060	Pier Cap #1	10	0	7/3/17 A	7/17/17 A	
Span 2						
1110	Form, Rein. & Pour - Span 2	10	10	8/31/17	9/14/17	0
1130	Curbs & Sidewalks - Span 2	5	5	9/29/17	10/5/17	0
Pier #2						
1055	Form, Rein. & Pour - Column #2	10	0	7/3/17 A	7/17/17 A	
1070	Pier Cap #2	10	5	7/18/17 A	8/7/17	0
Span 3						
1120	Form, Rein. & Pour - Span 3	10	10	9/15/17	9/28/17	0
1140	Curbs & Sidewalks - Span 3	5	5	10/6/17	10/12/17	0
Pier #3						
1065	Form, Rein. & Pour - Column #3	10	5	7/18/17 A	8/7/17	0
1075	Pier Cap #3	5	5	8/8/17	8/14/17	0

Figure 7.13 Comparison of Update No. 2 and Update No. 3 initial critical path activities.

A comparison of the planned and actual performance of Acts. 1060, *Pier Cap #1*, and 1055, *Form, Rein. & Pour—Column #2*, shows that they were performed as expected and, thus, resulted in no project delays. However, although Acts. 1070, *Pier Cap #2*, and 1075, *Form, Rein. & Pour—Column #3*, started as expected, they did not finish on July 31, 2017, as expected. The third update shows that these activities were now expected to finish on August 7, 2017, which is 5 workdays later than expected. Therefore, the slow progress of Acts. 1070, *Pier Cap #2*, and 1075, *Form, Rein. & Pour—Column #3*, was responsible for delaying the project 5 workdays or 7 calendar days.

However, as noted earlier, the third update does not show that the project experienced additional delay during July. Why not? The reason that the third update shows no additional project delay in July was that there was a schedule revision incorporated into the third update that mitigated or eliminated the project delay caused by the slow progress of Acts. 1070, *Pier Cap #2*, and 1075, *Form, Rein. & Pour—Column #3*.

A comparison of the initial critical activities in the second and third updates in Fig. 7.9 shows that the original duration of Act. 1075, *Pier Cap #3*, was reduced from 10 workdays to 5 workdays in the third update. The reduction of the original duration of this activity from 10 workdays down to 5 workdays mitigated the project delay caused by the slow progress of Acts. 1070, *Pier Cap #2*, and 1075, *Form, Rein. & Pour—Column #3*, in July.

Summary

Table 7.3 summarizes the schedule delay analysis from the baseline schedule through Update No. 3. The columns of the table display the *time period, critical activity, reason for delay/savings, discrete (delay)/saving in CDs, cumulative (delay)/savings in CDs*, and the *forecast completion date.*

The *time period* column identifies either the data date of the schedule or time period during which the listed activity is the initial activity on the critical path. The *critical activity* column is self-explanatory. The *reason for delay/savings* column describes the source of the delay.

The results of this analysis show that the late start and slow progress of Act. 1020, *Piles—Pier #1*, was responsible for 7 calendar days of project delay from May 20, 2017 through May 31, 2017. Then, the slow progress and late finish of Act. 1045, *Form, Rein. & Pour—Column #1*, was responsible for 4 calendar days of project delay from June 13, 2017 through June 30, 2017. Next, the slow progress and late finish of Acts. 1070, *Pier Cap #2*, and 1065, *Form, Rein. & Pour—Column #3*, was responsible for 7 calendar days of delay from July 18, 2017 through July 31, 2017. Lastly, when preparing Update No. 3, the contractor made a schedule revision, which consisted of reducing the duration of Act. 1075, *Pier Cap #3*, from 10 to 5 workdays. This resulted in a savings of 7 calendar days to the project completion date.

This analysis attributed the 18 calendar days of project delay from the May 8, 2017 data date of the baseline schedule through the August 1, 2017 data date of Update No. 3, to five activities, and a 7-calendar-day improvement was the attributed to the reduced duration of a sixth activity.

Table 7.3 Summary of delay from baseline schedule to Update No. 3.

Summary of delays (baseline to update no. 3)

Time period	Critical activity	Reason for delay/savings	Discrete (delay)/ savings in CDs	Cumulative (delay)/ savings in CDs	Completion date
5/8/17	**Baseline Schedule**		**–**	**–**	**10/16/17**
5/8/17–5/19/17	1000, *Mobilization*	Started and finished as planned, no delay	–	–	10/16/17
5/20/17–5/26/17	1020, *Piles—Pier #1*	Started late, 4-wd delay	(4)	(4)	10/20/17
5/27/17–5/31/17	1020, *Piles—Pier #1*	Progressed slower than expected, 1-wd delay	(3)	(7)	10/23/17
6/1/17	**Update No. 1**		**–**	**(7)**	**10/23/17**
6/1/17–6/5/17	1020, *Piles—Pier #1*	Finished as expected, no delay	–	(7)	10/23/17
6/6/17–6/12/17	1030, *Pile Cap #1*	Started and finished as expected, no delay	–	(7)	10/23/17
6/13/17 to 6/30/17	1045, *Form, Rein. & Pour—Column #1*	Started as expected, but finished late, 4-wd delay	(4)	(11)	10/27/17
7/1/17	**Update No. 2**		**–**	**(11)**	**10/27/17**
7/1/17–7/17/17	1060, *Pier Cap #1*, and 1055, *Form, Rein. & Pour Column #2*	Started and finished as expected, no delay	–	(11)	10/27/17
7/18/17–7/31/17	1070, *Pier Cap #2*, and 1065, *Form, Rein. & Pour Column #3*	Started as expected, but progressed slower than expected, 5-wd delay	(7)	(18)	11/3/17
8/1/17	**Update No. 3**	**Schedule Revisions**	**7**	**(11)**	**10/27/17**

Note that this analysis was performed without a focus on or knowledge of which party or parties were responsible for any of the delays identified. In this manner, the analysis remained objective. The next step in the analysis of delays would be to perform an in-depth evaluation of the actions of the parties in the context of the contract using the project documents that relate to the delayed work activities to determine the party responsible for each delay. This aspect of delay analysis will be discussed in a later chapter.

Critical path shifts

Most construction professionals would acknowledge that it is the rare construction project that is built exactly as planned. There are many reasons that the actual construction of a project deviates from the original plan. Some of those reasons include design changes, differing site conditions, significant changes in the character of the work, slow progress, better-than-expected progress, suspensions, unanticipated severe weather, etc. These factors affect the ability of the parties to complete the project within the original contract duration and are sometimes the cause of the project's late completion.

We should expect the project work and the critical path of the project to react to these factors. In addition to being responsible for delaying work on the critical path and, thus, delaying the project, these factors can and often are also responsible for causing shifts in the critical path from one work path to another in response to both the delays caused by the events and the efforts of the project team to mitigate these delays.

It is essential to identify precisely when a critical path shift occurs to ensure that project delays are properly assigned to the correct activity. The correct identification of the critical activity responsible for delaying the project is essential to ensure the correct determination of excusable and compensable delays. For example, correctly identifying when a critical path shift occurs can determine whether an owner assesses liquidated damages or grants a time extension and, if applicable, pays a contractor for its delay damages. Critical path shifts can be obvious but can also be difficult to identify.

What is a critical path shift?

To begin this discussion, recall the definition of the "critical path." As stated in Chapter 2, Float and the Critical Path, the critical path is the longest path through the schedule network. It is the planned completion date of this path that forecasts the completion date of the project. So, with this basic definition and understanding of the critical path, what then is a critical path shift?

Simply put, a critical path shift occurs when the critical path shifts from one work path to another. While this theoretically happens at a moment in time, in practice on a construction project it typically happens from one day to the next. For this to happen, a path that is shorter than the longest path on one day becomes the longest path on the next day. This will typically happen due to the actual progress of one path compared to the lack of progress on another, although shifts can also occur due to the way the logic of the schedule has been constructed. Obvious shifts in the critical path may be detected by comparing one schedule update to the next. Less obvious path shifts can be identified within an update through a comparison of the actual progress to the planned progress within the update period.

An example of a critical path shift between consecutive schedule updates is illustrated in Fig. 7.14.

The two screenshots in Fig, 7.14 depict a project that involves the replacement of a bridge. The project includes the construction of a temporary bridge, the demolition of the existing bridge, the construction of the new bridge, and, lastly, the demolition of the temporary bridge.

The August Update forecasts that the project will complete on March 30, 2017, and the critical path in the August Update consists of the construction of the temporary bridge, the shifting of traffic to the temporary bridge, the demolition of the existing bridge, the construction of the new bridge, and, finally, the demolition of the temporary bridge.

The September Update forecasts a project delay of 19 calendar days (March 30, 2017–April 18, 2017 = 19 CDs) compared to the August Update. Also, the critical path in the September Update is different from the critical path in the August Update. For example, the critical path in the September Update begins with procurement of structural steel for the new bridge, which was not critical in the August Update.

The analyst is tasked with figuring out which activity caused the 19 calendar days of delay experienced in September. The answer cannot be determined simply by looking at the August or the September Updates. Determining the right answer requires the analyst to dig a little deeper. This deeper evaluation requires the analyst to both fully understand why critical path shifts occur and how to determine the date that the shift occurs.

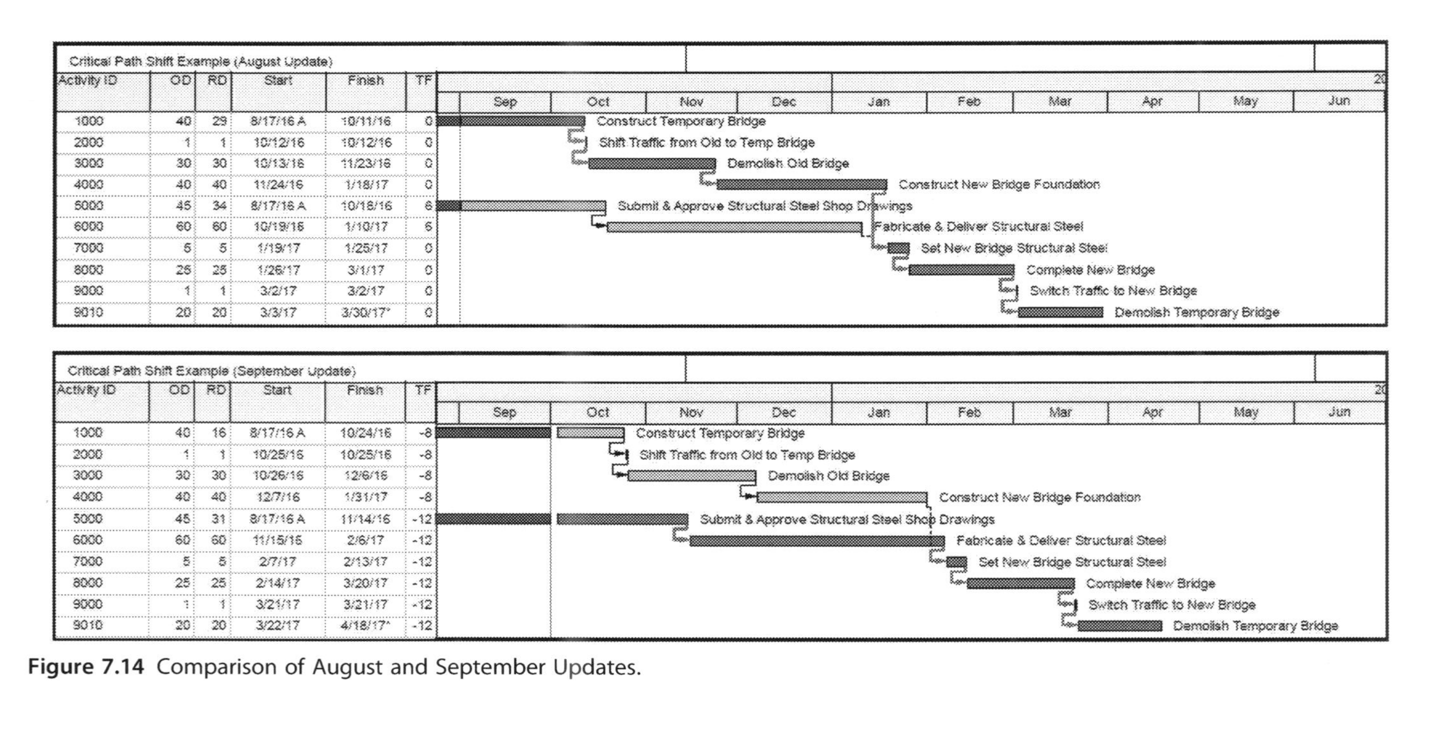

Critical Path Shift Example (August Update)

Activity ID	OD	RD	Start	Finish	TF
1000	40	29	8/17/16 A	10/11/16	0
2000	1	1	10/12/16	10/12/16	0
3000	30	30	10/13/16	11/23/16	0
4000	40	40	11/24/16	1/18/17	0
5000	45	34	8/17/16 A	10/18/16	6
6000	60	60	10/19/16	1/10/17	6
7000	5	5	1/19/17	1/25/17	0
8000	25	25	1/26/17	3/1/17	0
9000	1	1	3/2/17	3/2/17	0
9010	20	20	3/3/17	3/30/17*	0

Critical Path Shift Example (September Update)

Activity ID	OD	RD	Start	Finish	TF
1000	40	16	8/17/16 A	10/24/16	-8
2000	1	1	10/25/16	10/25/16	-8
3000	30	30	10/26/16	12/6/16	-8
4000	40	40	12/7/16	1/31/17	-8
5000	45	31	8/17/16 A	11/14/16	-12
6000	60	60	11/15/16	2/6/17	-12
7000	5	5	2/7/17	2/13/17	-12
8000	25	25	2/14/17	3/20/17	-12
9000	1	1	3/21/17	3/21/17	-12
9010	20	20	3/22/17	4/18/17*	-12

Figure 7.14 Comparison of August and September Updates.

Why does a critical path shift and what are the causes?

Conceptually, critical path shifts occur when the lengths of different work paths in a schedule network change with respect to one another. This change in the length of the schedule's work paths are typically caused by slower or better-than-expected progress and schedule revisions. Critical path shifts can be grouped into two categories: shifts caused by progress and shifts caused by schedule revisions. The way to properly determine when a critical path shift occurs in these different circumstances is illustrated in the following examples.

Critical path shifts caused by progress

Example 7.1. Typically, critical path shifts caused by progress result when one or more work paths make slower-than-expected progress or no progress at all. The classic example of this type of critical path shift can be demonstrated in the previously identified schedule update example. Therefore, let us again review the screenshots depicted in Fig. 7.14.

As discussed earlier, we know two things by comparing these two consecutive schedule updates. First, the critical paths in the updates are different. Second, the September Update shows that the project was delayed 19 calendar days (March 30, 2017–April 18, 2017 = 19 CDs) during the month of September 2017.

Digging deeper, we can determine that both work paths were delayed in September. The new critical path in the September Update has a total float value of −12 workdays and the work path that was critical in the August Update has a total float value of −8 workdays. This means that both work paths were delayed in September, but it is not clear just by looking at these screenshots of the update which of these work paths were responsible for the 19 calendar days of delay experienced in September 2017.

Before we get into the details of how to determine when the critical path shifts from the construction of temporary bridge in the August Update to the procurement of structural steel for the new bridge in the September Update, we should acknowledge the simplicity of the schedules used in this example.

By this we mean that all of the activities in this schedule are assigned to the same work calendar and, other than the end date constraint, there are no activity constraints to affect the float calculation. This means that the total float values of all of the activities in a particular path are equal and that we can use the progress achieved in September and workpath total float values to determine when the critical path shift occurs in September.

The first step in identifying when the critical path shifts from the construction of the temporary bridge to the procurement of the new bridge structural steel is to identify initial critical activities in the August and September Updates. The initial critical activity in the August Update was Act. 1000, *Construct Temporary Bridge*, and the initial critical activity in the September Update was Act. 5000, *Submit & Approve Structural Steel Shop Drawings*.

Next, we have to identify the status of both activities in the August and September Updates. In the August Update, Act. 1000, *Construct Temporary Bridge*, started on August 17, 2016, had a remaining duration of 29 workdays, and had a total float value of 0 workdays. Act. 5000, *Submit & Approve Structural Steel Shop Drawings*, also started on August 17, 2016 in the August Update, but had a remaining duration of 34 workdays and a total float value of 6 workdays.

In the September Update, Act. 1000, *Construct Temporary Bridge*, had a remaining duration of 16 workdays and a total float value of −8 workdays, while Act. 5000, *Submit & Approve Structural Steel Shop Drawings*, had a remaining duration of 31 workdays and a total float value of −12 workdays.

A comparison of these activities from both Schedule Updates shows that Act. 1000, *Construct Temporary Bridge*, made 13 workdays of progress in September, whereas Act. 5000, *Submit & Approve Structural Steel Shop Drawings*, only made 3 workdays of progress.

Using this information in conjunction with the project daily reports, we can create the as-built diagram shown in Fig. 7.15.

Fig. 7.15 includes the following information:

- The header identifies both the day of the week and the corresponding day of the month.
- The left-most columns identify the different work paths and the initial activities on the critical paths in the August and September Updates, cells identifying the activities' remaining duration and total float values in the August and September Updates, and a cell called "Work."
- The "Work" row contains an "X" to record when each activity made a day of progress during this update.

For example, we know that Act. 1000, *Construct Temporary Bridge*, made 13 workdays of progress during September by comparing the

	Day of Week	W	TH	F	S	S	M	T	W	TH	F	S	S	M	T	W	TH	F	S	S	M	T	W	TH	F	S	S	M	T	W	TH	F	S
	Date	31	1	2	3	4	5	6	7	8	9	10	11	12	13	14	15	16	17	18	19	20	21	22	23	24	25	26	27	28	29	30	1
PATH 1																																	
Const Temp Bridge	Work		X	X					X	X	X			X		X	X				X	X		X					X		X		
	RD =	29																														16	
	TF =	0																														-8	
PATH 2																																	
Sub & App SS SDs	Work		X	X				X																									
	RD =	34																														31	
	TF =	6																														-12	

Figure 7.15 As-built diagram of initial activities from both work paths.

remaining duration amounts for Act. 1000 between the August and September Updates. Then, the actual recorded days of progress for the Construct Temporary Bridge activity were based on a review of the daily reports that identified when the contractor actually constructed the Temporary Bridge.

Additionally, we know that Act. 5000, *Submit & Approve Structural Steel Shop Drawings*, made 3 workdays of progress during September by comparing the remaining duration amounts for Act. 5000 between the August and September Updates. Due to the nature of the work in Act. 5000, we cannot track the actual progress of submittal preparation by reviewing the daily reports as can be done for the construction activities because we do not know when the contractor actually worked on the submittal. Therefore, we have elected to assign the 3 workdays of progress for Act. 5000 on the first 3 workdays of the update period.

Because these two activities were the only activities on the project that made progress during the update period, we only need to track the progress of these activities. In Fig. 7.16, using the charted progress, we then calculate the changes in the remaining durations and total float values for each path, which, in this simple example, correlate to the length of each path. In this manner, we are able to identify the date on which the critical path shifted from the construction of the temporary bridge to the procurement of the structural steel during September.

Let us review in detail the change in remaining durations and the total float values of Act. 1000, *Construct Temporary Bridge*, as depicted in Fig. 7.16.

Starting with the first date of the August Update, Thursday, September 1, 2016, we know that Act. 1000, *Construct Temporary Bridge*, has a remaining duration of 29 workdays and total float value of 0. Act. 1000 made a day of progress on Thursday, September 1, 2016, and, thus, its remaining duration reduced from 29 to 28 workdays. Because this activity made expected progress on Thursday, September 1, 2016, its total float value does not change. Act. 1000 again made expected progress on Friday, September 2, 2016, and, in recognition of the progress made, its remaining duration was reduced from 28 to 27 workdays and its total float value remained the same at 0.

Saturday, Sunday, and Monday, which were September 3, 4, and 5, were nonworkdays because of a weekend and the Labor Day holiday. No progress was recorded or expected and, thus, there was no change in the remaining duration of the work or in the total float value.

	Day of Week	W	TH	F	S	S	M	T	W	TH	F	S	S	M	T	W	TH
	Date	31	1	2	3	4	5	6	7	8	9	10	11	12	13	14	15
PATH 1																	
Const Temp Bridge	Work		X	X					X	X	X			X		X	X
	RD =	29	28	27				27	26	25	24			23	23	22	21
	TF =	0	0	0				-1	-1	-1	-1			-1	-2	-2	-2
PATH 2																	
Sub & App SS SDs	Work		X	X				X									
	RD =	34	33	32				31	31	31	31			31	31	31	31
	TF =	6	6	6				6	5	4	3			2	1	0	-1

	Day of Week	F	S	S	M	T	W	TH	F	S	S	M	T	W	TH	F	S
	Date	16	17	18	19	20	21	22	23	24	25	26	27	28	29	30	1
PATH 1																	
Const Temp Bridge	Work				X	X		X					X		X		
	RD =	21			20	19	19	18	18			18	17	17	16	16	
	TF =	-3			-3	-3	-4	-4	-5			-6	-6	-7	-7	-8	
PATH 2																	
Sub & App SS SDs	Work																
	RD =	31			31	31	31	31	31			31	31	31	31	31	
	TF =	-2			-3	-4	-5	-6	-7			-8	-9	-10	-11	-12	

Figure 7.16 Example 1 identification of critical path shift.

However, because Act. 1000, *Construct Temporary Bridge*, did not progress on Tuesday, September 6, 2016, its remaining duration at the end of that day was the same (27 workdays) as it was at the end of the previous workday, Friday, September 2, 2016. The lack of progress achieved on Act. 1000, *Construct Temporary Bridge*, on Tuesday, September 6, 2016, also caused this activity and its work path's total float value to decrease 1 workday to −1 workday.

If we continue this tracking of progress and the lack of progress of these two activities, and identifying the changes to the activities' total float values between the August and September Updates, we find that Fig. 7.16 identifies two distinct time periods in September: (1) when Act. 1000 is critical and (2) when Act. 5000 is critical.

The first period, which was the time period within which Act. 1000, *Construct Temporary Bridge*, was critical, began on Thursday, September 1, 2016, and ended on Monday, September 19, 2016.

The second period, which was the time period within which Act. 5000, *Submit & Approve Structural Steel Shop Drawings*, was critical, began on Monday, September 19, 2016, and ended on Friday, September 30, 2016.

On Monday, September 19, 2016, both Acts. 1000, *Construct Temporary Bridge*, and 5000, *Submit & Approve Structural Steel Shop Drawings*, were concurrently critical and this marks the day when the critical path shifted from Act. 1000 to Act. 5000.

Fig. 7.16 shows that when the critical path shifted from Act. 1000 to Act. 5000, the critical path had a total float value of −3 workdays. Therefore, Act. 1000 was responsible for delaying the project for 3 workdays and Act. 5000 was responsible for the remaining delay to the project. These 3 workdays of delay translate to 5 calendar days of delay due to the project being extended over a weekend. This is clearly seen by reviewing the August Update's forecast project completion date of Wednesday, March 30, 2017, and comparing it to the September Update's forecast project completion date of Monday, April 4, 2016.

Therefore, of the 19 calendar days of delay experienced by the project in September, Act. 1000, *Construct Temporary Bridge*, was responsible for 5 calendar days of delay and Act. 5000, *Submit & Approve Structural Steel Shop Drawings*, was responsible for the remaining 14 calendar days of delay. Thus, in order to determine which activity or activities were actually responsible for delaying the project for 19 calendar days of delay in September, rather than just looking at the September Update critical path, it is necessary to investigate the actual progress of the work on a daily basis.

This example was simple and uncomplicated because the schedule only used one work calendar and there were only two activities that made progress in September. As such, we were able to track the progress along both work paths during the update, identify the exact day when both work paths were critical, and identify when the critical path shifted from construction of the temporary bridge to the structural steel procurement activities by tracking when the work was actually completed and how the activities' total float values and, thus, path lengths changed. Moreover, this example also shows that each day of activity delay usually equates to 1 day of delay (or more, if subject to a weekend effect) to the project completion.

In most instances, however, when a schedule contains multiple calendars, critical path shifts can occur instantaneously from one work path to another work path without both work paths being currently critical on a specific day. Moreover, lack of progress on critical paths in a multiple calendar schedule can cause additional critical path shifts and much more than a day-for-day delay.

The next example demonstrates why this occurs.

Example 7.2. In this example, we demonstrate how a critical path can shift instantaneously without the existing critical path and new critical path being critical on the same day as depicted in the previous example. This example is common on more complex projects that require the project schedule to contain more than one working calendar.

For Example 7.2, we are using a five-activity schedule, which is depicted in Fig. 7.17, to illustrate this critical path shift example as simply as possible.

This five-activity schedule consists of two independent work paths that both lead to Act. E, a finish milestone. One work path consists of Acts. A, B, and E. The other work path consists of Acts. C, D, and E. Note that there is a shaded time period during March and April 2017. This shaded period represents a nonwork period during which Acts. B and C cannot be performed.

The schedule data date is January 2, 2017, and it forecasts that the project will finish on May 30, 2017. This inability of Acts. B and C to progress in March and April is the reason Act. B is shown as not starting until May 1, 2017, despite its predecessor activity finishing on March 31, 2017.

This example schedule shows that, as of January 2, 2017, the critical path consists of the Act. A, B, and E work path, which is depicted by the *red* (gray in print versions) bars. It is important to note that the critical path in this example is identified using the longest path filter, which properly identifies the project's critical path as the A, B, and E work path. The key defining feature of the critical path is that it is the path of work responsible for forecasting when the project will finish, not necessarily the path of work whose activities have the lowest total float values.

Note that both Acts. A and C are "riding" the data date and are expected to either resume work or begin immediately.

So, what happens if neither Act. A nor Act. C makes any progress on January 2, 2017? To evaluate the effect of neither Act. A nor Act. C making progress on January 2, 2017, the example schedule is copied and the data date is changed to January 3, 2017. The result is depicted in Fig. 7.18.

Note that the remaining durations of all four activities remained the same in both Figs. 7.17 and 7.18, despite the fact that the data dates changed from January 2, 2017 to January 3, 2017, respectively. The unchanged remaining durations mean that no progress was achieved on

Critical path shift Example 2 (day 1)

Act ID	OD	RD	Start	Finish	TF
A	120	89	12/1/16 A	3/31/17	30
B	30	30	5/1/17	5/30/17	0
C	58	58	1/2/17	2/28/17	0
D	61	61	3/1/17	4/30/17	30
E	0	0		5/30/17	0

Figure 7.17 Example 2, Day 1 critical path.

Critical path shift Example 2 (day 2)

Act ID	OD	RD	Start	Finish	TF
A	120	89	12/1/16 A	4/1/17	61
B	30	30	5/1/17	5/30/17	32
C	58	58	1/3/17	5/1/17	0
D	61	61	5/2/17	7/1/17	0
E	0	0		7/1/17	0

Figure 7.18 Example 2, Day 2 critical path.

Acts. A or C on January 2, 2017, which were the initial activities of both work paths.

Also note that Acts. A and C's lack of progress on January 2, 2017, resulted in a delay to the project of 32 calendar days from May 30, 2017 to July 1, 2017, as depicted by Act. E. Additionally, the critical path shifted from Acts. A, B, and E on January 2, 2017, as depicted in Fig. 7.17, to Acts. C, D, and E on January 3, 2017, as depicted in Fig. 7.18, due to the lack of progress of Act. C on January 2, 2017.

It is somewhat surprising and perhaps a bit counter intuitive that the lack of progress on one day should result in a 32 calendar-day delay to the project. So, how did this happen?

As discussed earlier, Acts. B and C were unable to progress during the months of March and April due to a nonwork period calendar. In Fig. 7.17, although Act. A was forecast to finish on March 31, 2017, the nonwork period's effect on Act. B was such that it could not begin until May 1, 2017. This calendar effect created a total float value of +30 workdays for Act. A despite it being on the project's longest and, thus, critical path. In Fig. 7.17, because Act. C was planned to finish on February 28, 2017, the last workday before the March and April nonwork period, Act. D, which was not subject to the nonwork period, was able to begin on March 1, 2017, and finish on April 30, 2017, 30 calendar days before Act. B would finish.

It is important to acknowledge that it was not until January 3, 2017, that the analyst knows that both Acts. A and C made no progress. Then and only then would the analyst be able to determine that the lack of progress on Act. C combined with the inability of Act. C to work in March and April were responsible for both the shifting of the critical path to Acts. C, D, and E and for causing the 32-calendar-day project delay.

Critical path shifts caused by schedule revisions

Schedule revisions are often made to the project schedule during the preparation of schedule updates or to the project schedule during the preparation of a time extension request. Schedule revisions made in preparation of a time extension request may be the result of performing a TIA.

The first example of a critical path shift caused by a schedule revision depicts a critical path shift that occurs when activities are added to the project schedule, as may occur when preparing a Prospective TIA.

Example 7.3. This example utilizes the same schedule that was used in Example 7.1. Fig. 7.19 is the September Update that shows the project finishing on April 18, 2017, with the new bridge steel procurement path as the critical path.

In this example, the owner changed the requirements of an element of the temporary bridge on September 28, 2016. To model the change, the contractor and owner agree to the 15-workday fragnet that is depicted by Act. 0100, *CO #1 Change Temp Bridge Requirements*, to represent the time needed to incorporate the change, which is depicted in Fig. 7.20.

The contractor and owner also recognize and agree that the change order work has to be completed before the contractor can complete the construction of the temporary bridge, and agree to tie the completion of Act. 0100, *CO #1 Change Temp Bridge Requirements*, to Act. 1000, *Construct Temporary Bridge*, with a finish-to-start relationship. After this logic relationship is added and the September Update and scheduled, or recalculated, the result is depicted in Fig. 7.21.

Fig. 7.21 shows that the project delay resulting from the CO #1 fragnet activity is 10 calendar days, from April 18, 2017 to April 28, 2017. Additionally, note that the critical path shifts from the procurement of the new bridge structural steel to the construction of the temporary bridge.

This critical path shift example demonstrates how a project's critical path can shift due to the addition of work. The example that follows illustrates a critical path shift that results from logic changes made to mitigate previous delay.

Example 7.4. This example begins with the same schedule that was used in Example 7.1. Fig. 7.22 is the September Update that shows the project finishing on April 18, 2017, with the new bridge steel procurement path as the critical path.

Instead of assuming that Fig. 7.22 is the final version of the September Update, let us assume that Fig. 7.22 is the progress-only version of the August Update updated through the end of September. As such, Fig. 7.22 shows that the project will finish on April 18, 2017, and the critical path consists of the procurement of the new bridge structural steel.

As discussed in Example 7.1, from a comparison of the August Update to the September Update, we know that Act. 5000, *Submit & Approve Structural Steel Shop Drawings*, experienced slow progress in

Critical path shift Example (September Update)

Activity ID	OD	RD	Start	Finish	TF	Activity
1000	40	18	8/17/16 A	10/24/16	-8	Construct Temporary Bridge
2000	1	1	10/25/16	10/25/16	-8	Shift Traffic from Old to Temp Bridge
3000	30	30	10/26/16	12/6/16	-8	Demolish Old Bridge
4000	40	40	12/7/16	1/31/17	-8	Construct New Bridge Foundation
5000	45	31	8/17/16 A	11/14/16	-12	Submit & Approve Structural Steel Shop Drawings
6000	60	60	11/15/16	2/6/17	-12	Fabricate & Deliver Structural Steel
7000	5	5	2/7/17	2/13/17	-12	Set New Bridge Structural Steel
8000	25	25	2/14/17	3/20/17	-12	Complete New Bridge
9000	1	1	3/21/17	3/21/17	-12	Switch Traffic to New Bridge
9010	20	20	3/22/17	4/18/17*	-12	Demolish Temporary Bridge

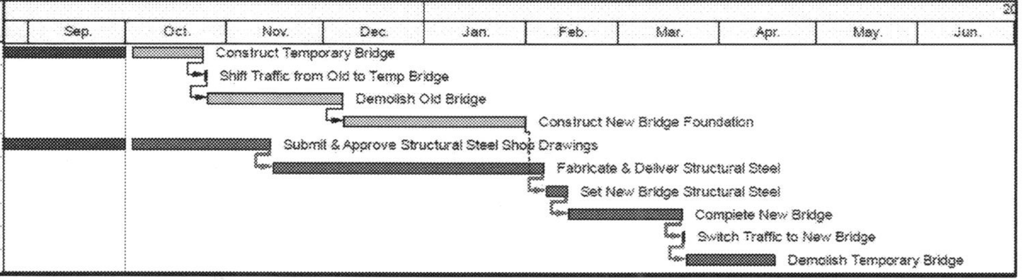

Figure 7.19 September Update.

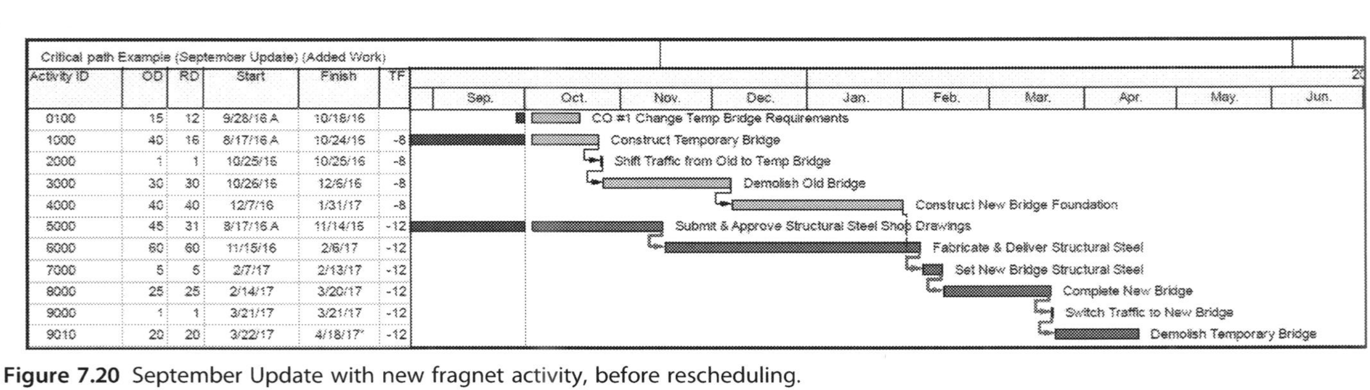

Figure 7.20 September Update with new fragnet activity, before rescheduling.

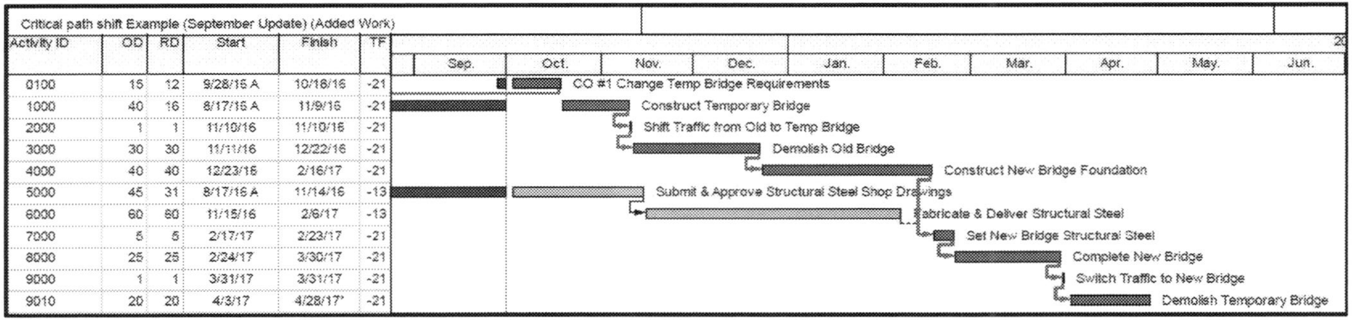

Critical path shift Example (September Update) (Added Work)

Activity ID	OD	RD	Start	Finish	TF
0100	15	12	9/28/16 A	10/16/16	-21
1000	40	16	8/17/16 A	11/9/16	-21
2000	1	1	11/10/16	11/10/16	-21
3000	30	30	11/11/16	12/22/16	-21
4000	40	40	12/23/16	2/16/17	-21
5000	45	31	8/17/16 A	11/14/16	-13
6000	60	60	11/15/16	2/6/17	-13
7000	5	5	2/17/17	2/23/17	-21
8000	25	25	2/24/17	3/30/17	-21
9000	1	1	3/31/17	3/31/17	-21
9010	20	20	4/3/17	4/28/17*	-21

Sep. Oct. Nov. Dec. Jan. Feb. Mar. Apr. May. Jun.

CO #1 Change Temp Bridge Requirements
Construct Temporary Bridge
Shift Traffic from Old to Temp Bridge
Demolish Old Bridge
Construct New Bridge Foundation
Submit & Approve Structural Steel Shop Drawings
Fabricate & Deliver Structural Steel
Set New Bridge Structural Steel
Complete New Bridge
Switch Traffic to New Bridge
Demolish Temporary Bridge

Figure 7.21 September Update with new fragnet activity, after rescheduling.

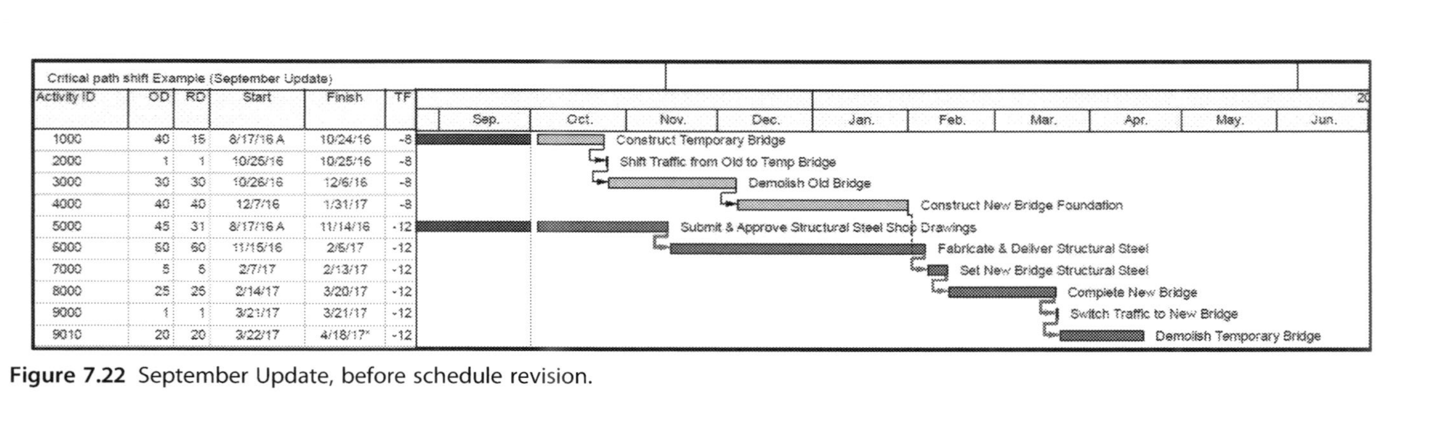

Critical path shift Example (September Update)

Activity ID	OD	RD	Start	Finish	TF
1000	40	15	8/17/16 A	10/24/16	-8
2000	1	1	10/25/16	10/25/16	-8
3000	30	30	10/26/16	12/6/16	-8
4000	40	40	12/7/16	1/31/17	-8
5000	45	31	8/17/16 A	11/14/16	-12
6000	60	60	11/15/16	2/6/17	-12
7000	5	5	2/7/17	2/13/17	-12
8000	25	25	2/14/17	3/20/17	-12
9000	1	1	3/21/17	3/21/17	-12
9010	20	20	3/22/17	4/18/17*	-12

Figure 7.22 September Update, before schedule revision.

September 2016 and caused some project delay. To mitigate some of this delay, the contractor changed the planned duration of the structural steel fabrication activity (Act. 6000, *Fabricate & Deliver Structural Steel*) from 60 to 47 workdays, as shown in Fig. 7.23.

Note that the reduction of the duration of Act. 6000, *Fabricate & Deliver Structural Steel*, from 60 to 47 workdays, caused the critical path to shift from procurement of the new bridge structural steel to the construction of the temporary bridge.

The schedule revision that is depicted in this Example 7.4 would be the type of schedule revision identified in the second step of a Contemporaneous Schedule Analysis.

A project critical path can shift from one work path to another for many different reasons during the project. As described above, critical path shifts can be grouped into two categories: shifts caused by progress and shifts caused by schedule revisions. Understanding when and why these types of critical path shifts occur is an essential step in properly assigning project delay to the responsible activity, change, or issue, which in turn enables the analyst to assign the project delay to the appropriate responsible party.

Correcting versus leaving errors

A delay analysis should rely on the contemporaneous project schedules as the basis of analysis to the maximum extent possible. By using the management tools employed by the parties during the project, the analyst is able to adopt the perspective of the project manager at the time of a particular delay and avoid applying the "wisdom" of hindsight. Reliance on the contemporaneous project schedules helps keep the analysis objective and guards against the analyst drawing erroneous or biased conclusions. For example, was the steel delivered late allowing more time to construct the foundations or was the steel delivery pushed back because the foundations were late? Analyses that stray far from the contemporaneous schedules, relying on after-the-fact creations, are often biased and unpersuasive.

Assessing the reliability of the CPM schedules to form the basis of an analysis of delays may be one of the most important decisions that the analyst has to make. While it is true that there may be sufficient cause to abandon the contemporaneous schedules as an analytical tool, this decision should be made carefully and with good reason. No schedule is

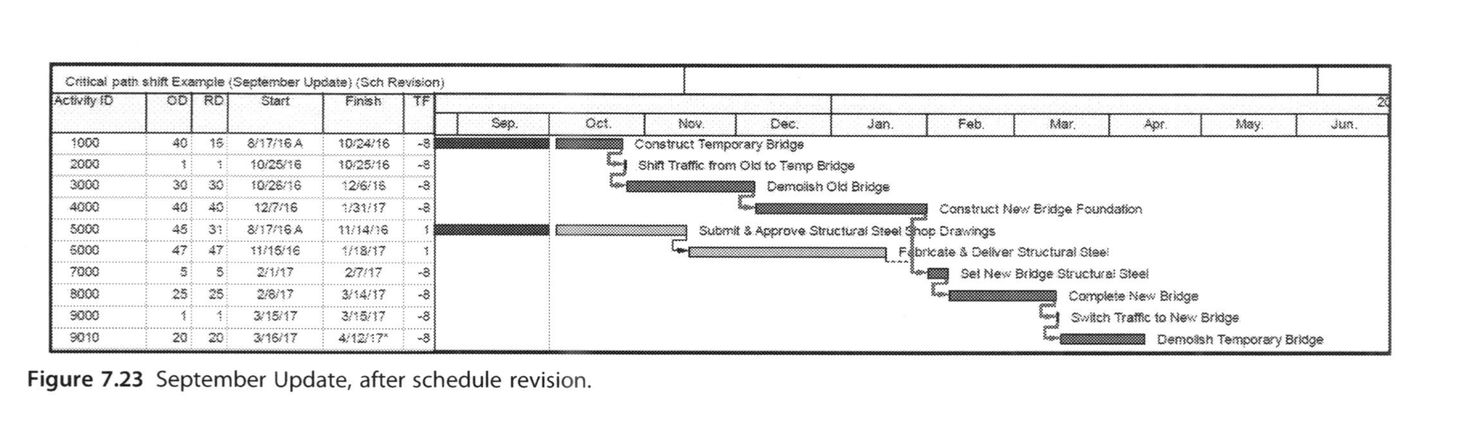

Figure 7.23 September Update, after schedule revision.

perfect. Most contain or omit logic that could be characterized as errors. But many of these are minor and have no effect on the project's critical path. And most errors are self-correcting, meaning that if the schedules are properly analyzed, the actual progress of the work will establish the correct answer.

To better understand this principle of self-correction, consider the following example. Presume that in a sequence of work along the critical path involving the installation of wall studs, rough-in wiring, and drywall, the scheduler omits the activity for rough-in wiring. The schedule presumes that drywall installation will begin at the completion of the stud installation—no time has been allotted for roughing in wire. One might argue that the schedule is flawed because the contractor cannot really complete all of the contract work on time. But what really happens? When the stud installation is actually completed, the drywall work fails to start. Instead, the contractor first roughs-in the wire. The contractor is assigned a critical project delay equal to the late start of the drywall work. And so the error self-corrects. Had the contractor included rough-in wiring in the original schedule, it would have done so in a way that did not increase the contract time. The same choice is available when the omission is discovered. Perhaps the contractor can perform the rough-in wiring concurrent with some of the wall stud and drywall work such that no delay occurs. Either way, the error in the schedule is self-correcting.

When is the schedule no longer good enough to allow for the reliable identification and measurement of delays? The contemporaneous CPM schedule should only be abandoned as an analytical tool when there is evidence that the schedule is not a reliable model of the contractor's plan or when the schedule contains such gross errors as to render its predictions useless. Often, a schedule that has been significantly revised to mitigate a large number of days of delay will begin to contain gross errors of the type that cause the schedule to model an unexecutable plan. In such cases, the plan being modeled by the schedule and the actual execution of the project depart, and the schedule is no longer a reliable tool to measure project delay.

A word of caution is appropriate. The proper use of advanced scheduling features should not be mistaken as scheduling errors. The mere existence of an additional calendar, a lag, or an activity constraint is not in and of itself an error. While the use of advanced scheduling features may make the task of analyzing the schedule to identify and measure delays more challenging, understanding how these features affect the calculation of the schedule allows the analyst to preserve the contemporaneous schedule as an objective analytical tool.

Also, the analyst must resist the temptation to alter the schedule by adding or deleting activities. Similarly, the analyst should resist changing the logic or durations to produce a schedule that seems more representative of the schedule that should have been used on the project. This practice can produce an erroneous analysis with potentially biased results.

If analysts note serious errors in the logic of the schedule, they should consider not accepting the contractor's schedule as a valid tool to measure the delays. The validity of the schedule is subjective; therefore, the analyst should always seek help from a qualified scheduling consultant before making this determination. If, indeed, the schedule does not reflect the reality of the job progress, or does not reasonably represent the contractor's plan for performing the work, then it may be wiser to abandon the schedule and perform a delay analysis using the as-built approach described in Chapter 8, Delay Analysis Using No Schedules.

Upon reviewing the CPM schedule, analysts may question the validity of the durations assigned to specific activities based on their knowledge of the project, estimating skills, or experience. However, if the reviewer does not know the specific resources that the contractor planned to apply to the work, the durations in question should not be dismissed as erroneous. After all, an experienced and creative contractor can devise the most expedient method to build the project, and this may well require less time than one would normally estimate. In the same vein, the contractor may decide to apply fewer resources to particular activities and have durations longer than one might normally estimate. Neither of these decisions on the part of the contractor makes the schedule incorrect. Without specific contract language constraining the contractor's sequence or imposing milestone dates, the execution of the project is the contractor's responsibility.

The analyst also has the option of performing the analysis of delay without correcting the errors and with the errors corrected to compare the results and describe the reason(s) for the differences.

One last thought about modifying schedules: While changes to schedule logic should be avoided, some analysis techniques require the addition of "dummy" logic to allow the schedule software to assist in the analysis of delays. Dummy logic is logic that has no net effect on the calculation of the schedule, but is required to allow the program to identify certain key aspects of, or milestones within, the schedule. For example, in order to identify the critical path on a complex schedule, it may be necessary to run the schedule through the longest path filter. In some scheduling software applications, this filter will determine the longest path from the first activity in the schedule to the last activity in the schedule. If the last activity is not the project completion milestone being analyzed, the filter may not identify the longest path to the completion milestone. In such cases, it may be necessary to add dummy logic to focus the filter on the path being analyzed. Such analysis techniques should not be confused with the types of logic changes that "correct" or alter the contemporaneous schedules in a way that reduces or limits the reliability of the results. Again, understanding how the advanced features of modern scheduling software affect the calculation of the schedule is essential in order to properly apply analytical techniques while preserving the contemporaneous schedule as an objective analytical tool. The use of scheduling software and other software tools in the schedule analysis process is discussed later in this chapter.

CONCURRENT DELAYS

The concept of concurrent delay is discussed ever more frequently in the analyses of construction delays. Concurrency is not only viewed from the standpoint of determining the project's critical delays, it is also being used as an argument to assign responsibility for delay damages associated with both critical path and noncritical path delays. Owners will sometimes cite concurrent delays by the contractor as a reason for issuing a time extension without additional compensation or even as the reason for denying a time extension. Contractors will sometimes cite concurrent delays by the owner as a reason why liquidated damages should not be assessed. Unfortunately, few contracts include a definition of "concurrent delay" and fewer address how concurrent delays affect a contractor's entitlement to additional compensation for time extensions or responsibility for liquidated damages.

An example of contract language that addresses a contractor's entitlement related to concurrent delays is provided in Fig. 7.24, which is taken from the New Mexico Department of Transportation's 2014 Standard Specifications for Highway and Bridge Construction.

> The Contractor is not entitled to an extension of Contract Time for any period in which a Non-Excusable Delay is concurrent with an Excusable Delay. When a Noncompensable Delay is concurrent with a Compensable Delay, the Contractor may be entitled to an extension of Contract Time but not entitled to compensation for the period the Noncompensable delay is concurrent with the Compensable Delay.

Figure 7.24 New Mexico DOT concurrent delay contract language.

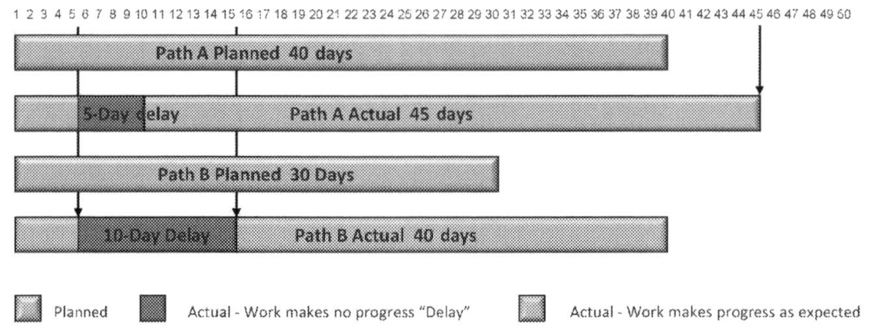

Figure 7.25 Concurrent delay example 1.

There is also a lack of understanding in the industry concerning the concept of concurrent delay. Simply stated, concurrent delays are separate delays to the "critical path" that occur at the same time. While this seems like a simple concept, some presentations of concurrent delays have significantly muddied the waters. So, the first question that should be asked is whether both of the alleged concurrent delays actually affected the critical path and, thus, the project completion date. Fig. 7.25 depicts an instance when two independent delays start on the same day and occur at the same time.

Fig. 7.25 shows a project that consists of two work paths (A and B). Path A has a planned duration of 40 days and Path B has a duration of 30 days. Path A is the project's critical path, because it is the longest path and forecasts when the project will finish, Day 40. Also, Path B has a float value of positive 10 days, which means Path B would need to be delayed 10 days, and Path A would have to progress as expected, before Path B would consume the available float on its work path, become the critical path, and start delaying the project.

Note that both work paths were delayed by separate delays that started on the same day, Day 6. The 5-day delay to Path A ended on Day 10 and the 10-day delay to Path B ended on Day 15. As such, Path A finished on Day 45 and Path B finished on Day 40, which means that the project finished 5 days late and was delayed 5 days.

It is clear that the 5-day delay to Path A was responsible for delaying the project, because despite being delayed 10 days Path B still finished by Day 40, the original completion date. So, the first step in every evaluation of concurrent delays is to determine whether or not both delays actually delayed the project's critical path.

The two delays, the 10-day delay to Path B and the 5-day delay to Path A are not concurrent delays. That is because only one of the delays is critical—the 5-day delay to Path A. In order to delay the project, the delay must occur on the project's critical path; therefore, to be concurrent delays, the delays must be concurrently critical.

Concurrent delays to separate critical paths

When considering concurrent critical delays, several situations should be considered. The first occurs when separate critical paths are concurrently delayed and, thus, responsible for delaying the project at the same time. For example, if shop drawings and bulk excavation are both on the critical path and are predecessors to the start of footing excavation and both predecessors are scheduled to finish on the same day, then both predecessors control the start of footing excavation. If the finish of bulk excavation was delayed 30 days, from June 1 to July 1, due to equipment failures and the finish of formwork shop drawings was delayed 30 days, from June 1 to July 1, due to a redesign, then the project was concurrently delayed by both the shop drawings and bulk excavation in June.

In this situation, the delay to bulk excavation is might be nonexcusable, and the delay to the shop drawings might be excusable. Most contracts do not specify which delay takes precedence, if any. Assigning 15 days of nonexcusable delay to excavation and 15 days of excusable, compensable delay to the shop drawings may be one way to apportion the delay. The existing case law in the jurisdiction of the project also may offer some guidance as to how this situation has been viewed from a legal perspective. In some jurisdictions, the occurrence of concurrent excusable and nonexcusable delays results in the contractor receiving a time extension, but no additional compensation as described above in Fig. 7.24. A carefully drafted contract that addresses this potential occurrence will avoid the disputes that may otherwise occur in this situation.

Another example of contract language that addresses this issue can be seen in some Suspension of Work clauses within various state departments of transportation, the relevant excerpt from which is depicted in Fig. 7.26.

These clauses provide for an equitable adjustment for suspended work only if there were no other causes of delay to the project. Applying the suspension of work clause shown in Fig. 7.26 would result in the contractor's receiving no time or additional compensation for the redesign because of its concurrent delay to the bulk excavation.

If both 30-day delays were excusable but only one was compensable, then once again the analyst must look to the governing case law. In this instance, it has been viewed in some jurisdictions that the noncompensable delay takes precedence, and the contractor would be issued a 30-day time extension but no additional compensation, as described above in Fig. 7.24.

A different situation occurs when there are initially concurrent delays, but one delay ends before the other delay as depicted in Fig. 7.27.

Looking at the example in Fig. 7.27, assume the owner-caused delay lasted 5 days and delayed critical activity A from Day 21 through Day 25. The contractor-caused delay lasted 10 days and delayed critical activity B from Day 21 through Day 30. In this situation, there were 5 days of concurrent delay from Day 21 through Day 25, when both delays simultaneously affected the critical path. The resolution of the 5 days of

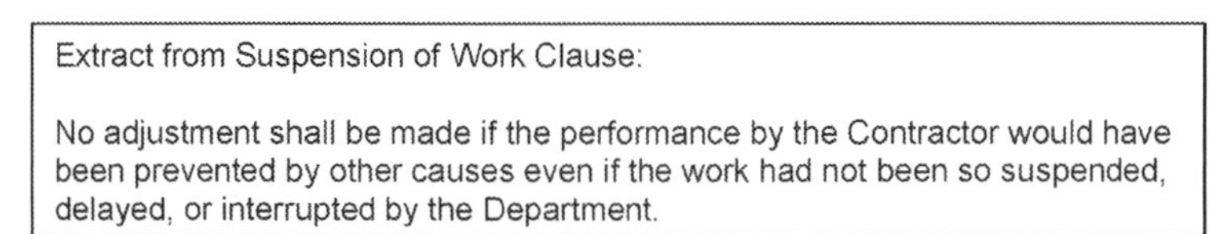

Extract from Suspension of Work Clause:

No adjustment shall be made if the performance by the Contractor would have been prevented by other causes even if the work had not been so suspended, delayed, or interrupted by the Department.

Figure 7.26 Example DOT Suspension of Work Clause.

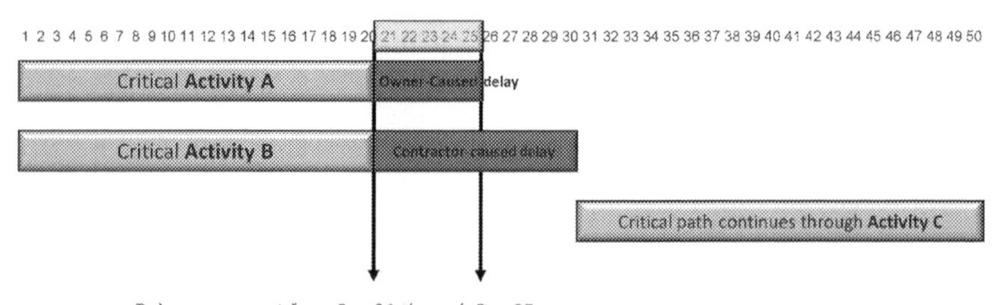

Figure 7.27 Concurrent delay example 2.

concurrent delay would be approached the same as previously described in that the contractor may be entitled to a time extension, but not the recovery of delay-related costs. The contractor, however, is not entitled to time or compensation of delay-related cost for the 5-day contractor-caused delay from Day 26 through Day 30.

However, if the delay durations are switched, as depicted in Fig. 7.28, the assignment of delay changes, as well.

In this situation, again there were 5 days of concurrent delay from Day 21 through Day 25, during which the contractor may be entitled to a time extension, but may not be entitled to the recovery of its delay-related costs. But the contractor would likely be entitled to both additional contract time and its delay-related costs from Day 26 through Day 30 when the owner-caused delay was solely responsible for delaying the critical path and, thus, the project.

Another situation occurs when there are two or more critical paths and the delay to one path starts before the delay to the other critical path. To evaluate this situation, we will alter the shop drawing and bulk excavation example. Bulk excavation work made no progress from June 1 to July 1 because of a strike. The preparation of shop drawings was suspended due to a design change from June 10 to June 25. The concept known as primacy of delay applies in this situation. The contractor's delay to bulk excavation began before the owner's delay to the shop drawings. Because of the delay to excavation, the contractor had more time to complete the shop drawings, and the delay to the shop drawings never became critical. Bulk excavation was responsible for a 30-day delay, and the contractor is not entitled to a time extension or additional compensation. The concept of primacy of delay recognizes that critical delays create float on other paths of work, and typically the contractor and owner can use this float.

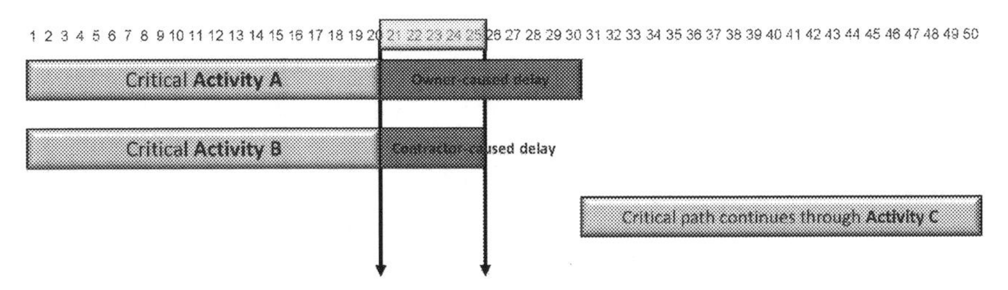

Figure 7.28 Concurrent delay example 3.

To demonstrate the primacy of delay, consider the following example, which consists of a project with two parallel critical paths of work. Fig. 7.29 shows Path A and Path B, each planned to finish in 40 days, but both actually finish in 50 days.

Fig. 7.29 shows us that both paths of work are scheduled to start the same day and finish on the same day, Day 40. Therefore, an analysis at the start of these paths would show that both paths are concurrently critical. Additionally, Fig. 7.29 also shows that both work path actually started on the same day and actually finished on the same day, Day 50. If we were to rely only on the information depicted in Fig. 7.29, it appears that both work paths were concurrently critical and because they both finished on the same day, it appears that both work paths were concurrently delayed.

However, this attempt to analyze concurrent delays with these limited facts and conclude that the project was concurrently delayed by both work paths is dangerous. To truly determine whether both work paths were concurrently delayed, it is necessary to drill down into the details of the actual work and timing of the delays that affected each work path.

As described earlier in the book, to properly identify critical path delays, it is necessary to evaluate the performance of the work along the critical path of the project from the beginning of the project to the end, tracking the critical path and identifying critical path shifts as they occur. Inherent in this process is the identification of concurrent delays.

To begin the evaluation of the potential concurrent delays to Paths A and B, we must start at the beginning of the project. We already know that both work paths were concurrently critical before the work began, but let us begin progressing both paths from the beginning to see what happens. Fig. 7.30 depicts the status of the project after Day 2 and shows that work on both paths progressed as expected resulting in no project delay through the completion of Day 2.

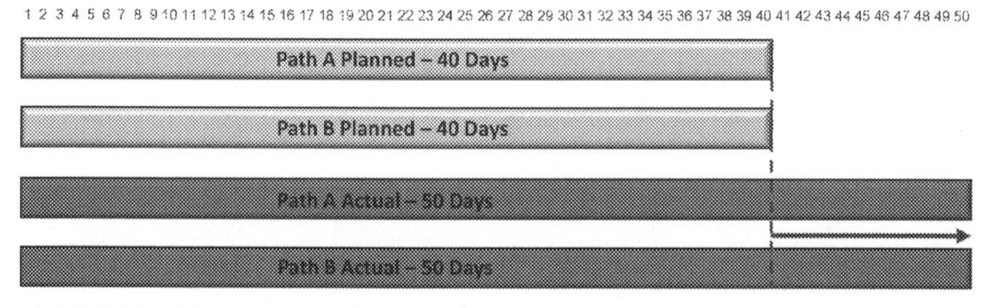

Figure 7.29 Concurrent delay example 4.

However, when we move forward in time one day, to the end of Day 3, we find the delay to Path B depicted in Fig. 7.31.

Fig. 7.31 shows that work on Path A made expected progress resulting in no delay to the Day 40 finish of Path A. However, note that a delay to Path B began on Day 3 that resulted in a delay to the finish of Path B to Day 41.

Moving further in time to completion of the delay to Path B, we see the following, which is depicted in Fig. 7.32.

Fig. 7.32 shows two things. First, Path A progressed as expected through the first 12 days, which did not delay Path A or the project. Second, the 10-day delay to Path B resulted in a 10-day delay to the completion of the project completion. The resulting forecast project completion date is Day 50. Therefore, the 10-day delay to Path B is responsible for delaying the project's completion date 10 days.

A consequence of the 10-day delay to Path B is that Path A now has 10 days of float with respect to the Day 50 project completion date.

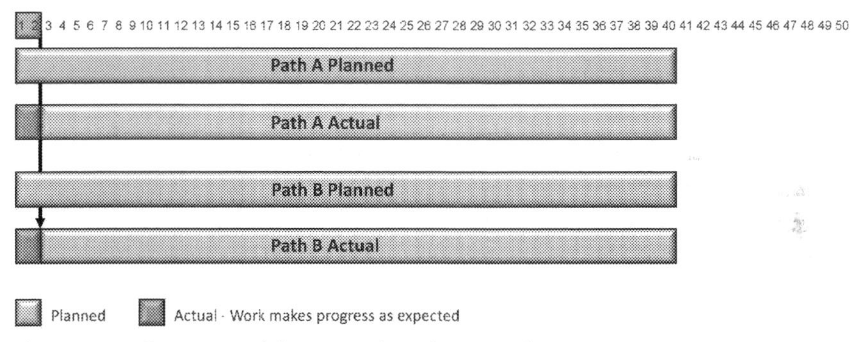

Figure 7.30 Concurrent delay example 4, Day 2 Update.

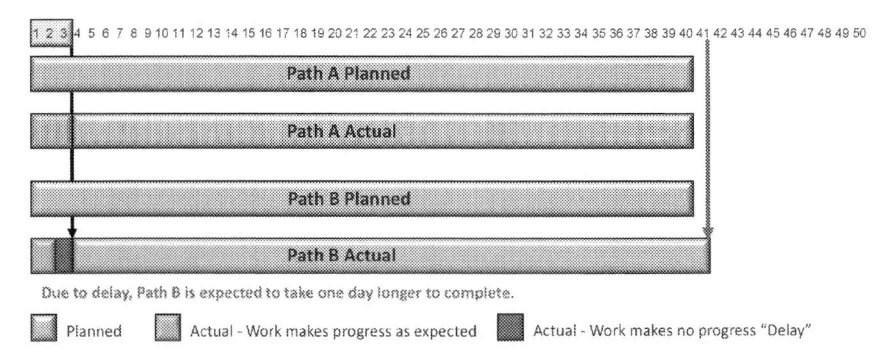

Figure 7.31 Concurrent delay example 4, Day 3 Update.

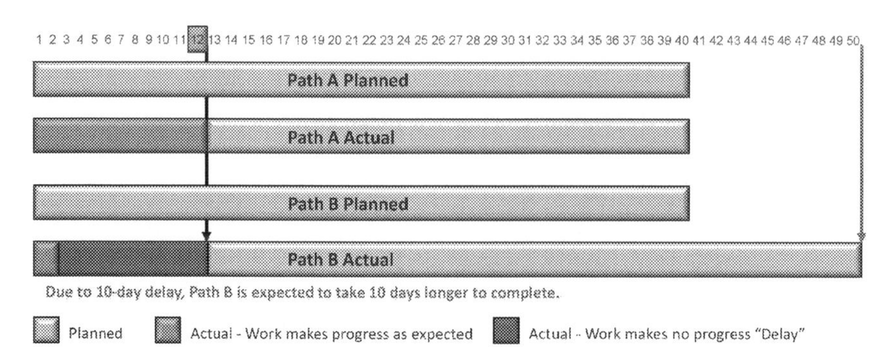

Figure 7.32 Concurrent delay example 4, Day 12 Update.

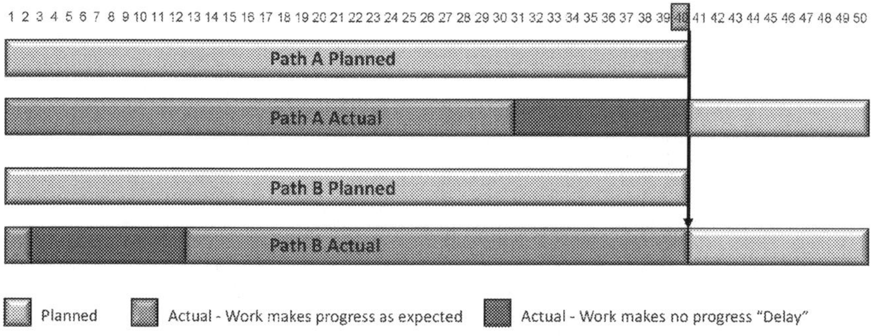

Figure 7.33 Concurrent delay example 4, Day 40 Update.

Moving further forward in time to when the delay to Path A occurred, we see the following, which is depicted in Fig. 7.33.

Fig. 7.33 shows us that Path B makes expected progress from Day 13 through Day 40 and that Path A makes expected progress from Day 13 through Day 30. Then, the delay to Path A starts on Day 31 and concludes on Day 40. As a result, Fig. 7.33 shows that as of the end of Day 40, the 10 days of float on Path A has been consumed or used up and the schedule forecasts that both Paths A and B will finish on Day 50. Because we already know that both Paths A and B actually finish on Day 50, we know that the remaining work on Path A and B finished by Day 50 and there were no further delays.

As stated above when discussing Fig. 7.29, if an analyst had solely looked at the start and finish dates of these two paths of activities, the analyst might conclude that they both delayed the project by 10 days. The more detailed analysis clearly shows that this is not the case. Instead, based on the chronological analysis of delays in Figs. 7.29–7.33, all 10 days of project delay were caused by Path B and no project delay was caused by Path A. In assessing delays during the same time frame, the analyst should perform the analysis on a day-by-day basis to correctly ascertain the exact activities that caused the delay and the correct magnitude of those delays.

Let us go back to our primacy of delay example whereby the bulk excavation work made no progress from June 1 to July 1, and the preparation of shop drawings was suspended from June 10 to June 25. There are some who advocate a theory of concurrent delay that would grant the contractor a 15-day, noncompensable time extension for the delay to the shop drawings. This theory of concurrent delay falls within the general category of a "but-for" argument. It basically argues that "but-for" my delay to the bulk excavation, I would have been delayed by the shop drawings. The major problem with this is that the "but-for" assumes that the two delays are unrelated and would have occurred regardless of the other. But the analyst really does not know that. It is entirely possible that the owner recognized the delay to bulk excavation and recognized that taking additional time on the shop drawing review would have no effect on the critical path of the project. The "but-for" argument is highly subjective and seldom persuasive. It is an argument, not an analysis.

The bottom line is that analyzing concurrent delays to separate critical paths is best done by using a technique to analyze delays that will identify the critical path of the project on every day of the project.

Concurrent delays to the same critical path

A delay to the project's critical path can have multiple causes. Assume that the first activity on the critical path of a highway project is the demolition of the substructure of an existing bridge. As of the schedule update of June 1, demolition was to be finished on June 15. If the start of demolition was delayed from June 1 to June 16 because of a union strike and concurrently the owner failed to obtain a permit, the two causes of delay are concurrent and run for the same period. The union strike is typically an excusable, noncompensable delay, and the lack of permit is typically an excusable, compensable delay. But recognize that this is not a concurrent delay but an example of concurrent causes for the same delay. Therefore, the analysis of the delay is simple: The critical path was delayed 15 days because of the late start of demolition of the existing bridge. The only task that remains is to determine which party is responsible for the delay. In this case, both the owner and the contractor share responsibility. The question of a time extension and of compensability is now a legal one and must be viewed in the context of the contract and the prevailing case law.

If both causes of the delay are concurrent but not of the same duration, then there may be an apportionment between the two causes. For example, if the causes both started on June 1 but the strike ended on June 10 and the Owner obtained the permit on June 15, then the first 10 days have a shared responsibility. The entitlement to a time extension and compensation must be determined from legal precedent for these 10 days. However, the remaining 5 days of delay appear to be attributable to the lack of a permit, and depending on the specific contract language, these 5 days may well be both excusable and compensable.

While application of the concept of apportionment appears simple, there are complications. First, there must be a clear definition of what delays are compensable and noncompensable. Second, the analyst must be able to show when the delays started and ended to allow an apportionment. The analyst must review all contemporaneous documents to ensure that the analysis reflects the status of the project when the delays were occurring. Proper research and documentation of what the parties knew as the delays unfolded will eliminate unsupported "but-for" arguments that advance predisposed conclusions or desired results.

An after-the-contract-completion-date concurrent delay argument

As stated earlier in this chapter, a contractor may wish to rely on a concurrent delay argument as a reason to convince an owner to not assess liquidated damages for a contractor-caused delay. An example of such an argument is when the owner is responsible for a noncritical delay after the project's contract completion date, which is illustrated in Fig. 7.34.

March Update

Act ID	OD	RD	Start	Finish	TF
A	60	0	2/1/17 A	3/31/17 A	
B	45	45	4/1/17	5/15/17	-35
C	55	0	2/1/17 A	3/27/17 A	
D	20	10	3/28/17 A	4/10/17	0
E	0	0		5/15/17*	-35

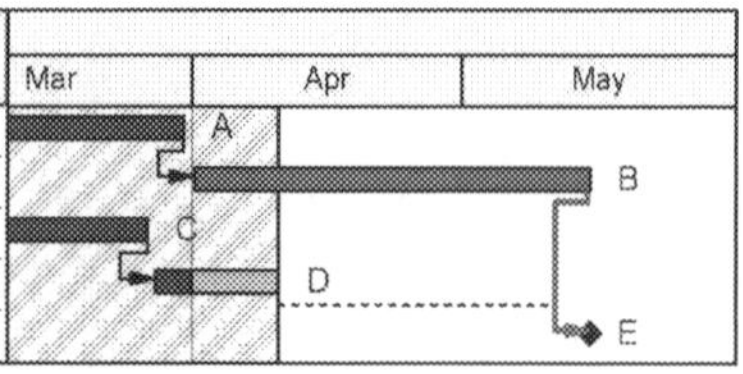

Figure 7.34 Concurrent delay example 5, March Update.

Fig. 7.34 depicts the project's March Update. The data date of the March Update is April 1, 2017, which is represented by the vertical *blue* (dark gray in print versions) line. All of the activities located to the left of the data date are completed or as-built and indicated as *blue* (dark gray in print versions) bars. The activities located to the right of the data date are planned. The project duration is represented by the shaded area on the bar chart portion of the printout and shows that the project's contract completion date is April 10, 2017. The example project includes two work paths. The first work path consists of Acts. A, C, and E, which is the project's critical or longest path, and which is forecasting that the project will finish on May 15, 2017. This path also has a total float value of −35 days because the project's contract completion date is April 10, 2017 and the May 15, 2017 forecast completion date is 35 days later than the April 10, 2017 contract completion date. The other work path consists of Acts. C, D, and E. Act. D, which is the last activity in this work path, is forecast to complete within the contract duration. In fact, Act. D is forecast to finish on April 10, 2017, which is also the project's contract completion date, and, as a result, Act. D has a total float value of 0 days. For the purposes of this example, let us assume that the contractor was responsible for the forecast late finish of work path A, B, E on May 15, 2017. In this example, the owner would assess 35 days of liquidated damages based on the contractor's delay to Act. B, which was the cause of the 35-day project delay.

Next, let us assume that during the month of April, Act. B progressed as expected, but Act. D did not. Fig. 7.35 depicts both the March and April Updates.

In the April Update, note that the data date moved to May 1, 2017, and the project's forecast completion date remained May 15, 2017. As such, the project was not delayed during the month of April. However, Act. D did not finish as expected on April 10, 2017. In fact, it made no progress in April and is now forecast to finish on May 10, 2017. As a result of Act. D's lack of progress in April, its total float value is −30 days. Let us assume that the delay to Act. D in April was caused by the owner.

In circumstances, such as those depicted in Fig. 7.35, the contractor might assert that, because the owner's 30-day delay to Act. D occurred after the contract completion date, the 30-day owner delay relieves the contractor of responsibility for 30 days of liquidated damages. This position would be taken despite the fact that this 30-day delay was not a delay

March Update

Act ID	OD	RD	Start	Finish	TF
A	60	0	2/1/17 A	3/31/17 A	
B	45	45	4/1/17	5/15/17	-35
C	55	0	2/1/17 A	3/27/17 A	
D	20	10	3/28/17 A	4/10/17	0
E	0	0		5/15/17*	-35

April Update

Act ID	OD	RD	Start	Finish	TF
A	60	0	2/1/17 A	3/31/17 A	
B	30	15	4/1/17 A	5/15/17	-35
C	55	0	2/1/17 A	3/27/17 A	
D	20	10	3/28/17 A	5/10/17	-30
E	0	0		5/15/17*	-35

Figure 7.35 Concurrent delay example 5, Comparison of March and April Updates.

to the project's critical path. In essence, the contractor is asking the owner for a 30-day time extension due to a noncritical delay.

There are at least two reasons why the contractor's request for a time extension should be denied. First, the contractor's delay was not critical. In accordance with the CPM theory and the primacy of delay concept discussed earlier in this chapter, the contractor's delay simply consumed float. The contractor cannot dispute that it was responsible for delaying the project 35 days well before the owner delayed noncritical work.

In this case, the contractor's position has many flaws, perhaps the most significant flaw being that it conflicts with the most basic tenet of CPM scheduling, which is that only delays to the critical path of the project will delay the project's completion. In recognition of this fact, nearly all well-written construction contract time extension provisions state that "the contractor is only entitled to a time extension when an excusable delay affects the critical path and delays the project completion date." Because the contractor is seeking a time extension for a noncritical path delay, the contractor is not entitled to a time extension or to relief from liquidated damages.

Other helpful software tools

As-built diagrams with Microsoft Excel

As-built information can be depicted in a number of ways, depending on the level of detail required by the analyst. As-built diagrams can be

ACTIVITY	MAY 2006																							
	1	2	3	4	5	6	7	8	9	10	11	12	13	14	15	16	17	18	19	20	21	22	23	24
Fly Forms - 4th Floor	x	x																						
Reinforce Slab - 4th Floor								x	x	x														
Pour Slab - 4th Floor											x													
Form Columns - 4th to 5th Floor													x		x									
Pour Columns - 4th to 5th Floor																x								
Strip Slab and Columns- 4th Floor																							x	
Fly Forms - 5th Floor																	x	x						
Reinforce Slab - 5th Floor																			x	x		x		
Pour Slab - 5th Floor																							x	
Form Columns - 5th to 6th Floor																								
Pour Columns - 5th to 6th Floor																								
Strip Slab and Columns- 5th Floor																								

Waiting for rebar fabrication
Received rebar
Only 10 ironworkers

Figure 7.36 As-built diagram.

plotted manually on a graph paper, but it is often more efficient to create as-built diagrams using Microsoft Excel (Excel). With Excel, the analyst can organize the information in the same manner as a graph paper, but, in Excel, the information can easily be reorganized to facilitate the analysis after all of the data has been entered, the data can be color coded to further facilitate the analysis, and the analyst can add notes from the as-built information into the individual cell or as a comment as the analysis is performed. Excel also allows the analyst to plot a larger amount of information than graph paper, customize the look of the printouts, and send as-built diagrams as a file.

Fig. 7.36 depicts an as-built diagram created in Excel that details as-built information for concrete work between May 1, 2006 and May 24, 2006.

Fig. 7.36 depicts a late start and slower-than-expected progress to the reinforcement of slab on the 4th floor. The shaded highlights denote nonwork days according to the contractor's schedule. While reading through the as-built information, the analyst noted that the owner's daily log on May 3, 2006, stated that reinforcement of the 4th floor slab could not start because the contractor was waiting on the rebar fabricator. The owner's daily logs also stated that on May 7, 2006, the contractor received the rebar and began reinforcement of the 4th floor on May 8, 2006. Including this level of detail in the as-built diagram will provide the analyst a clearer understanding of exactly when the delays are occurring to specific activities and will assist the analyst in determining the cause of delays to the critical path of the project.

Schedule analysis with CASE software

The analysis of project schedules can require a significant effort, depending on many factors. The analysis of a 3-year project with monthly updates could take the analyst weeks to complete, even for a relatively straightforward project. For a large project, with several thousand activities, the analysis would likely take additional time due to the increased number of near-critical paths that the analyst would need to evaluate against the critical path. The mathematical comparisons between the near-critical path duration and the critical path duration become increasingly difficult for schedules containing many leads/lags, constraints, multiple calendars, and/or lack actual start and actual finish dates. Although it is certainly possible to perform the mathematics of a schedule analysis manually, it is often more efficient and cost effective to evaluate numerous, complex schedules using software applications. Computer-Aided Schedule Evaluation (CASE) software, currently owned by Trauner Consulting Services, Inc., is one software application that automates the contemporaneous schedule analysis, often decreasing the time it takes to perform a schedule analysis by at least 50%. Why only 50%? Just as in the use of any other analysis software application (Primavera, Microsoft Project, Excel, Claim Digger, CASE, etc.), the analyst must still correctly interpret the results provided by the software and verify the results of the schedule analysis using as-built information and project correspondence.

Quantifying delays can be a very arduous process for an analyst to complete. However, with the software tools now available, analysts can greatly improve efficiency without sacrificing the quality of their analysis. Of course, proper training in the use of any software is essential because the software is only as effective as the analyst using it.

CHAPTER EIGHT

Delay Analysis Using No Schedules

USE OF CONTEMPORANEOUS DOCUMENTS FOR SEQUENCE AND TIMING

The preceding chapters discussed the performance of a delay analysis using a detailed Critical Path Method schedule and a bar chart. This chapter addresses the "worst case" situation: the project with no as-planned schedule or schedule updates. This is the most difficult situation in which to perform a delay analysis. Again, as the available information decreases, the analyst must make more assumptions, and the analysis becomes more subjective. However, while this type of analysis is difficult, it is not impossible.

When there is no schedule upon which to base an analysis of delays, it is usually because no schedule was prepared. Alternatively, it may be that a schedule was prepared, but the schedule was so flawed that it could not serve as the basis for a reliable analysis of delays. With regard to this latter circumstance, keep in mind that perfection is not the standard when determining whether a schedule can be used to analyze delays. No schedule is perfect and it is the rare project that is built exactly as scheduled. The analytical approaches discussed in Chapter 7, Delay Analysis Using Critical Path Method Schedules, are often self-correcting when it comes to schedule errors. Consequently, the existence of errors is not reason enough to abandon the use of the project schedules to evaluate project delays. The rejection of an available schedule as the basis of analysis is discussed at greater length in preceding chapters of this book.

Regardless of the reason why a schedule is not available to analyze project delays, the approach to analyzing delays without a schedule is essentially the same.

The first step to performing an analysis without schedules is to review the project document files to identify anything that might serve as a basis for determining how the parties planned to execute the project. There

Construction Delays.
DOI: http://dx.doi.org/10.1016/B978-0-12-811244-1.00008-2

are many potential sources, but the following is a useful guide for identifying the documents most likely to provide as-planned information regarding the planned sequence of the work or timing for the project work activities. The analyst should review the following:

- The contract documents for any specific required sequencing, phasing, or staging for the project work (Fig. 8.1).
- Correspondence between the general contractor and the owner or owner's representative for references to sequencing or timing, even if only for a portion of the project (Figs. 8.2 and 8.3).
- Subcontract agreements to look for any sequencing or timing dictated to the subcontractors by the general contractor concerning subcontract work.
- Correspondence with subcontractors for any discussion concerning scheduling, sequencing, or timing (Fig. 8.4).

Special Conditions of the Contract–West Street Bridge–Section 1.01, paragraph 5.a:
In staging its work the contractor must complete all abutment construction activities prior to paving of the approaches on both the east and west sides of the bridge.

Special Conditions of the Contract–West Street Bridge–Section 0.01, paragraph 3.b:
TIME FOR COMPLETION: All contract work must be completed by the middle of the 24th week after the Notice to Proceed.

Figure 8.1 Example contract language specifying work sequence.

June 15, 2007

John Lewis
Owner's Representative
Job Trailer
West Street Bridge
Podunk, New York 00001

Dear Mr. Lewis:

As we discussed yesterday during our meeting, we plan to work our construction operations on the West Street Bridge from an east to west direction. In other words, we will construct abutment #1 first, and then abutment #2. Our paving of our approaches will follow the same sequence.

Similarly, our operations on the piers will progress from east to west, starting with pier #1 and progressing through pier #3.

I trust this answers the question you raised, and you can notify the Township authorities accordingly so that they may affect the appropriate detours during the course of the project.

Sincerely,
Joe Super
Ace Construction

Figure 8.2 June 15, 2007 contractor correspondence.

August 20, 2007

Mr. Joe Super
Ace Construction
Job Trailer/West Street Bridge
Podunk, New York 00001

Dear Mr. Super:

Returned herein are the shop drawings submitted by Ace Construction for the reinforcing steel for the pier caps. As you will note, the drawings are approved as noted. Fabrication of the reinforcing steel can begin in accordance with the corrections noted.

I must note these drawings are being returned to you five weeks later than you requested. There are two reasons why the return of the shop drawings is five weeks beyond your requested date. First, you did not allow the engineer adequate time to review the drawings. As a minimum, the engineer requires three weeks to review, but you allowed only one week if they were to be returned within the time frame requested.

Second, the drawings were not prepared in accordance with normal shop drawing practices and consequently required much more time to review. As submitted, the drawings were extremely difficult to understand. While the engineer could have returned them disapproved and required resubmission in the proper manner, it was decided to take the extra time to correct them on the first submission in order to expedite the process.

If you have any questions, please feel free to call.

Sincerely,
John Lewis
Owner's Representative

Figure 8.3 August 20, 2007 owner correspondence.

June 25, 2007

Mr. Grey Ferrous
Ferrous Steel Erectors
Iron Street
Steeltown, New York 00002

Dear Mr. Ferrous:

In accordance with our previous conversations, Ace Construction will require the delivery and erection of the steel for the West Street Bridge to begin by the beginning of the 13th week of the project. Based on your present fabrication schedule, it appears that this should not be a problem.

Please be advised that time is of the essence on this contract. In the past, we have had problems with steel suppliers promising delivery dates and not adhering to them. Action such as this will not be tolerated on this project. Your delivery and erection of steel is critical to our timely completion of the project.

Sincerely,

John Lewis
Owner's Representative

Figure 8.4 June 25, 2007 contractor correspondence to subcontractor.

- Partial schedules produced during the project, which may describe the planned sequencing or timing for portions of the project (Fig. 8.5).
- Meeting minutes of any discussions concerning scheduling, particularly the preconstruction meeting minutes (Fig. 8.6).
- Daily log or diary entries by either the general contractor's personnel or the owner's representative (Fig. 8.7).

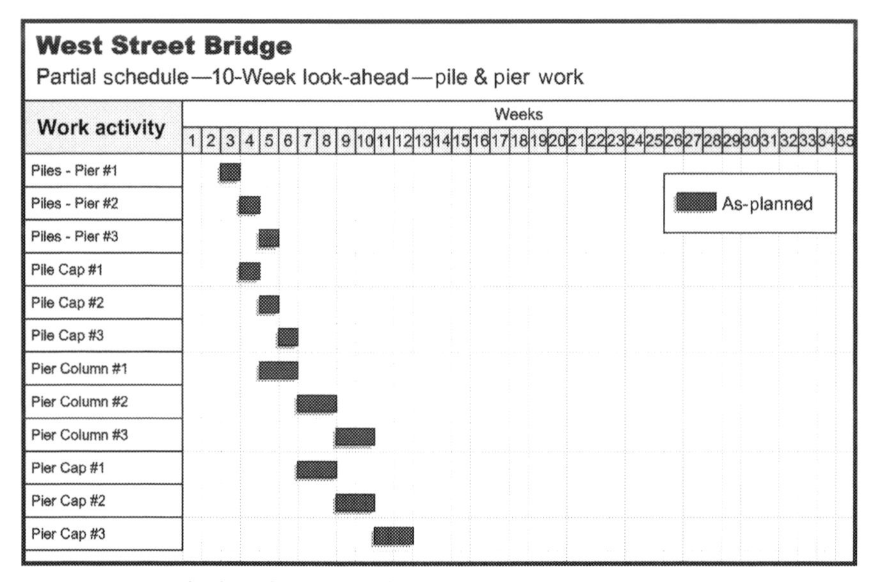

Figure 8.5 Partial schedule, 10-week look-ahead.

Preconstruction Meeting Minutes
(Excerpt)

Contractor noted that the forming, reinforcing, and placing of the concrete deck spans will take two weeks per span. Contractor will proceed from east to west constructing span #1 through span #4.

Figure 8.6 Preconstruction meeting minute excerpt.

- Purchase orders with suppliers that show planned dates for delivery of materials or equipment.

The analyst should use all the available project information to define some form of an as-planned schedule or at least to establish a general planned sequence for the work on the project for different parts or portions of the project.

Based on the sample information provided in Figs. 8.1–8.7, the analyst is unable to determine an as-planned schedule. However, the following information can be determined:

1. The contract required that all abutment work be completed before any paving was performed on the approaches. The contract also required that completion of the project occur by the middle of the 24th week (from Fig. 8.1).

Inspector's Daily Diary
(Excerpt)

During discussions today with Joe Super (Ace Construction), he noted that the abutment work would take about four weeks to perform for each abutment, and the paving of the approaches would take about two weeks for each side of the bridge. Joe noted that the existing abutment sills are acceptable as shown on the contract drawings. Therefore, they can accept the steel as is. The remainder of the work on the abutments can be performed either before the steel is in place or after it is set. He is going to try to have the abutments completed before he sets any steel, since the work will be easier that way. However, he is concerned about the lead time allowed for obtaining the epoxy material called for in the contract. He said that if necessary, he would set the steel so as not to delay the deck work which he feels controls his completion of the job.

Figure 8.7 Daily diary excerpt.

2. The contractor planned to work from east to west, or from Abutment #1 to Abutment #2, and from Pier #1 to Pier #3 (from Fig. 8.2).
3. The owner's representative noted that the review and approval of the shop drawings for the reinforcing steel in the pier caps had taken longer than required and were not returned to the contractor until 5 weeks later than requested (from Fig. 8.3).
4. The contractor planned on erecting the steel at the beginning of the 13th week of the project (from Fig. 8.4).
5. The contractor produced at least one, 10-week "look-ahead" schedule that addressed the work on the piers and described the specific sequence for this work, including the piles, pile caps, pier columns, and pier caps (from Fig. 8.5).
6. The contractor anticipated a 2-week duration for the forming, reinforcing, and placing of each of the concrete deck spans (from Fig. 8.6).
7. The contractor expected 4-week durations for each of the abutments and 2-week durations for each of the approaches (from Fig. 8.7).

Based on this information, the analyst can proceed with the analysis, despite the lack of a complete as-planned schedule.

Keep in mind that the goal is not to prepare an "alternative" or "after-the-fact" baseline or as-planned schedule for the project. Rather, the goal is to mine the project information in search of the contractor's original plan for construction. While the analyst may be tempted to create an after-the-fact schedule, reasoning that it will allow the analysis to be more precise, the opposite is true. To the extent a "planned" schedule is drafted as the basis of the analysis, it should be done so with only the information that can be gleaned from project documents, such as the examples outlined in this chapter, along with mandatory sequencing dictated by the physical construction attributes and conditions. More importantly, it should be used with the knowledge and understanding that it is incomplete and limited in terms of its precision.

USING AN AS-BUILT ANALYSIS TO QUANTIFY DELAYS

For projects that have no planned or updated schedules, the daily reports may provide sufficient detail to allow for a reliable analysis of delays, especially when used in conjunction with the planned information gleaned from the project documents. Daily reports allow the analyst to prepare an as-built diagram for the job. The as-built diagram for the example project based on the daily reports is depicted in Fig. 8.8.

In reviewing the as-built diagram, it is evident that the project was delayed seven-and-one-half weeks. It was to be completed by the middle of the 24th week, but was not actually completed until the end of the 31st week. Therefore, the analyst must account for at least seven-and-one-half weeks of net delay. The delay could, in fact, be greater if the contractor had planned to finish the project early, or if the contractor was able to mitigate some previous delays. However, there is no indication in the project documents to establish a plan to finish early. The as-built diagram also shows that the contractor did work from east to west as planned.

Based on the letter to Ferrous Steel Erectors (Fig. 8.4), the contractor planned to start steel erection at the beginning of the 13th week. The as-built diagram (Fig. 8.8) shows that steel erection began at the beginning of the 18th week. A 5-week delay can be identified for the start of the steel erection.

Following this conclusion, we would break down the activities that precede the start of steel erection. Using the 10-week "look-ahead" schedule from Fig. 8.5, which the contractor produced at the end of the second week of the project, a comparison of this schedule with the as-built diagram for this portion of the project shows that a delay occurred to the construction of the pier caps. A delay also appears in the construction of Pier Column #3, which took 3 weeks instead of 2 weeks to complete.

The comparison of the partial "look-ahead" schedule with the as-built portion is shown in Fig. 8.9.

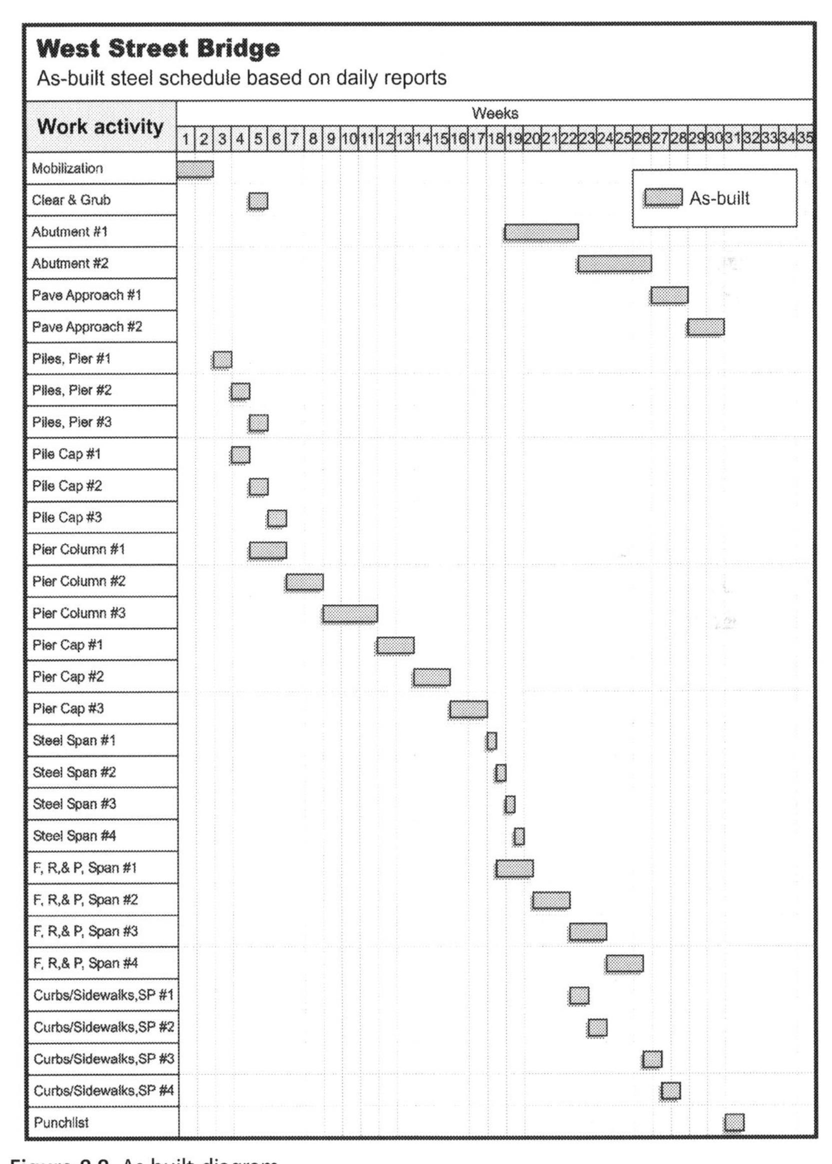

Figure 8.8 As-built diagram.

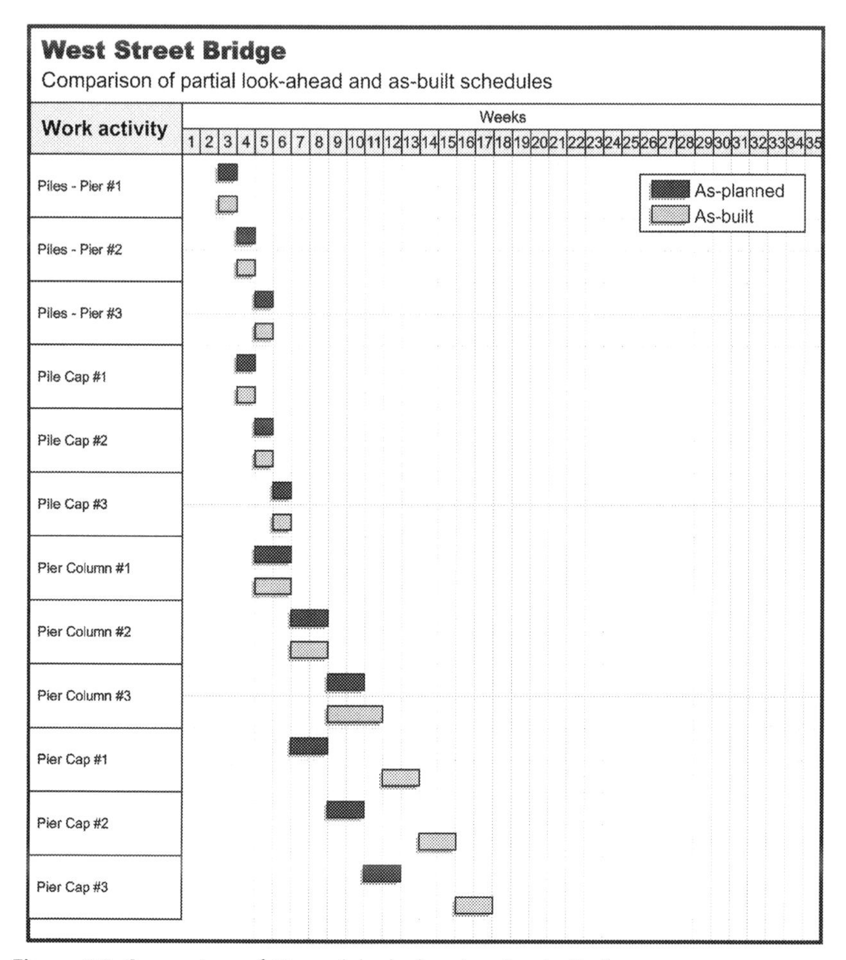

Figure 8.9 Comparison of 10-week look-ahead and as-built diagram.

Based on this comparison, it appears that the initial critical delay was the delay to the start of the Pier Cap #1. This may have been the result of the 5-week delay in approval of the shop drawings to which the owner's representative had referred in correspondence (Fig. 8.2). It does not appear that the delay in the completion of Pier Column #3 affected the completion of the project, as the flow of activities in the "look-ahead" schedule (and in the logical construction sequence) would have been in the stair-step fashion depicted.

The remainder of the work along the pier and deck path appears to have been performed in a logical, sequential fashion and does not indicate further delay. However, there is a gap between the completion of the curbs and sidewalks and the beginning of the punch list work. Based on the as-built diagram, the punch list work did not start until the abutments and approaches were completed. The abutment and approach work occurred in the sequence planned. However, it appears that the performance of the abutment and approach work delayed the project an additional two-and-one-half weeks, as this is the gap between the completion of the superstructure work and the start of the punch list work. Based on the documentation available, it can be determined that the abutment and approach work started later than planned, but was performed within the duration planned by the contractor.

A summary of the delays appears in Fig. 8.10. The delay of seven-and-one-half weeks is from:

1. Five weeks of delay to the start of the pier cap work, from start of Week 7, when Pier Cap #1 should have started, to the start of Week 12, when Pier Cap #1 actually started.
2. Two-and-one-half weeks of project delay due to the 13-week late start of the abutment and approach work.

At this point in the analysis, the analyst should again carefully review the project documentation to determine if other supporting information is available to further substantiate the delay analysis. Of particular interest in this case would be the need to confirm the assumption that the project delay is properly attributed to the late start of Pier Cap #1 and not the delayed start of abutment work. For example, if the project documentation shows that the start of the abutment work was delayed because the owner failed to procure the needed permits to allow the abutment work to start and that the permits were not going to be received until Week 11 or later, then the permit delay may be the reason that the contractor waited to start the pier cap work until all three columns were completed. In fact, doing so may have been prudent in that it avoided the mobilization of additional work barges that would have been necessary in the original plan. Keep in mind that, in the original plan, the pier and deck work may have been critical, but once the permit was delayed to the point where the abutment and approach permit path was now the longest

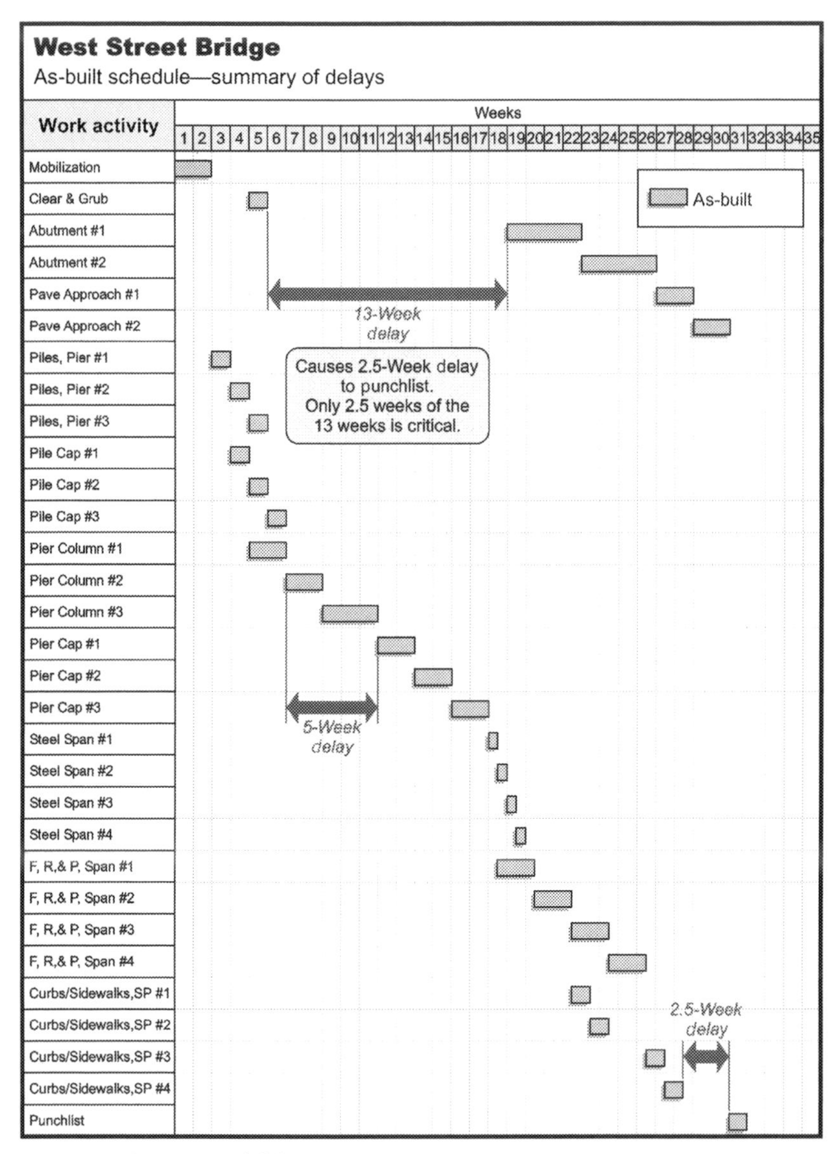

Figure 8.10 Summary of delays.

path, each additional day of delay would add float to the pier and deck path. Critical path shifts and the consumption and creation of float are discussed in detail in the previous chapters of this book. Despite the lack of a planned schedule in this case, the project still has a dynamic critical path that must be considered.

CHAPTER NINE

Other Retrospective Delay Analysis Techniques—Their Strengths and Weaknesses

Chapter 7, Delay Analysis Using Critical Path Method Schedules, described the Contemporaneous Schedule Analysis, which is the preferred retrospective delay analysis method. This chapter discusses other retrospective delay analysis techniques and their strengths and weaknesses.

There are several types of retrospective delay analysis techniques being used by analysts to identify and quantify critical project delays. While some may be appealing when first considered, their flaws become evident as the analyst further considers the assumptions upon which the technique is based. Ultimately, some techniques fail for obvious reasons. Others fail in more subtle ways. Some take an unbalanced view of project events. Others ignore important documents or information like the project schedule updates or as-built information.

There are several things to keep in mind when evaluating delay analysis techniques or methods. The following attributes may not encompass everything that must be considered, but they highlight important issues when evaluating alternatives:

1. To begin with, the analysis must be performed objectively. One way to achieve this objectivity is for the analyst to determine the source and magnitude of critical project delays without regard to the party responsible. For example, an analysis of the schedule may reveal that the late start of foundation excavation caused a critical delay to the project. This conclusion should be made independently from who caused this excavation work to start late. Determining the party responsible for a delay is a separate task that should not determine how a delay is analyzed.
2. Another way for the analyst to reinforce objectivity is to rely on the contemporaneous project schedules as the basis of analysis to the maximum extent possible. Analyses that stray far from the contemporaneous schedules are often biased and unpersuasive.

Construction Delays.
DOI: http://dx.doi.org/10.1016/B978-0-12-811244-1.00009-4

3. If the analyst is trying to identify and quantify all the delays to the project, then the analysis should account for all project delays (and time savings) throughout the duration of the project. In particular, every day of critical delay should be accounted for by the analyst. In the end, the sum of the critical delays and time savings identified by the analysis should equal the total net delay actually experienced by the project.

While there are numerous approaches used to analyze delays, care must be exercised throughout the process. The method used must incorporate the available contemporaneous information, recognize the dynamic nature of a construction schedule and the critical path, and avoid after-the-fact hypotheses.

SCHEDULE-BASED DELAY ANALYSIS TECHNIQUES

In the remaining sections of this chapter, we describe the following retrospective delay analysis techniques and identify their strengths and weaknesses.

- As-Planned versus As-Built Analysis
- Impacted As-Planned Analysis
- Collapsed As-Built Analysis
- Retrospective Time Impact Analysis
- Windows Analysis
- But-For Analysis

AS-PLANNED VERSUS AS-BUILT ANALYSIS

In every delay analysis, the analyst attempts to explain why the project finished late. When using the as-planned versus as-built technique, the analyst compares the project's baseline (as-planned) schedule to the project's as-built performance to discern the reasons for the project's late completion. This analysis is usually performed at a high-level using summary activities to depict both the project's baseline schedule activities and the actual performance of the work. AACE International's

Recommended Practice No. 29R-03, Forensic Schedule Analysis (RP-FSA), describes the technique as follows:

> *In its simplest application, the method does not involve any explicit use of CPM logic and can simply be an observational study of start and finish dates of various activities. It can be performed using a simple graphic comparison of the as-planned schedule to the as-built schedule. A more sophisticated implementation compares the dates and the relative sequences of the activities and tabulates the differences in activity duration, and logic ties and seeks to determine the causes and explain the significance of each variance. In its most sophisticated application, it can identify on a daily basis the most delayed activities and candidates for the as-built critical path.*

Fig. 9.1 is a graphical depiction of this analysis derived from AACE's RP-FSA.

Strengths and weaknesses of the As-Planned versus As-Built Analysis

One strength of the As-Planned vs. As-Built Analysis is that it is easy to understand and present. This is because the analysis is performed at such a high-level and does not get down into the details of how the project was planned and how it was actually built. Another strength is that the analysis can be performed with limited schedule and as-built information.

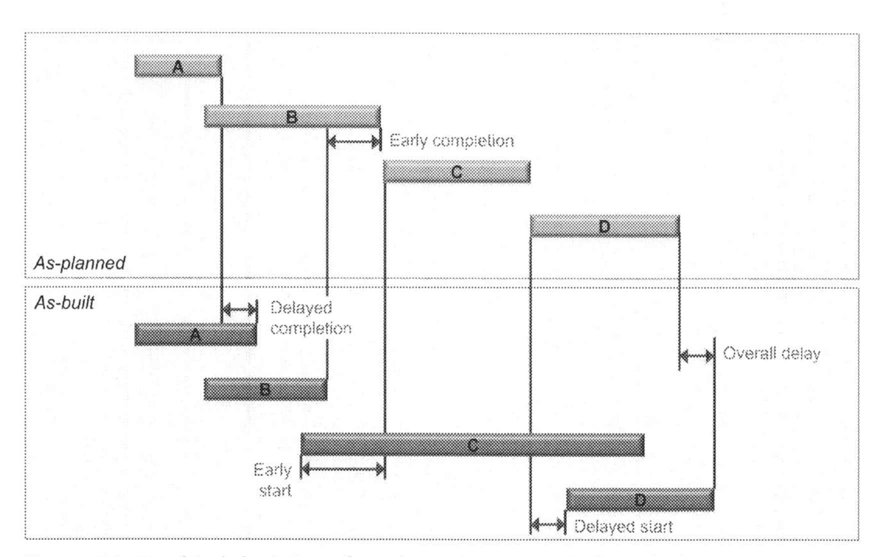

Figure 9.1 Graphical depiction of as-planned versus as-built analysis.

These strengths also point to a considerable number of weaknesses. For example, its simplicity is also its downfall. Because the baseline schedule is the only project schedule that the analysis relies on to establish the project's construction plan, the analysis automatically assumes that the project should have been built exactly as planned. However, if the project is built in a different sequence or order than planned, the analysis is unable to properly identify and measure the project delays because the revised plan does not match the baseline schedule that is being used as the basis for comparison. Additionally, this reliance on the project's baseline schedule may also limit the ability of the analysis to properly identify delays caused by the addition of work, because the added work did not exist in the baseline schedule. Another weakness that results from reliance on just the baseline schedule is that the analysis typically does not consider or account for shifts in the project critical path.

The description of the analysis technique excerpted from AACE's RP-FSA also does not specifically demand that the analyst identify the project's critical path or demonstrate that the identified delays were also critical path delays. The result is that this analysis may identify noncritical path delays as delays for which the contractor would be entitled to a time extension, even though these delays were not critical delays delaying the project completion date.

Also, this analysis technique's lack of detail limits its ability to identify and address concurrent delays.

IMPACTED AS-PLANNED ANALYSES

Some analysts prefer to identify and measure project delays using only the project's baseline or as-planned schedule. Based on this preference, these analysts insert delays that they believe were responsible for the project delays into the project's baseline or as-planned schedule. This method is known as an Impacted As-Planned Analysis.

In order to perform this analysis, the analyst must first identify each delay and, in doing so, will usually predetermine the party responsible for this delay. Then the analyst develops a "fragnet" for each delay that is to be analyzed and inserts this fragnet into the baseline schedule, usually all at once with all the other delay fragnets the analyst has identified for analysis. The analyst then reruns the schedule with all fragnets inserted and compares the result to the original baseline schedule. The difference in the scheduled project completion dates is the delay attributable to the delay fragnets inserted into the schedule. Note that for the analytical results of this approach to be useful, each delay fragnet the analyst inserts must have been caused by the same party, usually the owner. Otherwise, the results of the analysis would be inconclusive, because some of the delay could have been caused by the owner and some by the contractor with no quantitative differentiation. For the same reason, if the contractor is seeking additional compensation for all the delay shown by the analysis, then all of the delay fragnets inserted must be compensable delays. This type of analysis can be done using a scheduling program; however, it may also be done manually using bar charts.

The analysis can be performed by inserting an activity or activities to represent a single issue or delay, or multiple issues or delays in sequence, both of which will be described.

Single issue or delay—Impacted As-Planned Analysis

Fig. 9.2 represents a simple, four-activity excavation project that will be used to illustrate how an Impacted As-Planned Analysis with a single issue or delay is performed.

Fig. 9.2 depicts the project's baseline schedule and shows that the project consists of excavating four areas (A, B, C, and D) in sequence. Project completion was scheduled for the end of Week 4.

During the first week, while excavating in Area A, the contractor encountered rock, which the contractor believed to be a differing site condition. The contractor notified the owner of the rock immediately and the owner issued a stop work order for 1 week. At the end of the week, the owner agreed that the rock was a differing site condition, lifted the suspension, and authorized the contractor to remove the rock. The contractor removed the rock, which took another week. Once the rock was removed, the contractor resumed with the remaining planned excavation work in Area A.

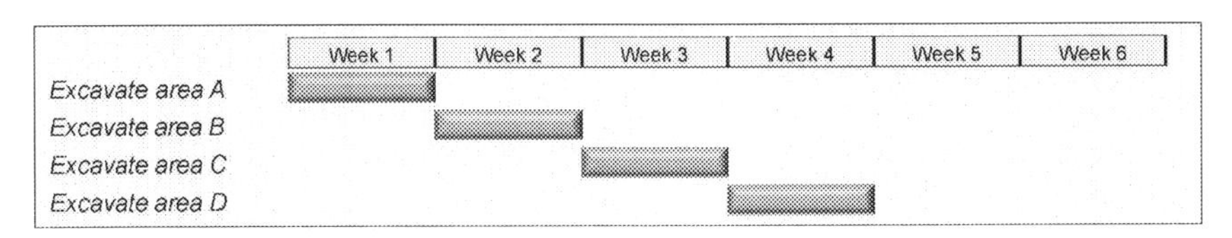

Figure 9.2 As-planned schedule.

Relying on this information, the analyst developed two activities, a 1-week long "Stop Work Order" activity and a 1-week long "Added Rock Excavation" activity. Then, the analyst inserted these two activities into the baseline schedule, splitting the Excavate Area A activity into two separate activities. The result is the impacted baseline schedule depicted in Fig. 9.3.

Using the impacted baseline schedule, the analyst concluded that the 2-week delay to Act. A caused by the addition of the "Stop Work Order" and "Added Rock Excavation" activities resulted in a 2-week delay to the project. Based on an initial review, the analyst's presentation makes sense. Its strength is that it is simple, straightforward, and relies on both the project's baseline schedule and known project facts.

However, this simplicity and the use of the baseline schedule is also its biggest weakness. By relying on the baseline schedule, the analysis is static and assumes that except for the added work, the project work progressed exactly as planned at the time of bid. Anyone with experience on construction projects knows that it is the rare project that is constructed exactly as-planned, particularly a project that is experiencing delays. The significant piece missing from this analysis is what actually happened, which again is ignored in the analysis because the tool used to measure the delay was the project's baseline schedule and the fragnet for the change. Fig. 9.4 depicts what actually happened on the project.

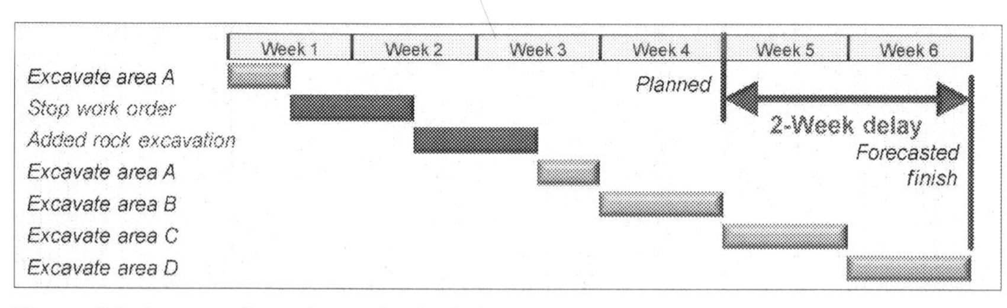

Figure 9.3 Impacted as-planned schedule.

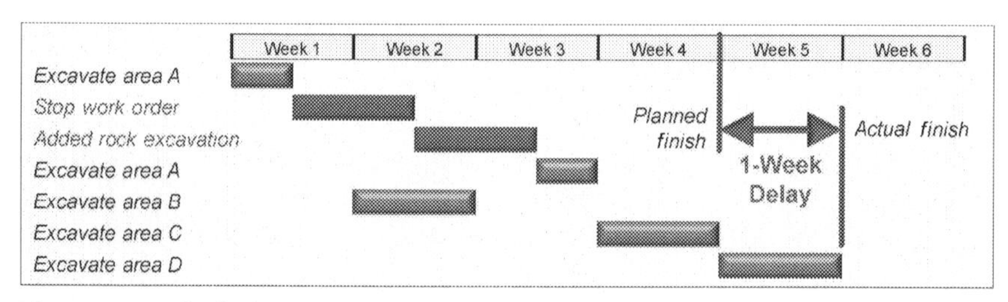

Figure 9.4 As-built diagram with delays.

Fig. 9.4 confirms that the 2-week delay caused by the "Stop Work Order" and "Added Rock Excavation" did, in fact, delay the Excavation in Area A, but the contractor acted prudently and mobilized its crew to Area B during Week 2 to complete the Area B Excavation during the 2-week delay to Area A. In other words, the contractor acted to mitigate some of the delay caused by the changed condition in Area A and the project was finished only one week late, not two, as represented in the impacted schedule.

Based on this example, one of the major weaknesses of an Impacted As-Planned Analysis is that it does not consider the actual project progress and events and, consequently, often overstates the project delay.

Note, also, that analysts performing an Impacted As-Planned analysis usually consider or depict only the delays caused by one party. The delay or delays that are inserted into the baseline schedule are typically not the delays attributed to the party performing the analysis. As such, the analysis is inherently biased since it does not consider delays caused by the other parties to the project. If only the delays caused by one party are considered, is it any surprise that only delays caused by that party are identified?

Note, also, that this analytical approach requires the analyst to identify, quantify, and develop fragnets for all delays attributable to a particular party, not just the critical delays. This can make for a lot of unnecessary analytical work. A final consideration is that the identification and quantification of delays and the development of fragnets is often subjective. This step in the analysis introduces an element of subjectivity that is undesirable.

Multiple issue or delay—Impacted As-Planned Analysis

To illustrate how the Impacted As-Planned Analysis is performed measuring the effect of multiple issues or delay, we will begin with the simple project depicted in Fig. 9.5. Note that the analyst is working for the owner for this example and the delays being identified are allegedly contractor-caused delays.

Fig. 9.5 shows that Activities A through F, representing Phase 1, must be performed in sequence, and are scheduled to occur based on the durations shown. At the same time, but not until after the completion of Activity B, there is another sequence of consecutive activities consisting of Activities G through I, representing Phase 2, with the respective durations shown. The total project duration is 35 days.

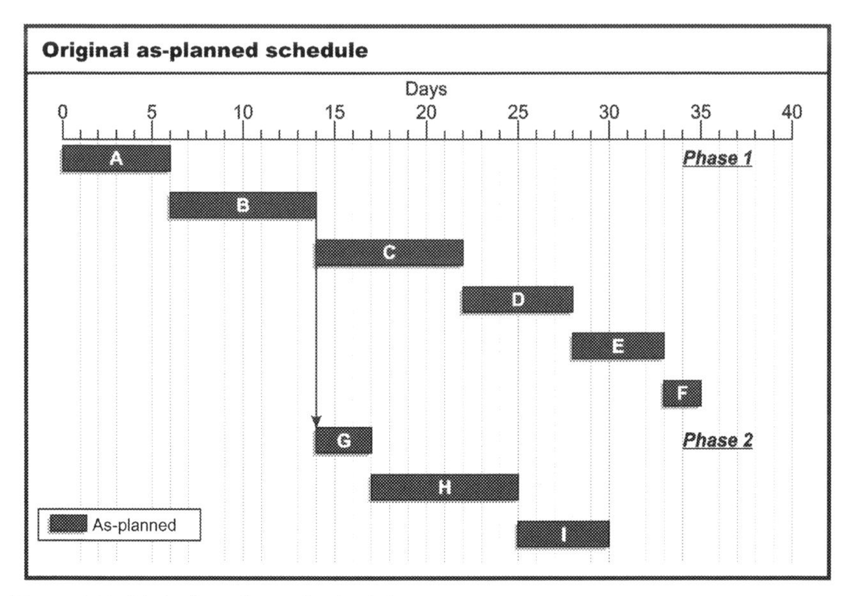

Figure 9.5 Original as-planned schedule.

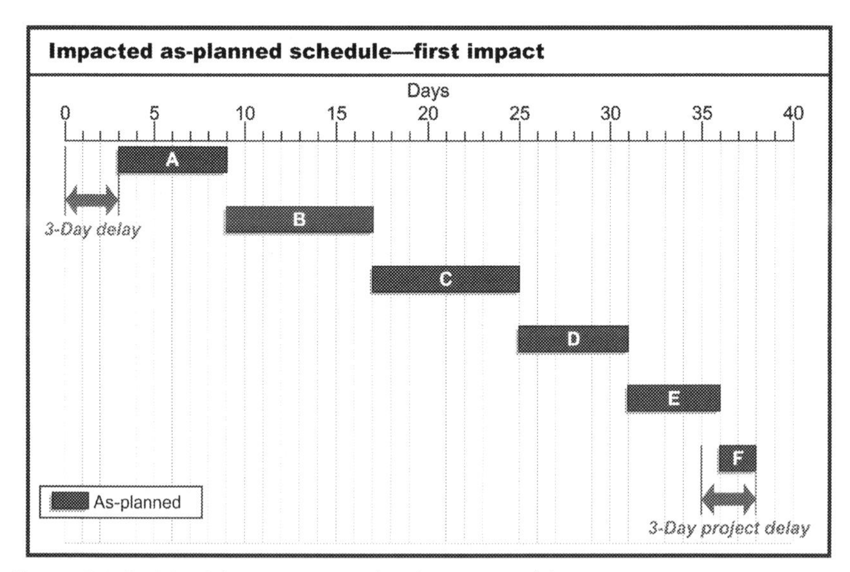

Figure 9.6 Activity A late start caused 3-day project delay.

In its Impacted As-Planned Analysis, the analyst alleges that three critical path activities (Acts. A, C, and F) were not performed as-planned due to events for which the contractor was responsible. These are the three events for which the analyst develops and inserts fragnets into the

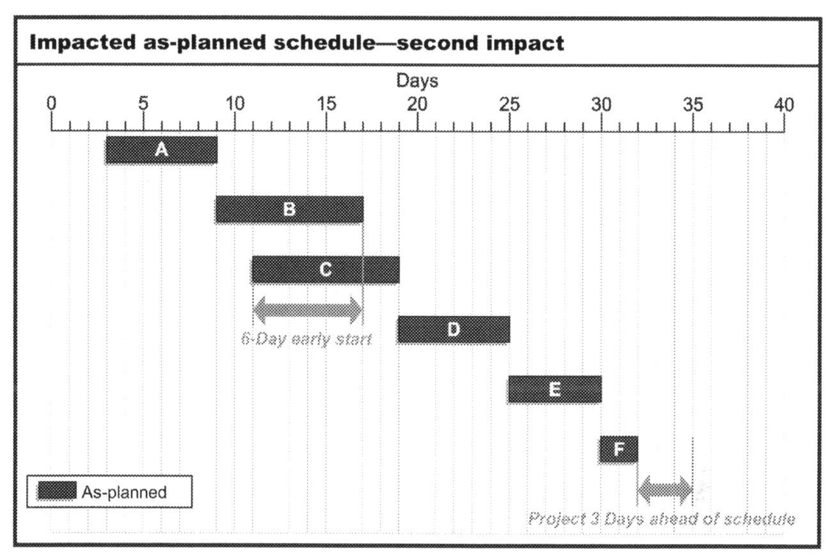

Figure 9.7 Activity C early start caused 3-day project savings.

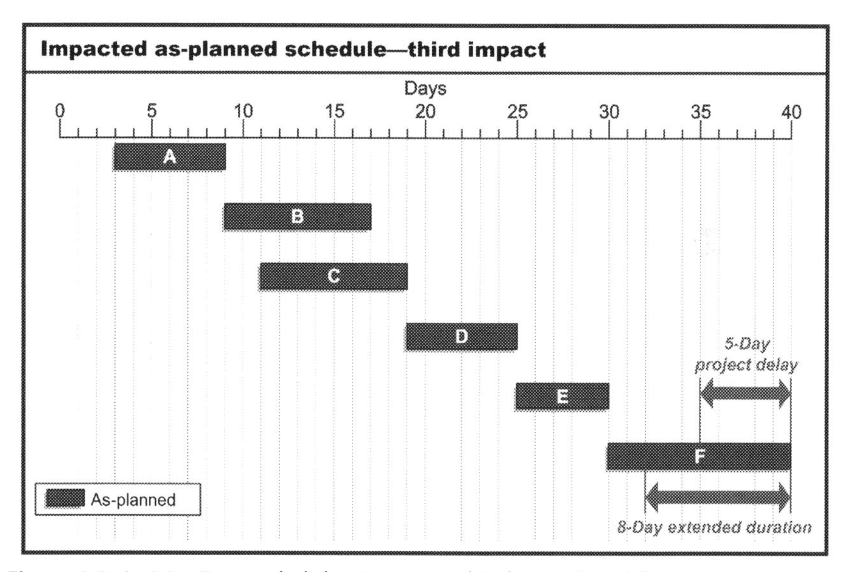

Figure 9.8 Activity F extended duration caused 5-day project delay.

baseline schedule to evaluate delays to the project. The analyst added the performance of each activity in sequence into the baseline schedule and obtained the following results:

The analyst concluded that that the project was delayed by 5 days, primarily because of the late finish of Activity F. Because all three of these events were determined by the analyst to be the contractor's responsibility, the analyst concluded that the net project delay of 5 days was the responsibility of the contractor. However, as is often the problem with an Impacted As-Planned Analysis, the analyst failed to consider what actually happened. In particular, the analysis did not consider owner-caused delays. As illustrated by the analytical approaches discussed in Chapters 5 and 7, Measuring Delays—The Basics and Delay Analysis Using Critical Path Method Schedules, instead of relying solely on impacting the baseline schedule with the three alleged contractor-caused events, the analyst should compare the baseline schedule to an as-built diagram of how the project was constructed, considering all the possible delays. Fig. 9.9 depicts the project schedule with all of the as-built information added for comparison.

Fig. 9.9 shows us that not only were the activities that the analyst identified completed later than planned, but additional work, represented by Activity J, was added to the contract. Note that Activity J finished on Day 40, which was the same day that Activity F actually finished. Also note that the analyst's Impacted As-Planned Analysis did not account for or consider the performance of Activity J because it was not included in

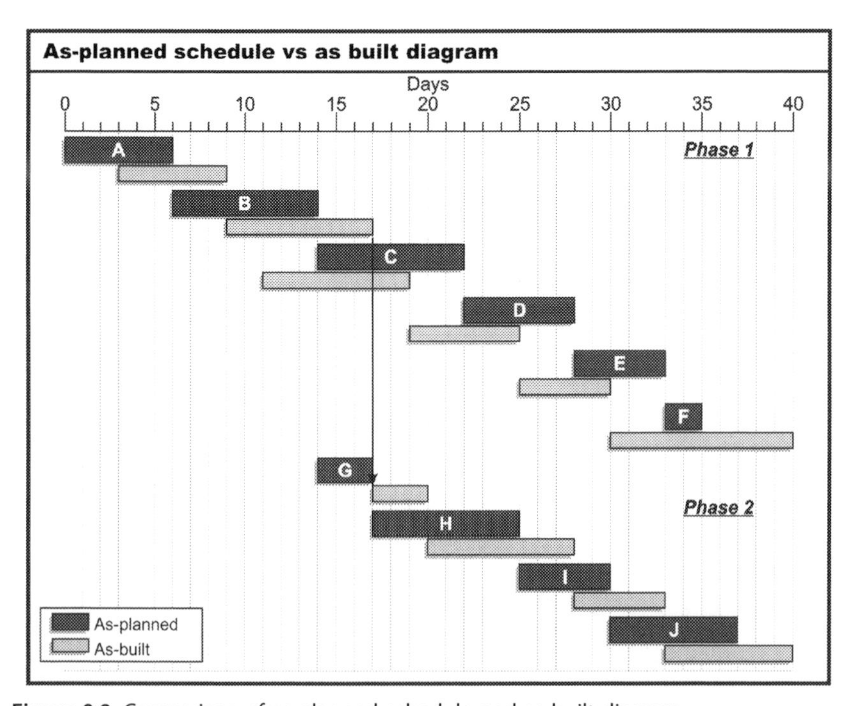

Figure 9.9 Comparison of as-planned schedule and as-built diagram.

the baseline schedule and because it was a change that was the owner's responsibility.

Additionally, we know that the extra work represented by Activity J was added to the project on Day 15. To ensure that we properly account for the project delay, if any, that was caused by the addition of the Activity J work, we updated the project schedule as of Day 15, which is depicted in Fig. 9.10. This update includes the addition of Activity J.

Fig. 9.10 shows us that despite the late start of Activity A, the early start and progress achieved on Activity C resulted in an improvement of 3 days to the completion of the Phase 1 path from Day 35 to Day 32, and a shift in the critical path from Phase 1 to Phase 2. Thus, by Day 15, prior to the insertion of new work, the Phase 2 activities were on the critical path and the project was 2 days ahead of schedule. However, with the addition of Activity J on Day 15, the project was now scheduled to finish on Day 40. The Phase 2 path remained critical for the rest of the project duration and completed on Day 40 with the completion of Activity J.

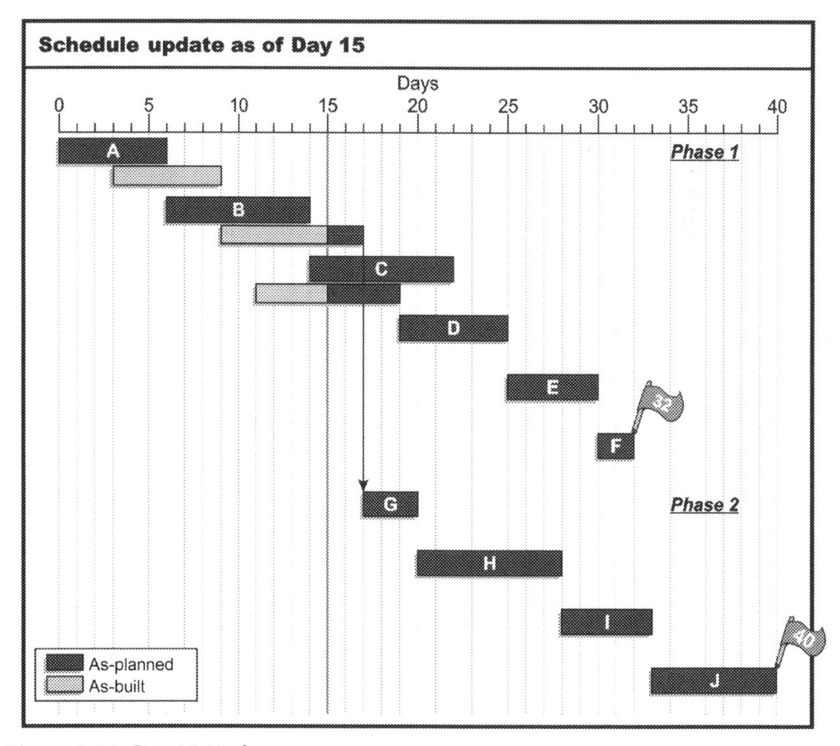

Figure 9.10 Day 15 Update.

Based on the proper analysis of the project, we see that, despite the fact that Activity F took longer than planned and also finished on Day 40, it was not the cause of any critical project delay. Rather, the entire 5-day delay to the project was caused by the addition of Activity J by the owner. Again, in this case, the Impacted As-Planned Analysis led to the wrong conclusion.

Strengths and weaknesses of the Impacted As-Planned Analysis

The strength of the Impacted As-Planned Analysis is the same as the strength of the As-Planned versus As-Built Analysis in that both are easy to understand and present.

However, the major weaknesses of this approach are that:

- The "impacted" schedule does not depict or consider the actual project events as they occurred.
- The decision as to what changes or impacts are to be modeled in the schedule is highly subjective and biases the ultimate outcome.
- It does not consider the dynamic nature of a construction project and the critical path.

Because the Impacted As-Planned approach does not update the schedule using actual as-built information or rely on the contemporaneous schedule updates, there is no comparison made between events depicted in the "impacted" schedule and actual project events. This technique assumes that the project would have progressed exactly as it was planned, except for the inserted fragnets. In most instances, this assumption ignores what really happened on the project and does not consider changes in sequence or changes in the critical path.

Additionally, the analyst cannot always identify from the project documentation all circumstances that affected the project schedule. Events that are never mentioned in any project documentation can and often do affect the schedule. For example, a specific activity might have a longer-than-planned duration. If the extended duration of this activity is not noted in the project documentation as an "impact" or a change, then its effects may not be measured. If all the delays that affected the schedule are not inserted into the impacted schedule analysis, then the results will be incomplete and unreliable.

In the methodologies described in the preceding chapters, an extended duration would be accounted for by updating the schedule with the actual as-built durations from contemporaneous documentation, such as the daily reports. In this way, every delay is "automatically" considered as part of the analysis. It does not depend on an analyst's decision to characterize the extended duration of the activity as a delay, develop an associated fragnet, and insert this fragnet into the schedule to quantify the delay that might have been associated with the performance of the activity.

Another major weakness of this analysis technique is that it is often a subjective or one-sided representation of the project delays. In most cases, when this technique is used, the analyst only inserts the delays caused by the party opposing the analyst's client. By inserting the delays of the opposing party, only, the results of the analysis are inherently biased against the opposing party.

Last, but not least, the reliance on only the baseline schedule does not consider the dynamic nature of construction projects or the critical path. As a consequence, the analysis does not consider the actions taken by the party to mitigate project delays. The result is that the Impacted As-Planned approach almost always overstates the actual project delays.

While this method is simple to present and understand, it almost always gives erroneous results. Also, it is usually easy to show that the results are erroneous. As a result, in addition to be inaccurate, it is not very persuasive. It also fails on a fundamental level. The analyst first identifies the delays without performing any analysis. The technique merely provides for the quantification of the delay created by the analyst. It is the tail wagging the dog. The purpose of a delay analysis is to both identify and quantify a delay, not just confirm the analyst's subjective evaluation of project events. Most qualified analysts consider the Impacted As-Planned approach to be unreliable and it is now rarely used.

COLLAPSED AS-BUILT ANALYSES

Another method of analysis is the Collapsed As-Built Analysis, also called the subtractive as-built. This approach is perhaps best described as the reverse of the Impacted As-Planned Analysis described in the previous section. It addresses one of the more significant weaknesses of the Impacted As-Planned Analysis in that it considers or accounts for how the project was actually built. However, it ignores the original project plan.

In the Collapsed As-Built Analysis, the analyst studies all of the contemporaneous project documentation and prepares a detailed as-built schedule or chart. Often, this as-built schedule is a CPM network that is input into a scheduling program with logic. However, it may also be drawn on a time scale, or done manually using bar charts.

It is important to note that when developing an as-built schedule in this manner, the analyst wants the as-built schedule to match how the project was constructed as closely as possible. For example, all of the activity start and finish dates in the as-built schedule should match when each of the work activities actually started and finished. In addition, the analyst must also insert logic relationships between the schedule activities. Typically, these logic relationships are added to ensure that the activities start and finish on their actual dates. However, unlike actual activity start and finish dates, which can often be determined or verified by using other documentation, such as daily reports, diaries, or photographs, the as-built schedule logic must be identified by the analyst. The determination of the logic of construction is much more subjective, since it is based not only on the mandatory logic of construction—the footing has to be constructed before the foundation wall that sits on the footing—but also preferential logic related to how many crews and how many pieces of equipment the contractor mobilizes to perform the work.

All schedules developed after-the-fact are subjective. The as-built schedule prepared for the Collapsed As-Built Analysis is no exception. This subjectively opens the door to bias.

Having prepared an as-built schedule, the analyst "subtracts" or removes delays. If the removal of these delays affects the "subtracted" schedule's end date, then the difference in days between the as-built and the collapsed as-built schedules' end dates is considered to be the delay caused by the delay or delays removed.

Note that for a scheduling program to be used to "collapse" the as-built schedule in this manner, the analyst must develop an as-built schedule with a data date of Day 0 to ensure that when it removes the alleged delays from the as-built CPM schedule, the end date will "collapse." If the data date was the date of final completion, then the scheduled completion date would not change as you removed delays from the schedule.

There are several ways to perform a Collapsed As-Built Analysis. Two of them are discussed in the paragraphs that follow.

Unit subtractive as-built

For simplicity, the first variation is referred to as a unit subtractive as-built approach. This method starts with the preparation of the overall as-built schedule and subtracts one "impact" at a time in an attempt to correlate a measurable number of days attributable to each "delay" removed. Once the alleged delays are removed, the argument is made that the project would have finished earlier were it not for the impacts removed.

Gross subtractive as-built

The second variation is referred to as the gross subtractive as-built approach. It is performed in two steps. This method starts with the preparation of the overall as-built schedule and removes all possible impacts that may have caused a delay to the project. These potential impacts are both owner and contractor caused. The resulting schedule and overall project duration supposedly represent the time the project would have taken if no problems had occurred.

During the next step in the process, the analyst adds specific problems back into the now "collapsed" as-built, one at a time. As the analyst reinserts each impact, the corresponding increase in the project duration or delay is attributed to the respective impact.

Strengths and weaknesses of the collapsed or subtractive As-Built Analysis

As noted above, one strength of the Collapsed As-Built Analysis, unlike the Impacted As-Planned Analysis, is that the analysis is based on what actually happened.

However, the overall analysis method, regardless of the variation used, has several major weaknesses. The three primary weaknesses are:

- This method requires the analyst to construct a detailed as-built schedule based on as-built information. This requirement, as described above, is extremely subjective and highly amenable to error and manipulation. With little effort, the analyst can create an as-built schedule that supports a predisposed conclusion.
- The method also requires the analyst to identify or determine specific "impacts" before performing any analysis. Not only is this subjective, it is impossible. The method forces the analyst to first reach a conclusion about what caused a delay, and then uses the analysis to prove its conclusion. This is the proverbial "tail wagging the dog." For example, the analyst is forced to add logic relationships between activities that never existed in the contemporaneous project schedule to force activities to start or finish on their actual dates. By doing so, float time

between the activities is improperly characterized as activity time or as a lag in a logic relationship.

- Lastly, the analysis' absolute reliance on only the project's as-built information ensures that the Collapsed As-Built Analysis suffers the same static perspective of the project schedule as the Impacted As-Planned Analysis. Where the Impacted As-Planned Analysis firmly set the analyst's perspective of the project from Day 0 and assumed the project was constructed exactly as planned, the Collapsed As-Built Analysis "freezes" the critical path based on the as-built critical path constructed by the analyst. The critical path then changes depending on the "subtraction" made, but these critical path "shifts" rarely match the shifts shown in the contemporaneous project schedules.

This method ignores the fact that the project schedule is dynamic and ignores the contemporaneous schedule update submissions that were used to plan and manage the project. It also ignores how the initial plan was depicted in the baseline schedule. In fact, if the essence of the delay is that it forced the contractor to execute the project in a different sequence than planned, because this analysis considers only what was built and not what was planned, it cannot reliably measure the delay.

RETROSPECTIVE TIME IMPACT ANALYSIS

The Retrospective TIA is similar to the Prospective TIA in that the performance of the analysis involves first preparing fragnets made up of the activities and logic used to model the delay. The fragnet is then inserted into the appropriate project schedule, the schedule is re-run, and the difference in the scheduled completion dates between the original schedule and the schedule with the added fragnet is used to quantify the associated critical project delay, if any.

However, unlike the Prospective TIA, which is more or less universally accepted as the approach an analyst should use to estimate the delay associated with a change that has not yet occurred, the Retrospective TIA is not. This is because it suffers from subtle but significant flaws that can cause the results to be unreliable and inaccurate.

For example, as discussed in Chapter 7, Delay Analysis Using Critical Path Method Schedules, in a perfect world and on a perfect project, the Prospective TIA should be performed before the changed or added work is started. On those projects, the parties are able to agree on both the cost and time needed to perform changed or added work before it is performed. This agreement includes agreement on the fragnet activities, the durations of these activities, the logic relationships among these and other schedule activities, and the update or version of the contemporaneous project schedule that will be used for the analysis.

Unfortunately, most Retrospective TIAs are performed well after the changed or added work is completed or even well after the project itself has finished. In these situations, the analyst typically performs the analysis without the agreement of the other party.

As with a Prospective TIA, most analysts focus on selecting the activities to be used in the fragnet, evaluating the logic relationships among the fragnet activities themselves and the rest of the project schedule, and selecting the appropriate schedule update for inserting the fragnet. However, when a TIA is performed retrospectively, the retrospective nature of the analysis introduces errors that undermine the objectivity, accuracy, and reliability of the analysis. These errors have three primary sources:

- The analyst relies on as-built durations for fragnet activities.
- The actual performance of parallel work paths is not considered.
- The analysis evaluates the delays caused by different parties differently.

Retrospective Time Impact Analysis Example 1

We will use the same tunneling example that was used to represent the Prospective TIA in Chapter 7, Delay Analysis Using Critical Path Method Schedules, to show how a Retrospective TIA that relies on the as-built durations for the fragnet activities can overstate the project delay.

D-Tunneling was performing the tunneling and drainage piping installation for a large terminal expansion at an airport. In its contract with the airport authority, D-Tunneling received its notice-to-proceed on February 1, 2017, and was required to complete the project by June 5, 2017. Before starting construction, D-Tunneling created a baseline schedule reflecting its original plan for the work, which is depicted in Fig. 9.11.

D-Tunneling's baseline schedule showed that the project's critical path consisted of the Tunnel A work and the project's scheduled completion date was June 5, 2017.

On the afternoon of March 6th, TBM #2 encountered rock at Station 75 + 00 that was harder than the geotechnical report indicated in

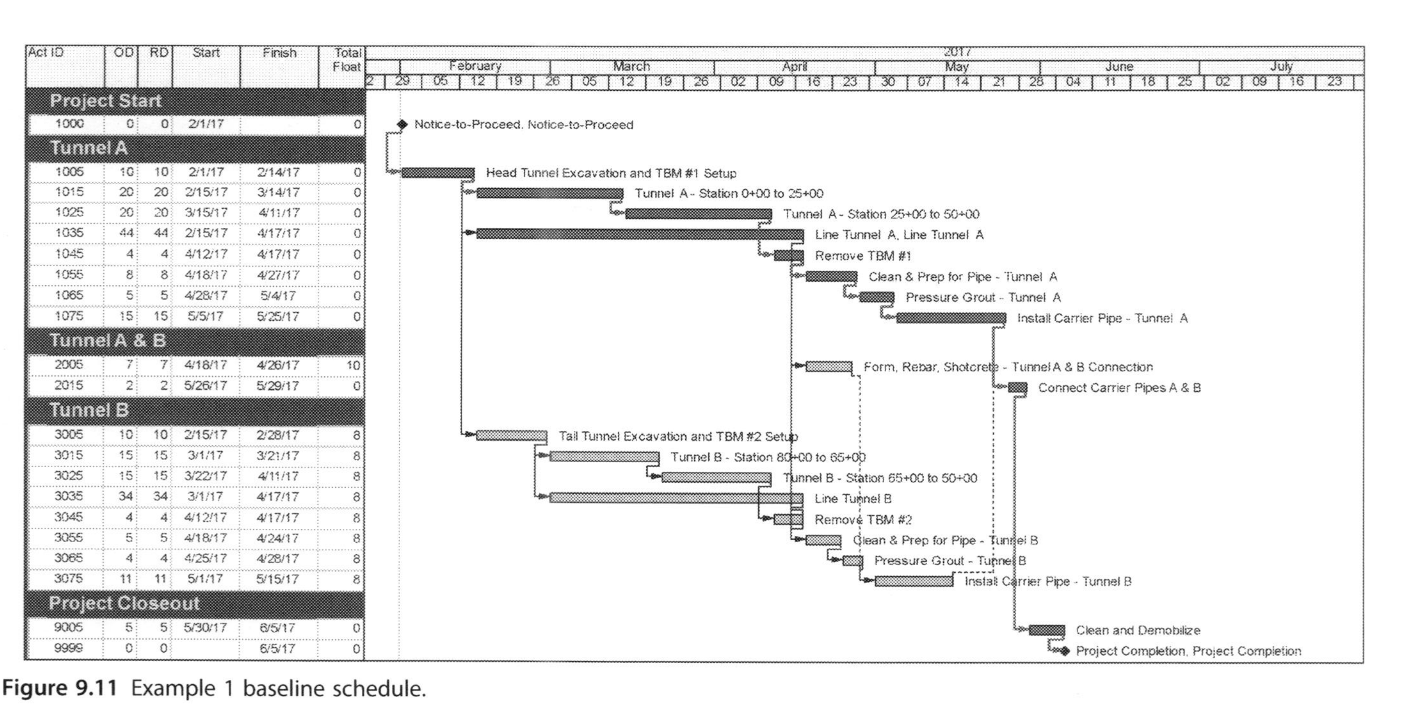

Act ID	OD	RD	Start	Finish	Total Float
Project Start					
1000	0	0	2/1/17		0
Tunnel A					
1005	10	10	2/1/17	2/14/17	0
1015	20	20	2/15/17	3/14/17	0
1025	20	20	3/15/17	4/11/17	0
1035	44	44	2/15/17	4/17/17	0
1045	4	4	4/12/17	4/17/17	0
1055	8	8	4/18/17	4/27/17	0
1065	5	5	4/28/17	5/4/17	0
1075	15	15	5/5/17	5/25/17	0
Tunnel A & B					
2005	7	7	4/18/17	4/26/17	10
2015	2	2	5/26/17	5/29/17	0
Tunnel B					
3005	10	10	2/15/17	2/28/17	8
3015	15	15	3/1/17	3/21/17	8
3025	15	15	3/22/17	4/11/17	8
3035	34	34	3/1/17	4/17/17	8
3045	4	4	4/12/17	4/17/17	8
3055	5	5	4/18/17	4/24/17	8
3065	4	4	4/25/17	4/28/17	8
3075	11	11	5/1/17	5/15/17	8
Project Closeout					
9005	5	5	5/30/17	6/5/17	0
9999	0	0		6/5/17	0

Figure 9.11 Example 1 baseline schedule.

the contract documents. In addition, the rock was considered "mixed face" and contained a mixture of rocks with varying compositions and hardnesses. As a result, D-Tunneling stopped work on Tunnel B while the issue was being investigated and resolved.

After the project was completed, D-Tunneling submitted a time extension request to the airport authority. D-Tunneling's time extension submission included the following explanation of the delay and requested a time extension of 24 calendar days from June 5, 2017 to June 29, 2017:

1. The airport authority/engineer was responsible for all soil borings and site testing in the contract.
2. On March 6, 2017, D-Tunneling had encountered "mixed face" rock at Station 75 + 00 that was harder than the tolerance limits identified by the engineers' soil boring tests in the contract. As a result, D-Tunneling had to shut down its TBM #2 operations in Tunnel B. D-Tunneling has and will continue work in Tunnel A, as planned.
3. The engineer has completed additional testing of the area and determined that the out-of-tolerance rock exists between Station 75 + 00 and Station 69 + 00.
4. D-Tunneling has purchased a new cutterhead that was delivered on April 3, 2017. The cutterhead was assembled in 4 workdays and D-Tunneling resumed tunneling in Tunnel B on April 7, 2017.
5. Using the above timeline to calculate the delay resulting from the differing site condition, D-Tunneling has inserted a fragnet into its March 6, 2017 Update, containing the following changes:
   a. Activity 3015 was given an actual finish of March 5, 2017, reflecting only the work completed between Station 80 + 00 and Station 75 + 00.
   b. Activity 3015A, Order/Ship/Rec. New Cutterhead for TBM #2, was added to the schedule to represent the time it took to receive the new cutterhead. Activity 3015A was given a duration of 21 workdays, which resulted in the calculation of a start date of March 6, 2017, and a finish date of April 3, 2017.
   c. Activity 3015A1, Assem. Cutterhead and Return to Station 75 + 00, was added to the schedule to represent the time it took to assemble the new cutterhead. Activity 3015A1 was given a duration of 4 workdays, which resulted in the calculation of a start date of April 4, 2017, and a finished date of April 7, 2017.
   d. Activity 3015B, Complete Tunnel B—Station 75 + 00 to 65 + 00, was added to the schedule to represent the remaining tunneling

work after the new cutterhead was assembled. Activity 3015B was given a duration of 13 workdays.

e. Logic was revised to incorporate the changes identified in items **a–d**. All logic changes shown are finish-to-start relationships with no leads or lags. No other logic relationships in the schedule were changed.

Fig. 9.12 depicts D-Tunneling's March 6, 2017 Update with the fragnet inserted and the schedule rerun. Note that Fig. 9.12 forecasts a completion date of June 29, 2017.

In essence, to perform its analysis D-Tunneling took the actual performance of the extra work items and inserted them into its March 6, 2017 schedule update to calculate the time extension to which it believed it was entitled.

In its evaluation of D-Tunneling's time extension request, the airport authority compared D-Tunneling's March 6, 2017 Update with the added fragnet to the project's as-built schedule. The as-built schedule is depicted in Fig. 9.13.

A comparison of this as-built schedule to D-Tunneling's Retrospective TIA analysis established the following:

- D-Tunneling did encounter harder rock on March 6, 2017.
- The new cutterhead did arrive on the project site on April 4, 2017.
- The new cutterhead was assembled and installed on TBM #2 on April 7, 2017.
- D-Tunneling resumed tunneling in Tunnel B on April 10, 2017.

However, the airport authority also noticed that the project actually finished on June 23, 2017, not June 29, 2017. It also noted that some of the activities that followed the delay event were performed differently than planned. Therefore, D-Tunneling's Retrospective TIA did not account for how the project was actually completed.

When compared to the Prospective TIA example in Chapter 7, Delay Analysis Using Critical Path Method Schedules, we know that D-Tunneling's time extension request in this example failed to consider the fact that the owner and contractor shifted some of the tunneling work that was to be completed by TBM #2 to TBM #1. This shifting of work from one TBM to the other was not considered in D-Tunneling's time extension request and accompanying Retrospective TIA.

Ultimately, the airport authority acknowledged that the differing site condition at Station 75 + 00 was, in fact, a critical delay to the project, but granted D-Tunneling a time extension of 18 calendar days from June

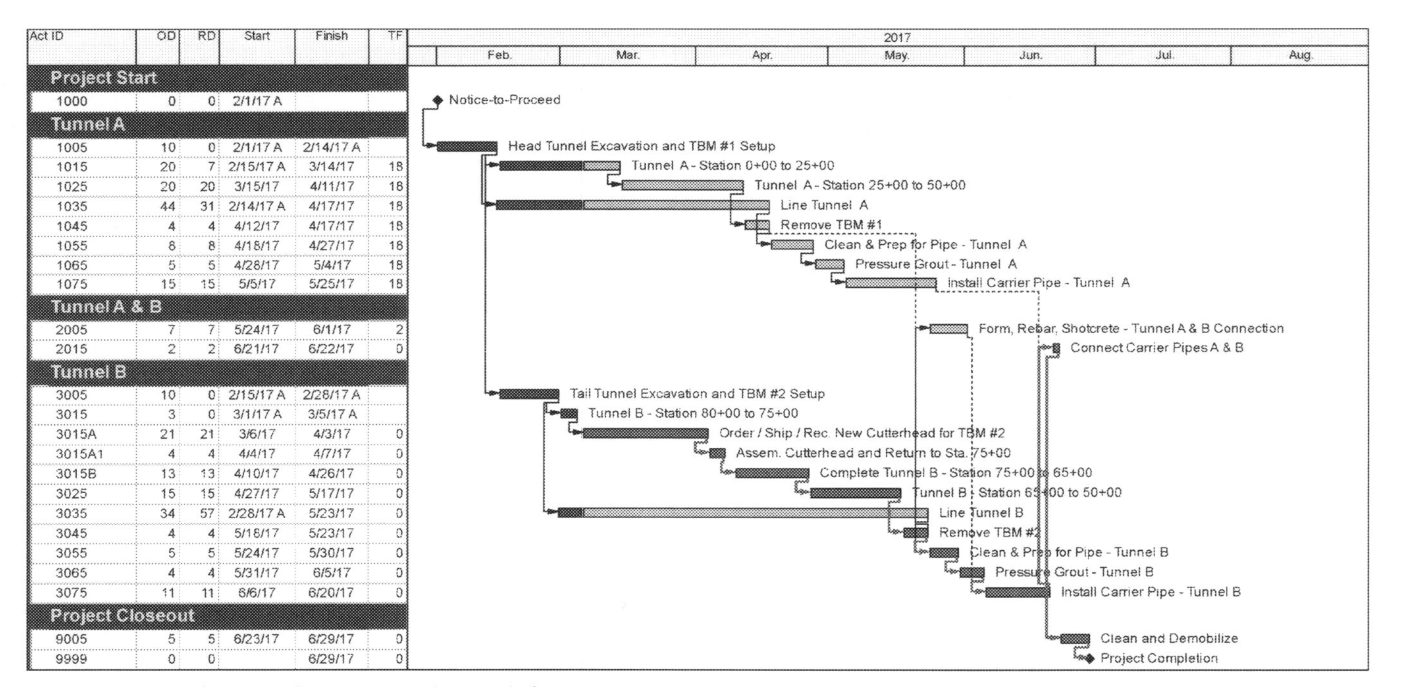

Figure 9.12 Example 1 March 6, 2017 Update with fragnet.

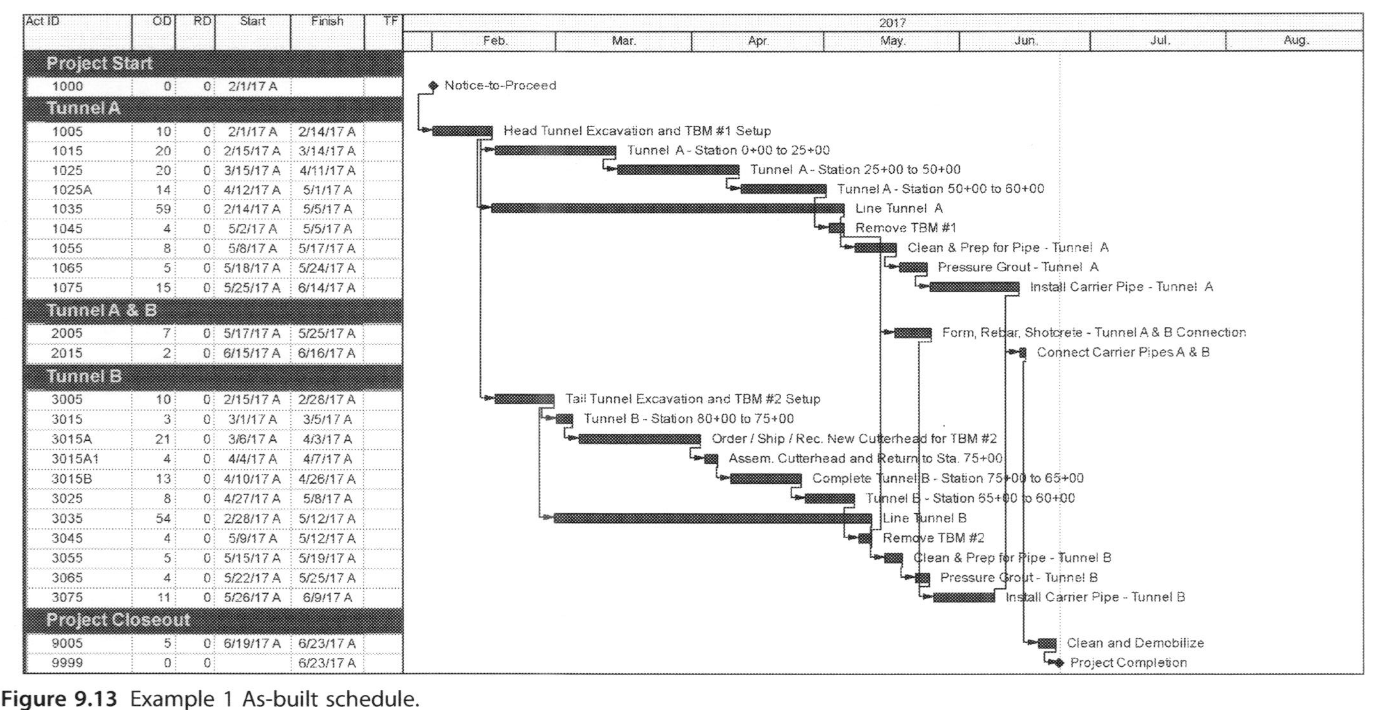

Act ID	OD	RD	Start	Finish	TF
Project Start					
1000	0	0	2/1/17 A		
Tunnel A					
1005	10	0	2/1/17 A	2/14/17 A	
1015	20	0	2/15/17 A	3/14/17 A	
1025	20	0	3/15/17 A	4/11/17 A	
1025A	14	0	4/12/17 A	5/1/17 A	
1035	59	0	2/14/17 A	5/5/17 A	
1045	4	0	5/2/17 A	5/5/17 A	
1055	8	0	5/8/17 A	5/17/17 A	
1065	5	0	5/18/17 A	5/24/17 A	
1075	15	0	5/25/17 A	6/14/17 A	
Tunnel A & B					
2005	7	0	5/17/17 A	5/25/17 A	
2015	2	0	6/15/17 A	6/16/17 A	
Tunnel B					
3005	10	0	2/15/17 A	2/28/17 A	
3015	3	0	3/1/17 A	3/5/17 A	
3015A	21	0	3/6/17 A	4/3/17 A	
3015A1	4	0	4/4/17 A	4/7/17 A	
3015B	13	0	4/10/17 A	4/26/17 A	
3025	8	0	4/27/17 A	5/8/17 A	
3035	54	0	2/28/17 A	5/12/17 A	
3045	4	0	5/9/17 A	5/12/17 A	
3055	5	0	5/15/17 A	5/19/17 A	
3065	4	0	5/22/17 A	5/25/17 A	
3075	11	0	5/26/17 A	6/9/17 A	
Project Closeout					
9005	5	0	6/19/17 A	6/23/17 A	
9999	0	0		6/23/17 A	

Figure 9.13 Example 1 As-built schedule.

5, 2017 to June 23, 2017, not the 24-day time extension requested from June 5, 2017 to June 29, 2017.

This example highlights a few of the deficiencies with the Retrospective TIA.

The first deficiency in this example relates to the fragnet inserted into the schedule by D-Tunneling. The fragnet modeled the added work in Tunnel B, but did not incorporate the shift of tunneling work from Station 50 + 00 to 60 + 00 from TBM #2 to TBM #1. This is a common deficiency of the Retrospective TIA. It considers the work that was added, but not how the parties reacted to or acted to mitigate the delay.

The second deficiency is more subtle. Note that all of the uncompleted work, which is located to the right of the March 6, 2017 data date, represents D-Tunneling's best estimate as to how much more time it needs to complete the remaining project work. D-Tunneling's March 6, 2017 Update, with the fragnet added, is comparing the as-built durations for the added work to the planned durations for work after March 6, 2017. Comparing as-built fragnet activity durations to the planned durations of existing project scope is an apples-to-oranges comparison. The delay forecast by the March 6, 2017 Update with the fragnet added is comparing what "actually" happened, as it relates to the added work, to what was "planned" to happen for the other project work. Said another way, the forecast delay is predicated on the remaining project work being performed actually as-planned, which is often a very poor assumption.

Additionally, analysts need to keep in mind that the as-built durations potentially include both the effects of the owner's change and inefficiencies that are the contractor's responsibility. If an owner grants a time extension based on a fragnet that was developed using as-built durations, then the owner may be granting the contractor a time extension and paying for delays that were caused by the contractor's own slow progress. The following example illustrates this problem.

Retrospective Time Impact Analysis Example 2

Looking again at Fig. 9.12, recall that this figure showed that the progress in Tunnel B using TBM #2 was stopped on March 6, 2017, due to D-Tunneling encountering a differing site condition. D-Tunneling submitted a time extension request after the project was finished requesting an additional 24 calendar days of contract time. This 24-calendar day time extension request was based on inserting the fragnet discussed in

Retrospective TIA Example 1, which included time to procure and install a new cutterhead.

In Example 2, to evaluate D-Tunneling's time extension request, the airport authority again compared D-Tunneling's March 6, 2017 Update to the project's as-built schedule. This as-built schedule is different from the as-built schedule used for Example 1. The Example 2 as-built schedule is depicted in Fig. 9.14.

When comparing Figs. 9.12 and 9.14, the airport authority noticed that the project actually finished on July 6, 2017, not June 29, 2017 as predicted by the March 6, 2017 Update with the fragnet added.

The airport authority's evaluation of the as-built schedule showed that the project's July 6, 2017 late completion was caused by the late completion of Tunnel A, not the delay to the Tunnel B operations. This was evident by the 22-day extended duration of Activity 1015, Tunnel A—Station 0 + 00 to 25 + 00, which actually took 42 days to complete compared to its planned duration of 20 days. While not conclusive, the airport authority recognized that this observation needed to be investigated.

In the March 6, 2017 Update with the fragnet added (Fig. 9.12), Activity 1015, Tunnel A—Station 0 + 00 to 25 + 00, was expected to finish on March 14, 2017; however, the as-built schedule showed that it actually finished 31 calendar days late on April 14, 2017. The airport authority's contemporaneous project documents showed that the reason Activity 1015, Tunnel A—Station 0 + 00 to 25 + 00, finished late on April 14, 2017, was that TBM #1 broke down in March and it took D-Tunneling nearly a month to repair it and place it back in service.

Based on the evaluation of the project's as-built schedule and the fact that D-Tunneling's progress in Tunnel A was delayed by an equipment breakdown, which is an unexcusable delay, the airport authority rejected D-Tunneling's request for a 28-calendar day time extension.

This example shows why a Retrospective TIA is a flawed approach that may not accurately portray what really delayed the project. In this example, D-Tunneling's time extension request did not consider the actual progress of the Tunnel A work and, thus, underreported the remaining duration of the Tunnel A work in comparison to what actually happened. In a Prospective TIA Analysis, future progress is assumed to be as planned because the work is still planned. But retrospectively, it is incorrect to ignore what has actually happened.

The underreporting of the Tunnel A durations in this example resulted in Tunnel B delays being critical and project delay being

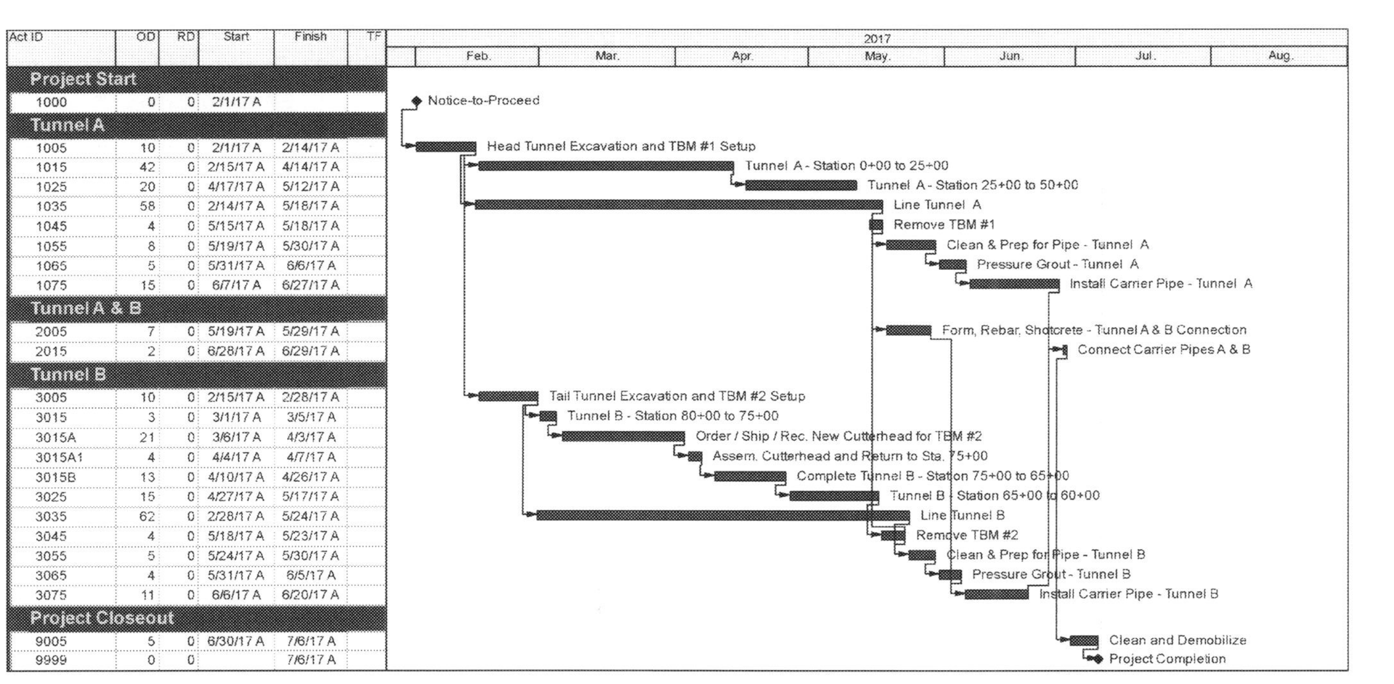

Figure 9.14 Example 2 as-built schedule.

erroneously assigned to the differing site condition in Tunnel B. If an accurate remaining duration had been shown for Tunnel A, the Tunnel B work would not have been critical and would not have resulted in a delay to the project. That is the case in the Contemporaneous Schedule Analysis presented in Chapter 7, Delay Analysis Using Critical Path Method Schedules. This example illustrates the problem with comparing the as-built information for Tunnel B to the as-planned information for Tunnel A.

Retrospective Time Impact Analysis Example 3

Some analysts take the Retrospective TIA to the extreme. Instead of just inserting the fragnet into one schedule update to demonstrate the contractor's entitlement to a time extension, some analysts attempt to make their analysis appear more reliable or acceptable by analyzing the project delays chronologically through the project from beginning to end using the project schedules. In doing so, they insert a fragnet or fragnets every month to represent an owner issue or delay and, then, attempt to measure the progress the contractor made that month to account for any potential contractor-caused delay. The first flaw with this approach is that the analysis is based on project schedules that did not exist during the project. Said another way, by changing the contemporaneous project schedules, the analyst is judging the decisions made by the project team using schedules that were not used by the team to build the project.

Another weakness with this approach is that the analysis method is inherently biased against the owner. To demonstrate this weakness, let us walk through the following example. Fig. 9.15 depicts the Month 1 update of a simple, five-activity schedule. All five activities are linked with finish-to-start logic relationships, and the project is forecast to finish in Month 5.

To perform this type of the Retrospective TIA, the analyst inserts an activity representing the owner delay into the Month 1 update, which is depicted in Fig. 9.16.

Fig. 9.16 shows that when the alleged owner delay activity is inserted into the Month 1 update, linked to the schedule as a predecessor to Activity B, and the schedule update is recalculated, the result is a critical path shift and a 2-week project delay. Note that Activity A is no longer on the critical path because its completion is not driving or responsible for determining the start date of Activity B. This first step measures the project delay resulting from the owner's alleged delay.

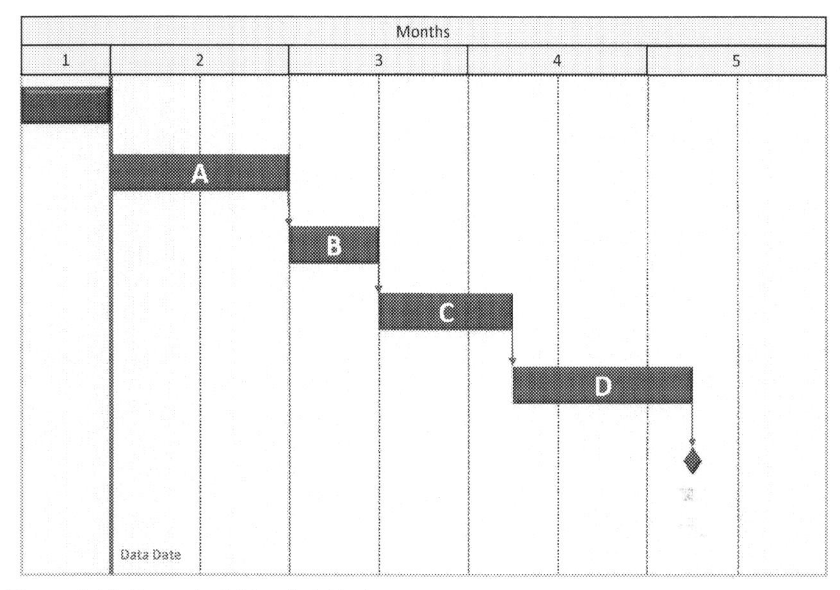

Figure 9.15 Example 3 Month 1 Update.

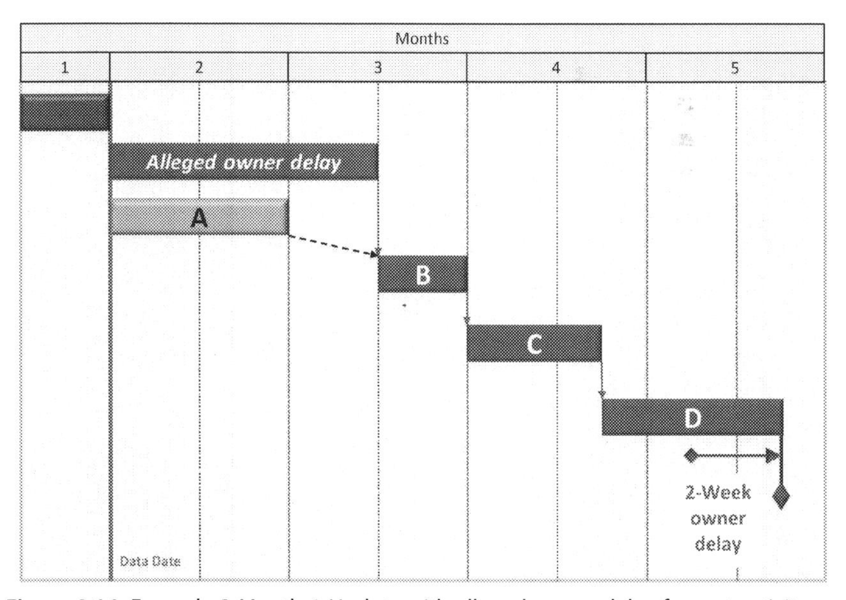

Figure 9.16 Example 3 Month 1 Update with alleged owner delay fragnet activity.

Then, the analyst attempts to measure the project delay resulting from the contractor's progress in Month 2, as depicted in Fig. 9.17.

To begin with, Fig. 9.17 shows that update's data date was moved from the end of Month 1 to end of Month 2. Because the alleged owner delay had a duration that was based on its as-built duration, it progressed as expected in Month 2. Activity A made slower-than-expected progress and did not finish in Month 2 and is now expected to finish in Month 3.

The first thing to notice was that the progress or lack of progress achieved in Month 2 did not result in additional delay in Month 2. By inserting the project delay caused by the alleged owner delay before the progress achieved in Month 2 is acknowledged or considered, the two weeks of project delay shown in the Month-2 update is attributed to the alleged owner delay.

If the progress achieved in Month 2 was evaluated before the insertion of the alleged owner delay, then 1 week of project delay would be assigned to the slow progress of Activity A and the remaining 1 week of project delay would be assigned to the alleged owner delay. Thus, the assignment of project delay would be different in Month 2. More to the point, if analyzed properly, as described in earlier chapters, the analyst could determine if the delays were initially concurrent or if one delay had primacy over the other.

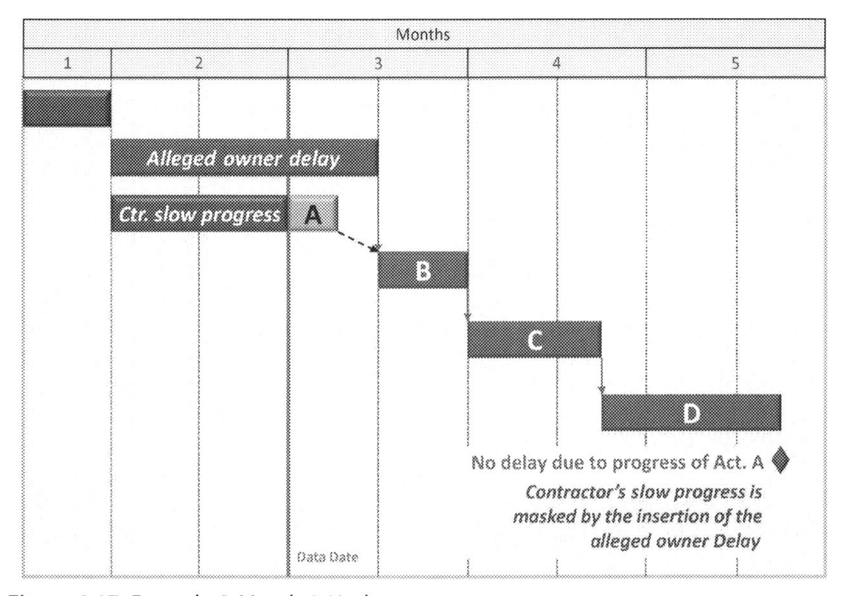

Figure 9.17 Example 3 Month 2 Update.

Analysts performing a Retrospective TIA in this manner will always be inserting a fragnet representing the owner's alleged delay before evaluating the project delay caused by the contractor's progress or lack of progress. The problem with this approach is that the owner delay is always given precedence over the contractor, making this analysis method biased against owners.

Strengths and weaknesses of the Retrospective Time Impact Analysis

A strength of a TIA, whether it is performed prospectively or retrospectively, is that it is specifically designed to measure project delay caused by added or changed work that was not included in the project schedule. At a minimum, the examples show the following weaknesses:

- The activities, logic, and durations used to develop the fragnets are as-built, not as-planned. This results in an apples-to-oranges comparison between a delay depicted using as-built information to a schedule that still shows the remaining project work as planned. It also means that the analysis does not consider the actions of the project team to mitigate delays.
- The fact that fragnets are inserted into updates before the contractor's performance during the update period is considered means that the analysis is always biased to emphasize owner delays over contractor delays. Note that while the analyst performing a Retrospective TIA evaluates the owner's delays by inserting a fragnet, contractor delays are never evaluated this way. This means that the responsibility for a delay has to be determined before the delay is inserted into the schedule for analysis and that delays caused by the owner are evaluated differently and, because delays are inserted, more harshly than contractor delays.
- The development and insertion of fragnets is inherently subjective.
- The analysis presumes that all of the remaining work in the schedule was performed precisely as planned, despite the fact that this is seldom, if ever, the case.

Another weakness of the Retrospective TIA relates to the nature or cause of the delay. Some delays may be caused by the addition of work. These kinds of delay are more amenable to "modeling" by the addition of a fragnet into the schedule update to represent the added work. Suspensions are different. The delay caused by a suspension is not the result of the addition of work. Suspension delays occur because work cannot occur. Despite this difference, the Retrospective TIA treats these causes of delay the same. For each type, the fragnet shows the delay being "added" to the schedule.

The problem with treating the addition of work the same as a suspension is subtle, but consider the following example.

The owner contracts with a contractor to build a new hospital. Early on, because it had happened on another hospital project, the contractor asks the owner if it is considering a modification to the electrical rough-in capacity in the basement based on a change to the specified radiation therapy equipment. The owner does not respond.

For reasons unrelated to the radiation therapy equipment, the project experiences a 1-year delay because the concrete subcontractor's work was defective and had to be demolished and replaced. The subcontractor eventually abandoned the site before completing the footings and foundation walls.

Ultimately, because of the extensive project delays, the owner opted to purchase the next generation of radiation therapy equipment. This required revisions to the basement electrical rough-in. To evaluate this change, the contractor's analyst prepared a fragnet. The fragnet began with the RFI activity. The activity related to the owner's answer to the RFI activity was given a 1-year duration. This activity was followed by activities related to the change to the electrical rough-in. This fragnet was inserted into the first schedule update. The schedule was rerun. The result was that there was a substantial project delay attributed to the owner's alleged delayed response to the RFI.

It is true that the owner took a year to respond to the contractor's question regarding electrical rough-in related to the installation of the radiation therapy equipment. However, though the time that passed between the day the RFI was asked and the day the RFI was answered was potentially a suspension, it was not "activity time" as characterized by the analysis. It was float created by the contractor's concrete subcontractor's poor performance and abandonment of the job. The analyst's incorporation of this float into the analysis fragnet was an inappropriate characterization of float as "activity time." Had the subcontract performed as planned, there would have been no need to change the radiation therapy equipment and no need to modify the electrical rough-in in the basement.

In summary, the Retrospective TIA, though widely used, is problematic on many levels. If used, it should be used with great caution.

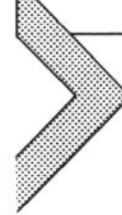

WINDOWS TECHNIQUES

The "Windows" Approach

Though many analysts classify their analysis technique as a "windows" analysis, there is actually substantial variation with regard how this analysis approach is actually executed. The only common feature among the many approaches to conducting a "windows" analysis is that the analyst chooses to analyze delays in isolated time periods, or "windows" during the project. (The term "windows," has nothing to do with the Microsoft products of the same name.) When comparing many of these "windows" analyses, both the criteria for selecting the "window" or time period and the delay analysis technique used to identify and measure the critical project delays within these "windows" differ.

When encountering a delay analysis that uses the "windows" approach, there are at least two questions that an analyst must answer when assessing the strengths and weaknesses of the delay analysis technique.

1. How were the time periods or "windows" established?
2. How were the delays identified and measured within the "window"?

The time periods or "windows" can be established in many ways. Selection of the time period can range from using the actual submission of the project schedule updates to establish the "windows" to arbitrarily choosing them. When using the project schedules to identify the time periods, the analyst should rely on the frequency of the schedule updates to determine the intervals.

Some analysts believe it is acceptable to base the "window" on the time period during which a particular controlling work activity is critical to determining the window. Allowing the controlling critical activities to determine the time periods means that each "window" should analyze a time period during which a specific work activity is the initial activity on the critical path. Performed in this manner, a "window" would begin when a specific work activity first becomes critical, either by its predecessor finishing or because the critical path has shifted. Similarly, the "window" should finish when the specific work activity is no longer critical, either when it finishes or the when the critical path shifts to another work activity. Realistically, selecting the "window" based on an activity or activities being critical still requires the analyst to determine throughout the specific time period if that activity or those activities remained critical throughout every day of the period. Consequently, the analyst still must go through a day-by-day analysis to determine what is critical. Therefore, the optimum objective selection of time frames for a "window" would be the contemporaneous schedule updates that existed on the project.

If you encounter a "windows" approach, you should require the analyst to clearly explain the reasoning for selecting these "windows." Anytime that an analyst makes a decision that is not directly supported by the contemporaneous project documents, such as the project schedules, or is not grounded in the facts, then the results become more subjective.

The other question that needs to be answered is: "How were the delays determined within the "window"?" The analyst should clearly explain how the critical project delays were identified and measured during the "window." Analyst should focus their attention on the critical path because only delays to the critical path can delay the project. Therefore, the analyst must demonstrate that the delay identified actually delayed the contemporaneous critical path. If the analyst cannot credibly show that the delay actually delayed the critical path during the "window," then its findings should be questioned.

The delay analysis method used to identify and measure the project delay in each "window" could be one of the techniques discussed in Chapter 7, Delay Analysis Using Critical Path Method Schedules, or in this chapter. Consequently, in addition to explaining the basis for choosing the window, the analyst must also address the efficacy, reasonableness, and objectivity of the method used to identify and measure the project delays in each "window."

It is interesting to note that if an analysis that the analyst chooses to call a "windows analysis" is performed correctly, then it is no different than the contemporaneous delay analysis explained in other sections of this book. Candidly, a "windows analysis" is just "window dressing." It has no analytical meaning and deserves no special significance or recognition.

BUT-FOR SCHEDULES, ANALYSES, AND ARGUMENTS

Sometimes analysts rely on the "but-for" approach to establish and, more often, to refute the cause of delays. In a "but-for" argument, the analyst takes the position that, regardless of what occurred, there were other delays that would have had virtually the same effect or would have been responsible for the same amount of delay. For example: but for the owner's delay to the shop drawing process, the contractor would have been late anyway, since he had not mobilized his subcontractor on time.

This approach actually sprouts from the concurrent delay argument. Use of this approach normally results from an analyst's comparison of the entire as-planned schedule with the overall as-built schedule. In this comparison, the analyst determines that several activities experienced delays. The analyst then identifies and argues that, e.g., an owner-caused delay is offset by an apparent contractor-caused delay. This analysis fails to recognize the fact that the original owner-caused delay provided the contractor with float on other activities, which, therefore, no longer had to be accomplished in the originally scheduled time frame. For example, Fig. 9.18 is a bar chart for a project showing the comparison between the original as-planned schedule and the actual as-built schedule. The contractor asserts that an activity was delayed because of a change by the owner. The owner responds that "but-for" its change, the contractor would have delayed the project. However, in our example, the real situation is that, once the contractor was delayed by the change orders, it chose not to proceed as originally planned and it chose to "pace" its work, since the work could be done more efficiently by waiting until the change order work was accomplished.

Contractors sometimes use a "but-for" analysis when there is no contemporaneous schedule to support their delay position. The contractor may insert the owner's delays into the original project schedule and then argue that "but-for" these delays, it could have finished the work much sooner. The difference between the actual completion date and the "but-for" schedule is then the measure of the delays caused by the owner.

When using the "but-for" approach, the party making the argument may even go so far as to admit that it is responsible for some of the delay. Purportedly, this adds credibility to the analysis. ("If I admit that some of the delays are my fault, then I am being fair in my evaluation.")

As acknowledged earlier, the "but-for" approach relates back to the concept of concurrency discussed earlier in this book. In the discussion of concurrency, we also addressed the primacy of delay. Reflecting back on that concept, we reiterate that any analysis must walk through the project in time, day by day, if possible, and, in that manner, determine the critical

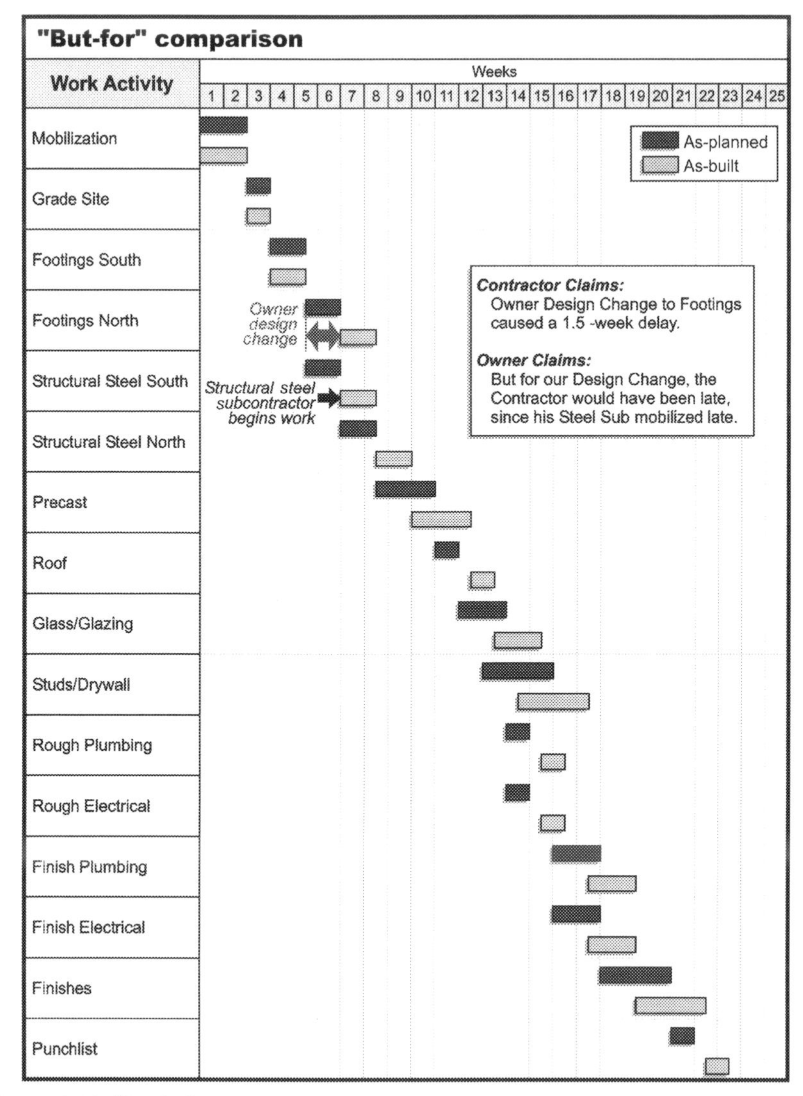

Figure 9.18 "But-for" comparison.

path at any given time. It must always be kept in mind that the schedule is not static; it is dynamic, changing over the life of the project. The schedule does not guarantee that the job will be built a certain way. It only provides a "road map" or plan for construction and for the durations of the activities. Therefore, any time an analyst "freezes" the schedule to measure delays, the analyst is asking for trouble. One cannot apply a static measure to a dynamic situation.

Strengths and weaknesses of the but-for analysis

A strength of this analysis technique is that it can be performed with very little contemporaneous project scheduling information. However, the many weaknesses of the technique, which include not using all of the project's contemporaneous schedules, ignoring the primacy of delay, ignoring the possibility that one party chose to pace its work based on earlier delays by the other party, and ignoring how the project was constructed from beginning to end, make the analysis unpersuasive and easily refuted.

NONSCHEDULE-BASED ANALYSES

Analyses based on dollars

Some analyses that are presented to support delay conclusions are based on the relationship between dollars and time. There is a widely held, but mistaken belief that the dollar value of the work performed is directly related to the time progress of the job. Some arguments incorporating this method are as follows:

> *Contractor: "I was able to complete 90% of the work in ten months on the job. Then, because of the Owner delays in inspection and punch list, it took four months to complete the last 10% of the work."*
>
> *Owner: "During the period of the alleged delay to the contractor, he was able to perform 25% of the dollar value of his work on the project. The dollar value of work accomplished during that period reflected the same rate of progress as that before and after the delay. Thus, the owner's change did not really delay the project work."*

While both of these arguments may appear plausible and logical from an initial quick reading, there is no quantifiable linear relationship between time progress and dollars on any project. The dollar value at each stage of the project depends on the nature and cost assigned to the specific activities completed. For instance, some high dollar value items of work may be performed in a short period of time. Whereas, other low dollar value items of work may take more time. The dollar value method does not identify and track the activities on the critical path of the project, which was identified earlier as the only reliable method for determining delays on the project.

Some public owners have structured their contract documents to reflect the idea that time and dollar value have a linear relationship. For instance, some owners grant time extensions based on the dollar value ratio of a change order to the original contract amount. For example:

- Original Contract amount: $100,000
- Dollar value of change: $10,000 (or 10% of original Contract amount)

- Original Contract duration: 100 days
- Therefore, the time extension granted is 10 days (10% of original Contract time).

In fact, an owner could make a change that would increase the contract amount by ten percent that would not affect the duration of the project. Likewise, the owner could make a change with a minimal (direct) dollar impact that could significantly delay the project.

S CURVES

Some owners require contractors to provide "S" curves to show job progress. The contractor provides a bar chart of the major activities on the job and applies the dollar value from the schedule of values to the bar chart. By summing the dollars over time, an "S" curve is generated. Fig. 9.19 is an example of a typical "S" curve for a project.

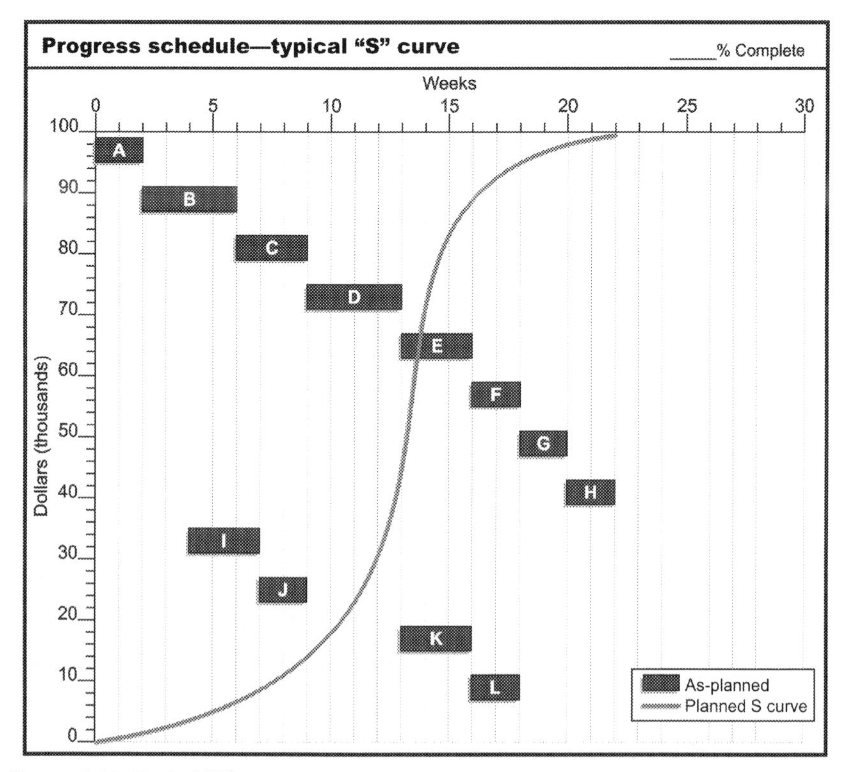

Figure 9.19 Typical "S" curve.

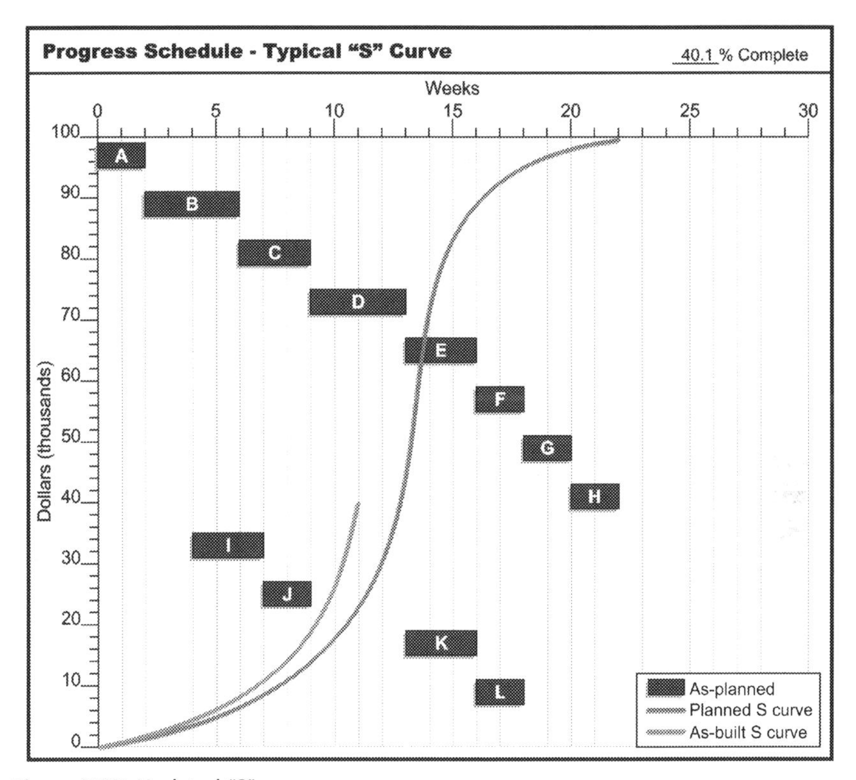

Figure 9.20 Updated "S" curve.

The contractor must then submit an updated "S" curve with each monthly pay request. The monthly submission will also have an entry for "progress to date," recorded as a percentage (see Fig. 9 20). This update would be used to measure the contractor's progress in time; however, this is very often misleading. "S" curves developed from a contractor's total billing dollars do not measure the time progress of the project, but merely show the progress of billings over the course of time. There are many reasons why we do not recommend the use of this "S" curve for determining job progress. For example, the original planned "S" curve might not be accurate due to "front end loading," or other factors. It should also be noted that the updated "S" curve information might be misleading because of payments for stored materials and equipment.

The contractor could overbill or "front end load" its bid items to recover its costs as early as possible and, therefore, reliance on the "S" curve could portray more "progress" than had actually occurred. Because they are unreliable, time-to-dollar relationships should not be used to establish physical progress or delays on a project.

CHAPTER TEN

The Owner's Damages Due to Delay

When a project is delayed, the owner, contractor, subcontractors, and other parties engaged in the project may all incur unplanned costs that are a direct result of that delay. The determination of whether these additional costs can be recovered from another party is based on the contract, the applicable laws, and a determination of the party or parties responsible for the delay. This chapter addresses the unplanned costs that the owner may incur and the quantification of these should be recoverable.

The general concept for recovery of owner's delay costs is similar to the concept that applies to contractors or other parties. The recovery of these delay damages serve to place the owner in the same position it would have been in had the contractor performed as required by the contract. In addition, the legal standard regarding damage calculations is easy to state, but sometimes difficult to implement. Damages need not be calculated with absolute certainty, but they may not be based on speculation.

In the broadest sense, the owner's damages are determined based on either the actual costs incurred or a damage amount liquidated in the contract. Both of these categories of damages are discussed in this chapter.

ACTUAL COSTS

Absent a contract provision indicating otherwise, the owner may be entitled to recover the costs it incurred as a result of the contractor's failure to complete the work within the time (or times) established by the contract. The recoverable costs would typically be the actual costs incurred by the owner as a result of the contractor-caused delay. The following is an incomplete, but a representative list of costs the owner might incur due to the contractor's delay:

- Added costs to provide project inspection services over the project's extended duration.

Construction Delays.
DOI: http://dx.doi.org/10.1016/B978-0-12-811244-1.00010-0

- Added costs to provide for continued construction services from the design consultant.
- Added costs to provide continued project oversight and administration by the owner's staff or by the construction manager hired by the owner.
- Added costs to maintain temporary or existing facilities longer than anticipated.
- Added costs related to renting space while waiting for the project to be completed.
- Added costs related to extending insurance coverages.
- Added expenses related to continued storage costs incurred to store the owner's fixtures and other building contents while the building is being renovated.
- Added expenses incurred because new fixtures and other building contents have to be stored while waiting for the new facility to be completed.
- Lost earnings because the facility cannot be rented, sold, or used for the purpose it was built.
- Costs to the public because the facility has not been completed; e.g., the cost incurred by the trucking industry because trucks are stuck in traffic jams caused by the ongoing construction (known as road-user costs).
- Additional moving expenses; e.g., the cost incurred to "double handle" FF&E that must be put into storage rather than being installed directly into the completed facility. (FF&E are movable furniture, fixtures, or other equipment that have no permanent connection to the structure of a building.).
- Cost escalation related to increased costs of labor or materials due to inflation.
- Financing costs.
- The cost of finding replacement facilities because the project was not completed by the contracted date.

When attempting to recover its actual costs, the owner has the legal burden to substantiate these costs. Properly substantiated costs are clearly identified and segregated from other costs, calculated correctly, and documented sufficiently to establish that they were actually incurred. For many owners, substantiating these costs can be a daunting task, particularly if adequate cost-related documentation has not been accumulated and organized during the course of the project. Owners should seek out qualified counsel and hire appropriate expertise to ensure that delay costs are properly calculated and documented. The items listed in Fig. 10.1

Considerations for Documenting and Preparing an Owner's Delay Damages

1. The owner should have its construction manager or design consultant maintain **separate records** that reflect the costs of meetings, inspection, site visits, and other ongoing management and oversight services that are the result of the project's extended duration.
2. The owner should track and accumulate **increased moving costs**, such as the escalation of moving costs and added storage expenses.
3. If the owner had to pay **additional rent** to occupy its previous facilities longer than planned, those costs should be carefully tracked and documented.
4. If the owner's staff is involved with the ongoing administration of the project, then each staff member should **maintain a timesheet with a detailed description of tasks performed**, meetings attended, telephone calls, emails prepared, and other actions performed to provide continued support to the project.
5. Depending on the nature of the project, the owner may have increased costs associated with or providing temporary or alternate lodging or services, **these considerations are particularly important to residential and educational projects**. All such costs should be separately tracked, documented, and justified.
6. If the project is financed, then the owner may be exposed to **additional interest and other financing costs or fees**. These costs must be tracked, documented, and calculated correctly.
7. The owner may face costs from **other follow-on contractors** who have been delayed by the delay caused by the first contractor. These costs would then be a part of the owner's actual delay costs.
8. If the project is a revenue generating facility, the owner may claim the **lost earnings from not being able to operate.**

Figure 10.1 Considerations for documenting and preparing an owner's delay damages.

provide a description of some of the documentation that the owner should maintain in order to establish its entitlement to delay damages.

The last item of costs listed in Fig. 10.1—lost earnings—may be difficult to demonstrate. Depending upon the nature of the project, the courts and boards may look upon lost earnings as being somewhat speculative and, therefore, not easy to reliably quantify. However, this does not mean that owners should not seek reimbursement of lost earnings. Instead, owners should be realistic about the challenge of proving these. The chance of recovering lost earnings may be enhanced, for instance, if the owner is able to show a measurable difference in the production rates of an existing facility compared to the increased capability of the replacement facility. Also, an owner may find that it will increase its chances of recovering lost earnings through a liquidated damage provision.

Fig. 10.2 is an example of an owner's calculation of actual costs sustained when the contractor completed the project 150 days late and no time extensions were granted for excusable delay.

In general, actual costs are more difficult to recover, because, by their nature, they can be difficult for an owner to demonstrate; particularly a public owner. For example, consider how difficult it might be for a public owner to establish the added cost of staffing a delayed project. The cost of salaries and benefits might be possible to identify and document, but the

Statement of Actual Damages – ABC Corporation

1. Additional Design Services

   Based on the records of the ABC Corporation's design firm, the A/E made 22 more site visits than the number originally scheduled. The cost of these 22 additional site visits was $10,560 as tabulated and documented in Appendix A.

2. Moving Costs

   Because of the delay, ABC Corporation was forced to renegotiate its contract with its moving company. The cost of the renegotiated contract was $6,520 higher than the original moving contract. The original and renegotiated moving contracts documenting the price increase are provided in Appendix B.

3. Rental Costs

   ABC Corporation was forced to stay in its existing facilities for an additional five months. The rental costs during this period were $617,000. This added rental expense is documented in Appendix C.

4. Staff Costs

   ABC Corporation's in-house construction staff remained active on the project for an additional 150 days. The staff costs based on salaries and benefits were $61,845. A tabulation of the added staff costs, time sheets, proof of salary and benefit costs, and the company's audited administrative overhead rate are provided in Appendix D.

5. Temporary Lodging

   ABC Corporation's plant start-up engineer moved to the project site based on a promised completion date by the contractor. The contractor missed this promised date by 35 days. The lodging and subsistence paid to the start-up engineer during this 35-day period totaled $2,265. The receipts for the expenses incurred by the start-up engineer are included and tabulated in Appendix E.

6. Finance Cost

   ABC Corporation financed this project at an interest rate of .5% over prime. Because of the delay, ABC Corporation recorded additional financing costs of $792,000. ABC Corporation's loan agreement with its lender and detailed calculations of the interest expense are provided in Appendix F.

7. Claims by Follow-on contractor

   ABC Corporation received a delay claim from its start-up contractor. The claim was settled for the amount of $29,000. The start-up contractor's claim and the settlement agreement are provided in Appendix G.

8. Lost Earnings

   Because of the delay, ABC Corporation was unable to benefit from the increased production capability of the new facility. Based on sales records and production records related to expenses, the lost earnings totaled $2,675,000. Appropriate sales and production records are provided in Appendix H.

Figure 10.2 Statement of owner's actual damages.

cost of the owner's administrative overhead might be much more challenging. Given the difficulties associated with quantifying its actual costs, many owners estimate these costs before the project is advertised for bid. The estimated cost of late completion is then "liquidated" in the contract as part of a liquidated damages clause.

LIQUIDATED DAMAGES

Liquidated damages are determined prior to the execution of the contract. The exact amount of the liquidated damages is specified in the contract. A typical contract clause is provided in Fig. 10.3.

Sample Liquidated Damages Clause

Should the contractor fail to complete the project within the time allowed by the contract, including time extensions issued by executed change orders, then, for each calendar day of delay, the owner has the right to withhold the amount of $500 as liquidated damages. These liquidated damages are compensation to the owner for costs the owner may incur due to the contractor's delay, and are not to be construed as penalties.

Figure 10.3 Example 1 liquidated damages clause.

Fig. 10.3 presents only one example of a liquidated damages clause. Such clauses can be much more extensive and elaborate. However, this example contains language common to most liquidated damages clauses. The owner should seek the assistance of a qualified counsel in structuring the wording of the clause and should carefully compute the damages it may reasonably sustain as a result of a delay.

One might ask why an owner would specify liquidated damages amount in advance as opposed to seeking recovery for actual damages if a delay occurs. The answer is that liquidated damages are desirable when it is difficult or impossible to accurately calculate and document the actual damages that the owner would incur in the event of a delay, particularly for public projects. Projects such as highways and transit systems have a value to the public, but the cost to the public of late completion is challenging to calculate. Rather than rely on recovery based on difficult calculations and documentation that is time consuming to gather and organize, owners prepare reasonable estimates of their delay costs before the project is advertised and then "liquidate" this estimated cost in the contract. This liquidated damage is assessed against the contractor when the contractor delays the project beyond the date (or dates) established in the contract. If accurately estimated, the amount assessed then serves to cover the damages the owner incurs because the project is completed late.

Many owners believe that the inclusion of a liquidated damages clause has the added benefit of acting as a deterrent to lateness. In other words, fear of having to pay liquidated damages "motivates" the contractor to complete the project on time. In reality, the purpose of a liquidated damages clauses is not to deter lateness, although they may have that effect. In the absence of a liquidated damages provision, a contractor would be exposed to the owner's actual damages. This exposure would also serve as a deterrent to lateness. The primary advantage provided by the liquidated damages clause is that the amount is known with certainty. Also, having the amount liquidated in the contract does make it easier for the owner to assess these damages as soon as the contract completion date has passed, even before the project is completed, which may support the theory that a liquidated damage provision motivates the contractor to complete on time.

Generally, the inclusion of a liquidated damages clause does not affect contractors' bids. For one thing, contractors recognize that without a liquidated damages clause, they are still liable for actual damages should they finish late. However, most contractors view substantial liquidated damage amounts as increasing project risk. The greater the liquidated damage amount, the greater the risk. Still, if the owner has allowed plenty of time to perform the contract work or the risk of delay is small, even a high liquidated damages amount (e.g., $30,000 per calendar day) may have no effect on the contractor's bid price.

When a high liquidated damage amount is coupled with an extremely short or unreasonably short contract duration, contractors may increase their bids to address this risk. In these circumstances, contractors may price this risk in their bids by assuming that some amount of liquidated damages will be assessed. This might also have happened if there was no liquidated damages clause, but owners should consider the possible bid impact that results from extremely aggressive contract durations.

A potential benefit of a liquidated damages clause arises when a contractor falls behind schedule during the project. The liquidated damages clause allows the contractor to determine whether acceleration efforts will be cost effective. For example, if a contractor is behind the schedule by 10 days on a project and the liquidated damages are $300 per day, the potential exposure is $3000. If the cost of accelerating the work to make up the 10 days is $7000, then the cost-effective decision is to finish late. However, one consideration of such a decision is that a contractor that intends to complete a project late may be in breach of its contract with the owner. Any contractor considering such a choice should consult with an experienced legal counsel. In addition, the contractor and its subcontractors may incur added costs due to the late completion of the project, as well. For these reasons and many others, a contractor may still decide to accelerate its work even if the cost of the acceleration exceeds the cost of the potential liquidated damages assessment.

Estimating liquidated damages

The list of costs that the owner might incur due to the contractor's delay presented at the beginning of this chapter can be used to identify the cost categories that should be included in the owner's estimate of the liquidated damages amount. While other costs may be included in the liquidated damages estimate, this list is a good starting point that will lead an owner to the types of costs to consider.

A note of caution: Liquidated damages are specific to each project. Some owners use standard tables for liquidated damages that may not reasonably reflect the damages they will sustain if a delay occurs on their project. For instance, many State Departments of Transportation have liquidated damages clauses similar to the one shown in Fig. 10.4.

While the use of standard tables is convenient, the owner should ensure that the amounts in the table are valid and appropriate for the particular project. Often, such tables approximate the administrative costs that the owner will incur if the project extends beyond the contract completion date. However, owners should also recognize that liquidated damages do not necessarily bear a direct relationship to the contract amount. Two different projects of equal value can have very different potential damages. An owner who uses a standard table to figure liquidated damages may risk either understating the damages and thereby shortchanging itself or overstating the damages, which may

Sample Liquidated Damages Clause

Should the contractor fail to complete the project within the time allowed by the contract, including time extensions issued by executed change orders, then, for each calendar day of delay, the owner has the right to withhold the Daily Charge amount shown in the table below as liquidated damages.

These liquidated damages are compensation to the owner for costs the owner may incur due to the contractor's delay, and are not to be construed as penalties.

For Contract Amounts More Than	To and Including	Daily Charge
$ 0	$ 100,000	$ 300
100,000	500,000	550
500,000	1,000,000	750
1,000,000	2,000,000	900
2,000,000	5,000,000	1150
5,000,000	10,000,000	1350
10,000,000	--------	1400

Figure 10.4 Example 2 liquidated damages clause.

expose the owner to a legal challenge by the contractor. Fortunately, from the perspective of enforceability, the liquidated damages amounts provided in tables of this type are almost always low. Nevertheless, it is important that owners recognize that each project is different and will have different estimated delay costs.

When do liquidated damages begin and end?

The contract should clearly specify when the assessment of liquidated damages will begin and end. With regard to when the assessment of liquidated damages begins, the contract will typically allow the assessment of liquidated damages from the date established in the contract for completion of the contract work, plus any authorized extensions. Authorized extensions are typically time extensions mutually agreed to by the owner and the contractor or, sometimes, unilaterally issued by the owner.

Some liquidated damages provisions will tie the start of liquidated damages to the contract substantial completion date, rather than the overall contract completion date. Although this term is often defined, the date of substantial completion is usually understood to be the date when the project can be used for the purpose it was intended. For example, for a highway or bridge project, the substantial completion date might be the date the highway or bridge is open to traffic. Some landscaping or cleanup work might still remain, but the project is sufficiently complete to be used by the traveling public. Again, in the absence of a contract definition for the term, the date of substantial completion is often synonymous with the date of beneficial use or beneficial occupancy. Like substantial completion, these dates are also usually defined as the date from which the project can be used for its intended purpose.

In addition to the question of when the assessment of liquidated damages begins, there is also the question of when the owner can begin enforcing this assessment. Must the owner wait until the contract date has passed or can the owner begin enforcing the assessment as soon as the project schedule predicts that the project will finish late? A careful reading of the liquidated damages clauses in Figs. 10.3 and 10.4 suggests that the actual collection of liquidated damages would not start until after the contract date has passed.

If an owner wants to be able to collect liquidated damages sooner, then the liquidated damages clause should be written accordingly. As an example, many owners now limit their withholding of retainage. If the project schedule shows the project finishing after the established contract completion date, an owner might want to withhold retainage (or more retainage). This retainage amount (or added retainage amount) might be determined by multiplying the liquated damages daily rate by the number of days the project is scheduled to finish late. Owner's wanting to use this approach should make sure their contracts support such an assessment or contact qualified legal counsel.

At the other end, where the assessment of liquidated damages ends may be less clear. Many liquidated damages clauses are written to allow the assessment of liquidated damages through the date the contract work is fully completed. However, liquidated damages are supposed to be based on the owner's costs of delay. The contractor will likely argue that the owner's most significant costs end at substantial completion, not the completion date of all of the contract work. For example, the contractor might argue that the owner's delay costs ended when the owner was able to move into and operate its new warehouse, not when the landscaping was installed the following spring. It is common for a contractor to make this assertion, and unless the owner's estimate of its liquidated damages clearly shows that the costs of delay upon which the liquidated damage amount is based continue all the way through to the completion of all of the contract work, the contractor may be able to limit the assessment to the project's substantial completion date.

Also, even if the contract refers to substantial completion, but lacks a clear definition of precisely what needs to be complete in order to achieve that milestone, a dispute may develop as to when the assessment of liquidated damages should end. Therefore, when a liquidated damage provision is used, the owner should take care to clearly define the beginning and end points of their assessment.

Application to project milestones

Liquidated damages clauses can also be written to apply to milestone dates or events during the project. For instance, liquidated damages may be linked to the completion of work phases, such as building close-in or the completion date for a section of the project. The amounts of these milestone liquidated damages may be separate from the liquidated damages amount that applies to project completion. For instance, a project may involve the construction of several buildings. In the contract, the liquidated damages clause may specify separate damages for the completion of each building, as well as a liquidated damages amount for the overall completion of the project. A highway construction project may specify liquidated damages for the completion of each bridge and for project completion. When multiple liquidated damages are specified, the contract should clearly state if and when these may be assessed simultaneously. For liquidated damages that are assessed simultaneously, the owner must take care that the estimates used for these amount do not duplicate any of the estimated cost components such that the assessment of damages in the specified fashion would amount to a penalty.

Hourly fees

For some projects, liquidated damages are specified on an hourly basis. On certain critical highway projects, the owner may specify hourly liquidated damages for failing to open portions of the roadway to traffic at set times for each day of the project. It should be noted that this type of hourly fee is most likely based on the cost to the traveling public, known as road-user costs, and not necessarily on the owner's delay-related costs. These fees are sometimes known as "Lane Rental" fees. Properly calculated road-user costs have been found to be a reliable measure of delay damages on public road projects.

Graduated damages

When justifiable, liquidated damages may be graduated. For example, the liquidated damages may be $1000 per day up to a certain date or for a defined number of days, and then may increase to $1500 per day for delays beyond the date or in excess of the initial number of days. These graduated liquidated damages should reflect the owner's increased damages as the delay continues.

Alternatively, liquidated damages may be assessed at one rate, usually a higher rate, until the contractor achieves substantial completion. A second rate, usually lower, based only on the owner's ongoing project oversight expenses, might then be assessed successfully until all the project work is completed. Again, the contract should clearly state when and for what periods these damages will be assessed.

Bonus or incentive clauses

It is sometimes asserted that liquidated damages must also have a corresponding bonus or incentive. This is not true. There is no requirement that the owner offer a bonus or incentive merely because the contract includes a liquidated damages clause. Said another way, the lack of a bonus or incentive does not justify a challenge to the liquidated damages clause.

The owner may, in fact, include a bonus or incentive clause in the contract for early completion. If a bonus or incentive is included, it does not have to match the amount of the liquidated damages. The bonus can be higher or lower, and can have limitations. For example, the bonus in a contract could allow $1000 per day for each day the contractor finishes the project earlier than the specified contract completion date, up to a limit of $50,000. Alternatively, the owner may allow a bonus for early completion that increases or decreases over time. For example, the owner may offer a bonus of $1000 per day for early completion up to 50 days, and for every day that the project is finished early in excess of 50 days, the bonus may be increased to $1500.

The bonus is computed from a contract-specified date. If the contract-specified date is extended by a change order, the bonus may be computed from the new, later date. In some instances, the benefit that the owner will realize from early completion may evaporate after a certain calendar date. In such cases, the bonus date may be associated with "no-excuse" language that limits the contractor's entitlement to time extensions related to the bonus date. On projects with these types of provisions, the bonus date is often fixed and may not be extended. As an example, on one highway project, the contract clearly stated that the bonus date would not be extended for weather-related delays, even if the delays were unusual or extreme. The contractor challenged the enforceability of this provision and lost.

Such clauses can be difficult to write and should be drafted by qualified counsel. Furthermore, in order to minimize the potential for disputes, the owner should make an extra effort to ensure that all bidders understand the intent of the bonus clause.

Enforceability

One of the owner's major concerns when using a liquidated damages clause is whether it will be enforceable. A reasonable amount of case law exists, and with proper guidance by counsel, the owner should be able to structure a clause that will be upheld.

If a contractor completes a project late and is assessed liquidated damages by the owner, it is possible that the assessment may be challenged. There are two basic approaches that a contractor may use to challenge the assessment of the liquidated damages. First, he or she may attack the propriety of the assessment by disclaiming responsibility for the delay. Second, the contractor may claim that the specified amount of the liquidated damages is excessive and, consequently, is actually a penalty as opposed to a reasonable estimate of the owner's delay damages.

If a contractor challenges responsibility for a given delay, it must show through a delay analysis that the delays to the project were excusable delays and, therefore, warranted a time extension. If the delay analysis establishes that the assessment of the liquidated damages is inappropriate, the contractor may be granted relief from the damages. Similarly, the contractor may attempt to show only partial responsibility for delays to a project, arguing that the owner also caused some concurrent delays. As previously discussed, if it can be shown that delays were also caused by the owner, then the contractor may be granted relief from the assessment of some or all of the liquidated damages.

The second approach used to challenge liquidated damages is based on the magnitude of the damages specified. The contractor may argue that the amount specified was excessive and was in effect a penalty rather than a reasonable estimate of the owner's actual costs due to the delay. Some owners may feel that it does not matter whether the amount reflects a penalty or a loss, since the damages were clearly specified in the contract that bears the contractor's signature. However, in construction contract law in the United States, penalties in a construction contract are not enforceable. If it is found that the amount specified was too high, it may be judged as a penalty and not a liquidated damage. In such cases, the courts may not enforce the clause and may or may not allow the owner to seek recovery of its actual delay damages. For this reason, most knowledgeable attorneys carefully avoid the use of the word penalty anywhere in the contract. Judges have been known to disallow clauses merely because the word "penalty" was used in the contract wording.

High estimates

When the contractor challenges the amount of liquidated damages, the owner must substantiate the validity of the damages. This does not mean that the owner must demonstrate that actual damages are comparable to the liquidated damages specified in the contract. The issue that must be decided is whether or not the estimate of liquidated damages was reasonable at the time it was prepared. In other words, when the contract was drafted, given what was known at that time, was the estimate a reliable measure of reasonably anticipated costs? Therefore, it is in the owner's best interest to maintain the documentation used to estimate the liquidated damages amount.

If it is determined that the owner's estimate was not reasonable or did not reasonably approximate the liquidated damages amount specified, the clause may not be enforceable. For example, if the owner's estimate showed potential damages of $4500 per day, but the liquidated damages amount specified in the contract was $10,000 per day, then the amount specified might very well be determined to be a penalty and, therefore, not be enforced.

Low estimates

Many times, the amount of liquidated damages specified in a contract is too low. The owner's damages are often greater than the specified liquidated damages. Can the owner recover its damages when they are greater than the specified liquidated damages? In most cases, the owner is limited to the liquidated damages amount specified. There are very few exceptions where an owner can recover more than the liquidated damages amount. The argument is that the owner wrote the contract and calculated the damages and is not entitled to collect more than the specified amount.

The liquidated damages clause is sometimes referred to as the "owner's sword" and the "contractor's shield." It is viewed as the owner's sword because the owner can use the clause to prod a contractor to strive for timely completion. It is viewed as the contractor's shield because the liquidated damages amount typically caps the amount that the owner can recover in the event of a contractor delay.

CHAPTER ELEVEN

The Contractor's Damages Due to Delay

Even after a contractor establishes its entitlement to the recovery of additional compensation, the change or claim can remain unresolved due to the contractor's inability to properly calculate and present its delay costs. Once entitlement is established, the contract should be used as a guide to determine what costs are recoverable, how these costs should be calculated, and what documentation is required to support these calculations. Problems that may prevent resolution of the change or claim may include overstated or incorrectly calculated costs, inclusion of costs that are not delay-related, costs that are not adequately supported, or costs that are not allowable under the terms of the contract.

The calculation of costs attributable to delay must be properly supported by the relevant contract provisions and the applicable facts supporting the change or claim, whether these come from the project documents, cost records, testimony, or other evidence. They must also be consistent with standard approaches used in the industry and accepted by the courts.

In the following sections, we will explore the types of costs that may be incurred by a contractor as a result of project delays, and how they are typically calculated. This chapter concludes with guidelines for the presentation and recovery of delay costs and some examples of typical delay cost calculations.

TYPES OF DELAY COSTS

Delay costs can be summarized within the broad categories presented in Fig. 11.1, some of which may not be recoverable under the contract or the law. A more thorough review of these cost categories follows in this and the next few chapters.

However, because the possibilities and permutations of each of these cost categories is virtually unlimited, these categories are not addressed in

Construction Delays.
DOI: http://dx.doi.org/10.1016/B978-0-12-811244-1.00011-2

Extended and increased field costs (Addressed in this Chapter 11)	These costs include the additional labor, material, and equipment costs resulting from project delays. These costs are typically quantified and supported by actual measurements of increased units and/or rates
Home office overhead (Addressed in Chapter 12)	When a project is delayed, the contractor may experience unabsorbed or unrecovered company-wide overhead costs. Generally when recoverable as delay costs, the contactor's home office costs, which support all projects, are allocated to the delayed project and recovered based on a daily rate applied to the compensable days of delay
Other categories of delay costs (Addressed in Chapter 13)	A change proposal or claim for delay costs may include other related costs such as the costs associated with noncritical delays, legal and consulting costs, lost profits, lost opportunity costs/bonding impairment, interest, and other costs
Inefficiency or lost productivity costs (Addressed in Chapter 14)	Depending on the nature of the delay, a contractor may also experience some measure of inefficiency. The resulting increased labor and equipment costs may be included within the proposed delay costs
Acceleration costs (Addressed in Chapter 15)	When the labor force is accelerated to mitigate delays, the resulting increased labor and equipment costs may also be included within the proposed delay costs

Figure11.1 Delay cost categories.

every detail. Rather, an example of the general methods is presented so that the reader may apply the same principles to each specific situation.

The next few sections in this chapter address the costs incurred on the project due to delays. These costs are comprised of field costs that are either extended or increased. Extended field costs are those time-related costs that are incurred over a longer period of time because they are required to support the project for an extended period. Extended field costs are discussed in more detail later in this chapter. Increased field costs are those costs in the field that, but for the delay, would otherwise not have been incurred. Generally all of these costs are incurred as labor, equipment, or material costs.

LABOR COSTS

Labor costs are incurred either in the performance of the work or as field overhead to support the work. Because delays generally increase the time of performance rather than the amount of work, delay costs related to labor are usually in the form of extended field overhead labor. This labor is typically associated with project management and superintendence and general support. Extended field costs are addressed in more detail later in this chapter.

However, the cost of labor may increase even when the amount of work has not increased. This typically happens when more labor is required to produce the same amount of work, typically defined as a loss of productivity or inefficiency, or when additional labor becomes necessary to complete the work faster, which may result in additional charges for overtime, shift work, or travel costs to add labor in a limited labor environment. Delay costs arising from inefficiency and acceleration are addressed in later chapters.

A contractor may also incur additional labor costs due to a delay when additional tasks become necessary to accomplish the work, such as in the case when a subgrade needs to be repaired because the delay allowed it to degrade, when the hourly cost of the labor increases due to either skill level or escalation, or when labor is forced to be idle.

Idle labor

When a project is delayed or suspended, the contractor's workers may be at the project site, but may be unable to perform the work. To recover the cost of idle labor, the contractor must be able to demonstrate that its workers were on the site, but unable to perform any work. The owner would appropriately question why the contractor was unable or unwilling to shift those workers to other tasks or other projects, or lay them off. The contractor should be prepared to substantiate an idle labor claim with project daily reports. If the owner does not retain such daily reports, there may be no basis for a challenge to the contractor's documented claim. For this reason both the owner and the contractor should diligently maintain documentation of labor activity on each day of the project.

Consider the following example:

A contractor is installing concrete pipe for a new storm sewer. While the crew is installing the sewer piping, it comes across a buried foundation from a structure that previously occupied the site. The existence of this structure was not known by the owner or the contractor and the contract provided that the owner would take responsibility for such unknown obstructions. The contractor's installation crew immediately stopped work and asked for direction as to how to proceed. There was no other sewer piping to be installed. The owner's engineer came to the site immediately and directed the piping installation around the obstruction. Within a few hours of stopping, the owner's pipe installation crew recommenced installation. The crew had been idle for four hours.

Even if the crew had not been idle, but had instead moved to another location on the site, the delay may have been reduced or eliminated, but the contractor may have incurred additional costs to move equipment and materials. Similar additional costs might also be incurred when the contactor returns to the original location to recommence the work that was stopped. If the contractor's work is disrupted many times this way, the contractor may also experience inefficiencies associated with these disruptions.

It should be noted that under most contracts, the contractor has an obligation to mitigate the costs when a delay occurs. Therefore, if possible, the contractor is expected to shift its labor to other work during a delay or be prepared to demonstrate why this was not possible.

Additionally, some construction contracts specifically address instances when the owner would agree to pay for idle labor. However, in those instances, recognition and payment of idle labor costs are triggered by the owner's issuance of a stop work order or other direction approving the need for the contractor to maintain its crew on the project site to resume work when directed to do so. Understanding the requirements and limitations in the contract regarding the contractor's recovery of idle labor costs is essential.

Union personnel

On certain projects labor is provided through signatory agreements that require the workers to be paid a certain minimal amount, such as for "show" time or a minimal number of hours per week. These agreements may also require additional nonworking personnel be at the project site while work is being performed. Some examples include a master mechanic, the shop steward, or certain levels of foremen. If there is a delay, these payments to personnel also represent extended or increased labor costs. The contractor should provide copies of the union agreements to substantiate these costs to the owner.

EQUIPMENT COSTS

A contractor's equipment costs on a project can also be affected by a delay. For example, delays may cause the equipment to be idle. The costs that can be recovered for idle equipment depends on the contract provisions. In some cases, the contract may address idle or standby equipment and allow for recovery at a reduced rate or provide no compensation at all. If the contract is silent on this issue, the contractor would likely seek reimbursement at the full rates of the equipment, reduced by any operating cost component, during the idle time. As in the case of idle labor, the contractor would need to identify specific equipment on the project in the daily reports to demonstrate exactly when this equipment was idle and the duration of the idle time.

The owner will generally question whether the equipment was idle because of the delay. It is often worthwhile for the owner to verify the contractor's use of equipment prior to and immediately following the period of the delay. If the equipment was idle both before and after the delay period, then the owner may appropriately question whether the contractor actually sustained any idle equipment costs for the idled equipment during the period of the delay.

When equipment is idle there is often a question related to the contractor's obligation to mitigate costs. For example, if the equipment remains idle for a prolonged period due to an ongoing delay, it may be less expensive to remove it from the site and remobilize it later when the work can progress. Such a decision should be considered when the delay is of a known or predictable duration. However, if the delay continues despite expectations that the delay will not be prolonged, the opportunity to mitigate the costs in this manner may not be readily apparent. The contractor should consult with the owner to discuss the propriety of such a decision when a delay appears to have the potential to continue for a prolonged period. The contractor should notify the owner of its intention to remove equipment from the site with its expectation that the owner will pay the demobilization and remobilization costs when such actions are warranted.

As with labor costs a contractor's equipment costs may also be subject to escalation. If a contractor is using rental equipment, the delay could shift the contractor's equipment into a time period with a higher rental rate. This situation can and should be documented with invoices.

Other costs can be associated with equipment in the case of a delay. For example, if a significant delay occurs to a portion of a project, specialized equipment originally planned for the work may no longer be available. The contractor-owned paving machine may have been assigned to a project for a particular period of time, but it is no longer available after the delay period. In such a case, the contractor may be forced to replace the owned equipment with more expensive rental equipment to meet the project's revised schedule. In another case, the contractor may have originally planned to perform mass excavation work using scrapers. But, because of the delay and the progress of adjacent work, the contractor now finds it must use loaders and trucks to excavate the material. The added cost to the contractor would be measured by comparing the cost of the two methods, the originally planned method versus the actual. In some respects this type of damage may have a productivity component, which is addressed later in this book. It should also be noted that when the performance of work on the project's critical path is less productive or inefficient and, as a result, takes longer than planned to complete, the extended duration of this work could cause delays to the project that result in additional delay-related costs.

MATERIAL COSTS

The most common additional material costs caused by a delay are storage costs and escalation costs. If a delay occurs after the purchase or fabrication of certain materials or equipment that are intended to be incorporated into the work, the installation of this equipment may be delayed as a result of the subject delay. This situation may necessitate the need to store these materials at some additional cost not contemplated in the original project plan. Such costs may include facility rentals, additional handling, and maintenance of the stored materials.

In some cases, a contractor may recognize that a delay is going to result in the purchase of certain materials after an anticipated price increase and that the cost to purchase and store these materials may be less expensive than purchasing them at a later date at the escalated price. In the case of an owner-caused delay, the contractor should notify the owner in advance of this course of action to ensure that the stored materials can be billed in the progress payments and that the storage costs will be reimbursed.

ESCALATION COSTS

Some delays can cause the contractor to experience escalation of its labor, material, or equipment costs. This may occur when the delay shifts these expenditures into a more expensive time period than that in which they were originally scheduled to be expended. To recover such costs, the contractor must demonstrate when the work that comprises these expenditures was planned to occur over time and compare this to how these expenditures were actually incurred.

In the case of labor, e.g., the comparison is made between the quantity of labor that would have been purchased before and after the price increase in the original schedule and the quantity of labor actually purchased before and after the price increase under the delayed schedule. The damage is the additional cost component of the quantity of labor that shifted from the period prior to the price increase to the period after the price increase. Material and equipment escalation is determined in the same manner.

FIELD OFFICE COSTS

One common delay-related cost that a contractor may claim is extended field costs. For example, when a project experiences a delay, the contractor's field staff and field equipment may be on site longer than originally scheduled. The discussion in this chapter addresses the major categories of extended field labor and equipment costs. This chapter explains how to calculate these damages to clearly illustrate the additional cost to a contractor due to a delay and the nature of its delay-related costs.

Extended field labor costs

Depending on the project-specific circumstances, when a project encounters a delay, the contractor would typically retain its supervisory team at the job site. Personnel such as the project manager, project engineer, superintendent, assistant superintendent, and administrative support positions fall into this category. These personnel represent a direct labor cost to the project.

In the context of field office overhead costs, the term "direct cost" merits some discussion as the characterization of costs as either direct or indirect is a matter of perspective. For example, in the context of the contractor's costs, the cost of the home office is an indirect cost that is not directly chargeable to any one project and the cost of the field office of a particular project is a direct cost that is directly chargeable to a project. However, in the context of the cost of a particular project, field office costs may be considered to be an indirect cost in that they are not directly chargeable to a particular aspect of the work.

In the simplest scenario, the contractor's cost to maintain staff on site during a delay is calculated by adding the daily cost of each staff member's salary, including burden, and then multiplying that sum by the number of days of excusable, compensable delay. Make sure that if the delay is expressed in calendar days that the costs are also expressed in cost per calendar day.

(Aggregate of each daily salary + burden)
× number of days of compensable delay
= Costs for extended field labor (supervisory personnel)

While this calculation is straightforward, it is important to note that the analyst also needs to address the propriety of the contractor's proposal to recover extended field costs for supervisory personnel.

To properly calculate the costs of extended field supervisory labor, the analyst should start with a review of the contractor's normal accounting procedures for these costs. For example, if the contractor routinely charges the project directly for the project manager, then claiming costs for the project manager is appropriate; of course, subject to the contract requirements. However, if the contractor routinely charges the project manager to home office overhead, then claiming this cost as a direct field cost may require further investigation. The owner would likely argue that the project manager's salary is a home office overhead cost and not an extended field cost. The same principle applies to the salary of any other field supervisory personnel extended on the project because of the delay. Home office overhead costs are addressed in the chapter that follows.

Extended field office costs

Extended field office costs will also consist of the daily cost of all of the support equipment and services that are necessary to maintain the project site. Such equipment will typically include office and storage trailers, portable sanitary facilities, office equipment, and security fencing. Extended field office services will typically include phone service, electrical services, drinking water supply, and security. The full list of such costs is too extensive to name here. The key to determining which costs are included in the daily rate is the time-related nature of the expenditure. If the expenditure is incurred on a recurring basis over time, e.g., the monthly rental of the security fencing, then it would be part of the extended daily field cost rate. If, however, the security fence was purchased for the project, such that the contractor does not incur an additional cost for having it on site an additional month, it would not be considered an extended field cost.

Extended daily field cost rate

The daily rate for field office labor and equipment is generally not linear over the life of the project. Rather, these costs typically follow a curve whereby they build up during the initial phases of the project, level off at their highest once the project is in full swing, and then gradually decrease as the project work nears the end. This begs the question of what rate to use when calculating extended field overhead delay costs.

The easy answer is to use the rate that was being incurred during the delay period. However, determining the precise rate at any given point in time is not always practical, especially when considering the exact timing of recurring payments with respect to the timing of particular delays. Because of this, a reliable way to determine the applicable rate is to develop an average monthly rate over a particular period and convert that to a daily rate. Costs are often looked at monthly as this is the typical basis upon which field overhead costs recur.

To determine the rate to apply to a particular delay, it is appropriate to break the project into phases. Typically, three phases are investigated, the start-up phase, the construction phase, and the completion phase, although more may be applicable depending upon the way the field overhead costs were incurred. The beginning and ending points of these phases should coincide with the trend of expenditures discussed earlier. Determining these points will require judgment, the objective being to arrive at a daily rate that best corresponds to the timing of the delay. For example, if there is a delay at the beginning of the project, the extended field office costs incurred as a result of this delay will be at the start-up rate, which may be less than when the project is in full swing. Similarly, if the delay occurs in the middle of the construction period, when the maximum resources are being applied, the rate will be greater than that occurring during the start-up or completion phases of the project and, thus, greater than an average rate developed for the entire project.

Both owners and contractors will often calculate one daily field office rate for the entire project, resulting in an amount that is more or less than the amount that should properly apply. When the project finishes late as a result of delays by the owner, contractors often make the mistake of requesting the field office costs incurred after the planned completion date, seriously undervaluing the costs that were incurred. These costly mistakes on the part of both owners and contractors can be avoided by properly breaking the project up into the various phases of overhead expenditures that occurred.

OTHER DELAY COSTS

Delays on a construction project may cause a contractor to incur a variety of costs that would otherwise not have been incurred. These items may include:

- need for temporary utilities
- use of temporary facilities
- purchase of extended warranties
- maintenance and protection of work during delays
- inefficiencies
- increased bond costs
- increased insurance coverages.

While these costs may not fit easily into one the general cost categories described in this chapter, it is important for the parties considering delay costs to carefully assess all of the additional costs to the project occasioned by the delay. In this way items such as these will not be overlooked.

GENERAL GUIDELINES FOR THE PRESENTATION AND RECOVERY OF DELAY COSTS

Overview of the cost presentation

The objective of the contractor's presentation of its delay costs in either a change order request or a claim should be twofold. First, the contractor needs to ensure that it has captured all of the costs associated with the delay, since it is unlikely to get a second bite at the apple. Second, the contractor wants to present the costs persuasively. The following paragraphs offer some suggestions aimed toward helping a contractor to meet these goals.

Before calculating costs, the contractor must carefully review all applicable facts supporting the proposal or claim in order to establish that it has been harmed and suffered a loss based on this harm. The calculation of costs must be based on the facts and merits of the situation, and not on unrealistic demands or expectations. To the extent possible, costs should be based on a cause-and-effect relationship with the change for which additional compensation is being sought. The fact that a contractor may have suffered a loss does not establish the contractor's entitlement to be paid for this loss. Too often, cost calculations are not realistic or adequately supported and a settlement becomes difficult to accomplish. A well-presented proposal or claim will lend itself to subsequent revisions if deemed necessary. The more supporting documentation accompanying the proposal or claim, the more likely a settlement may be negotiated.

Keep in mind that it is a contractor's responsibility to mitigate costs whenever practical in order to not create added costs greater than those reasonably justified under the circumstances. When in a delay situation, carefully document and review any and all efforts to mitigate the added costs of the delay when applicable. It is generally recommended to identify these efforts in correspondence to the owner. Correspondence addressing the need to maintain equipment on site, reduce or not reduce manpower, and other similar considerations will serve to strengthen the contractor's proposal or claim.

The contractor has to support its proposal or claim and related costs. The contractor's actual costs are generally presumed reasonable unless the owner can show that the costs were unreasonable. Once liability for delay has been established, the resulting costs do not have to be proven to an exact measure, but amounts should be reasonable and adequately supported, and should not be speculative.

Although this section focuses on establishing the cost of delay by identifying actual costs, the reader should be aware that other methods of recovery may be applicable under some circumstances. These methods are usually specified by the contract and legal counsel and may be applicable when pursuing reimbursement for costs based on wrongful termination, total cost, modified total cost, quantum meruit, or jury verdict approaches.

Basic guidelines for presenting delay costs

The following are basic guidelines for preparing and reviewing the costs associated with a delay:

1. The initial step in formulating the cost calculation, whether caused by delays or other reasons, is to carefully review and closely follow the contract provisions. A typical agreement will require both contracting parties to fulfill specific requirements and a certain measure of risk or loss can accrue to either party if the requirements are not satisfied. For example, the changes clause will likely provide rules for the recovery of delay costs, and the contractor's recovery of such costs may be limited to the terms set forth in the contract. It is essential that the costs proposed or claimed follow these specific rules, or recovery may be forfeited or reduced.
2. Avoid frivolous items and overstated proposal or claim amounts. Although some parties believe you must start high and negotiate down, frivolous or inflated proposals or claims cause a contractor to lose credibility and ultimately delay the process of resolution. Public contracts often require the contractor to certify the claim and may include clauses or be governed by laws that classify overstated claim amounts as false claims with harsh consequences. A well-documented proposal or claim should satisfy the audit requirements of the contract.
3. Support costs based on the verifiable facts, and not one's expectations anticipated prior to the award of the contract. If a project incurs a loss, the reasons for the loss that are being claimed with appropriate support and measurement, not the bottom line loss.
4. Summarize your costs within a format that will allow subsequent updates or revisions, especially if the quantitative measure is not fully available until a future date.
5. Avoid duplication of proposed or claimed costs and calculation or posting errors. Cross-check and double check the calculation as it evolves over time.
6. Remember that the responsibility for establishing entitlement to reimbursement for the costs of a delay fall on the party submitting the associated proposal or claim. The claimant must produce facts to establish that the costs were incurred as a result of actions of the other party, along with the appropriate measure of these costs.
7. There is always an obligation to mitigate costs. If certain costs could have been mitigated, these costs are unlikely to be recoverable.

8. Generally a claimant's actual costs are considered reasonable, unless it can be shown the costs were not reasonable. However, the fact that a cost was actually incurred does not automatically establish that costs were incurred as a result of the action of the other party.
9. The supporting documentation submitted with the proposal or claim may ultimately be used as evidence and, thus, should be prepared so as to be admissible under the jurisdictional rules governing the case.
10. The documents supporting your costs are generally more persuasive if prepared contemporaneously with the actual progress of the project, rather than documented after the fact.
11. Remember to establish a cause-and-effect relationship to the extent possible for each component of cost being proposed or claimed. Each party needs to show how its costs increased as a result of the other parties' actions or inactions.
12. When feasible, submit your proposal or claim for costs using the suggested format of the owner. This process will lessen the areas of disagreement with the owner about the format and structure of the proposal or claim, and keep the discussion focused on the content of the claim.

EXAMPLES OF DELAY COST CALCULATIONS

Labor escalation

Example 11.1. A project is scheduled to start on a specific date and finish 300 calendar days later. The contractor is using union labor contracted under a collective bargaining agreement that specifies pay increases at certain intervals. In this example, the pay increase for common laborers is to occur on Day 150. The rate before the wage increase is $25 per hour; it escalates to $27 per hour after the increase. In the third month of the project, the owner causes a 10-day delay to the project. The owner accepts responsibility for the delay so there is no question of liability or the magnitude of the delay. The project is extended by change order an additional 10 days, and the contractor must present the labor costs associated with the delay (See Fig. 11.2).

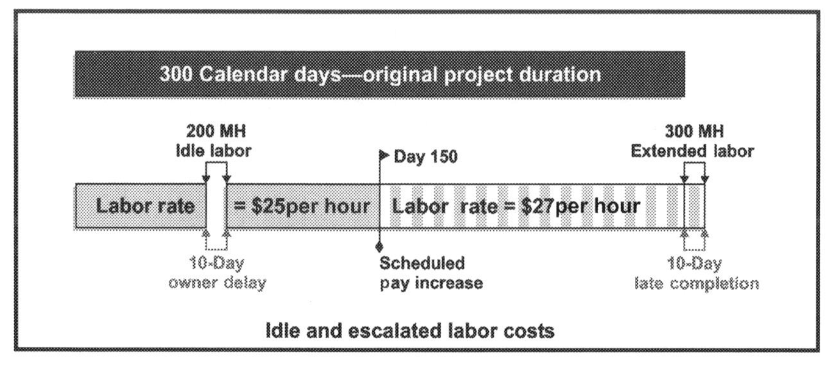

Figure 11.2 Idle and escalated labor cost graphic.

The contractor's records show that during the delay, there were 200 labor hours of idle time for common laborers. To figure delay costs, the contractor multiplies the number of hours by the wage rate:

$$200 \text{ hours} \times \$25.00 \text{ per hour} = \$5000.00$$

The contractor's records also show 300 labor hours for common laborers during the last 10 days on the job, which occurred in a time frame later than that originally planned. The contractor seeks compensation for the additional $2 per hour ($27.00–25.00) for the 300 labor hours.

$$\$2 \text{ per hour} \times 300 \text{ labor hours} = \$600$$

The contractor presented a total proposal of $5600; with $5000 for idle labor and $600 for extended labor.

The contractor's approach, however, is incorrect. To correctly determine the costs, the analyst must examine each component of the reimbursement being sought separately, as follows:

The idle labor request of $5000 was correct. Of course, the records provided by the contractor must validate the idle labor hours and substantiate that it was not possible to reassign the idle workers to other productive work activities or demobilize them from the project.

The request for $600 for the extended labor was incorrect. Had there been no delay, the 300 labor hours worked at the end of the project would have been worked only 10 days before the date they were actually worked. Therefore these hours still would have been worked at a rate of $27 per hour. There were no added costs associated with these last 300 hours. Instead the contractor should be looking at the 10-day time period immediately following the wage increase. The labor hours in this time frame were shifted from the less expensive time period into the more expensive time period. The contractor's certified payrolls showed that 400 labor hours were worked during the 10-day period immediately after the wage increase. Therefore the contractor would be entitled to an additional $800 for escalated cost of its labor due to the delay.

$$\$2 \text{ per hour} \times 400 \text{ labor hours} = \$800$$

Thus the contractor's costs were actually $5800 for the direct labor costs caused by the delay. The contractor would naturally add the appropriate burden to this amount to calculate the total additional labor costs attributable to the delay.

This simple example illustrates that the analyst should not estimate costs without first clearly understanding the nature of the delays. Hasty conclusions or conclusions based on unverified assumptions often produce erroneous calculations. A more realistic and complex project example might involve several trades under different collective bargaining agreements, each with different wage increases occurring on different dates. A wage increase might extend over a period of more than a year for some trades. The updated labor agreements may make new and different requirements for apprentices or supervisory labor. The contractor or owner must assess how the shift in the work performed by the labor force increased the labor costs or exposed the contractor to escalated wage rates. The example just given is further simplified by the fact that no additional work was added by the delay. It also assumes that once the delay ended, the work was performed in the same sequence as originally planned.

What happens when a delay alters the work sequence, or when a delay combines with other changes to affect the actual labor distribution on the job? In these situations, the contractor must substantiate the original planned distribution of labor compared with the actual distribution. The following example illustrates this approach:

Example 11.2. A project is scheduled for a duration of 300 calendar days. During the project the contractor maintains a resource-loaded CPM schedule to show the distribution of manpower planned over time. Fig. 11.3 shows the distribution of planned carpenter hours for the project taken from the original schedule. Fig. 11.4 shows the distribution of

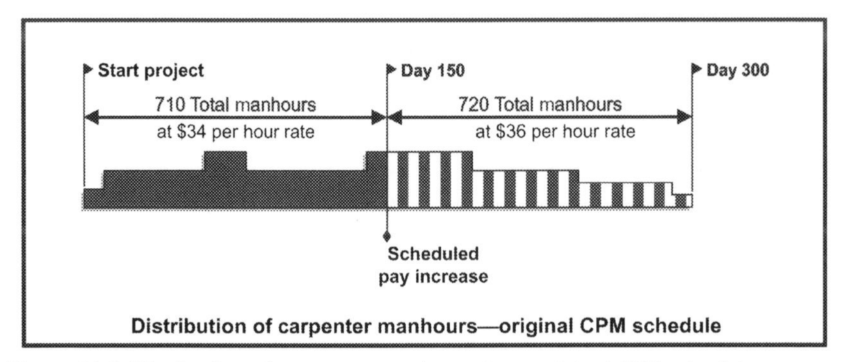

Figure 11.3 Distribution of carpenter manhours from original CPM schedule.

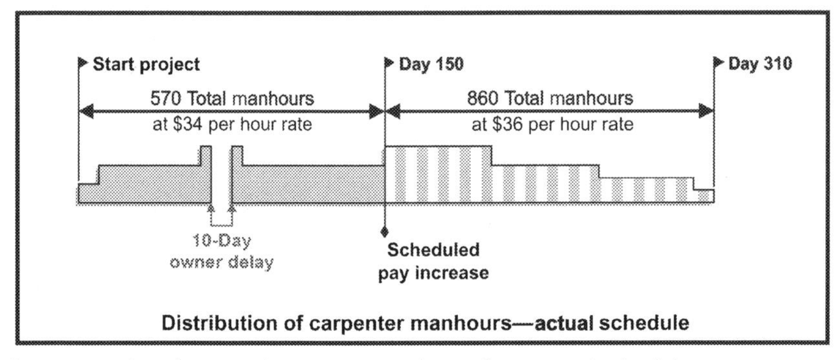

Figure 11.4 Distribution of carpenter manhours from actual schedule.

actual carpenter hours on the project, based on the delay caused by the owner.

As can be seen from these figures, the wage increase for the carpenters took effect on Day 150. In the original schedule, the contractor planned to use 720 carpenter labor hours after Day 150, at the higher labor rate of \$36 per hour. In the actual distribution of labor, the contractor used 860 carpenter labor hours during the higher wage rate period. Therefore 140 labor hours (860–720) were shifted into the period of the increased labor rate. The contractor should calculate \$280 (\$2 per hour × 140 hours) for the escalated labor cost of the carpenters.

$$\text{860 actual hours} - \text{720 planned hours} = \text{140 labor hours}$$
$$\text{140 hours} \times \text{\$2 per hour wage increase} = \text{\$280.00}$$

In this example, the actual distribution of carpenter labor was caused solely by the owner's delay. In a different situation, if the owner can demonstrate that this was not the case, then the delay costs should not include the escalated labor calculation.

Of course, many projects do not have a resource-loaded CPM schedule to allow a reasonably precise comparison between the planned and the actual expenditure of labor hours. Without the CPM schedule, the analyst can estimate the planned labor distribution from the project bar chart or other available project records. The analyst can compare the labor hours actually expended on the job (taken from the project daily reports) to the planned labor hour distribution. The following example illustrates this procedure:

Example 11.3. A contractor plans to perform the work in accordance with the bar chart shown in Fig. 9.4. No labor hours are shown in the bar chart. Due to an owner-caused delay, the contractor's work is performed later than originally planned. As a result, the contractor experiences escalated labor costs on the project after a rate increase (from $35 to $36 per hour) went into effect on Day 200 of the project.

Fig. 11.5 is the contractor's bar chart for the work as originally planned. Fig. 11.6 shows the actual bar chart for the project, reflecting the delays. It also shows the actual carpenter labor hours worked on each

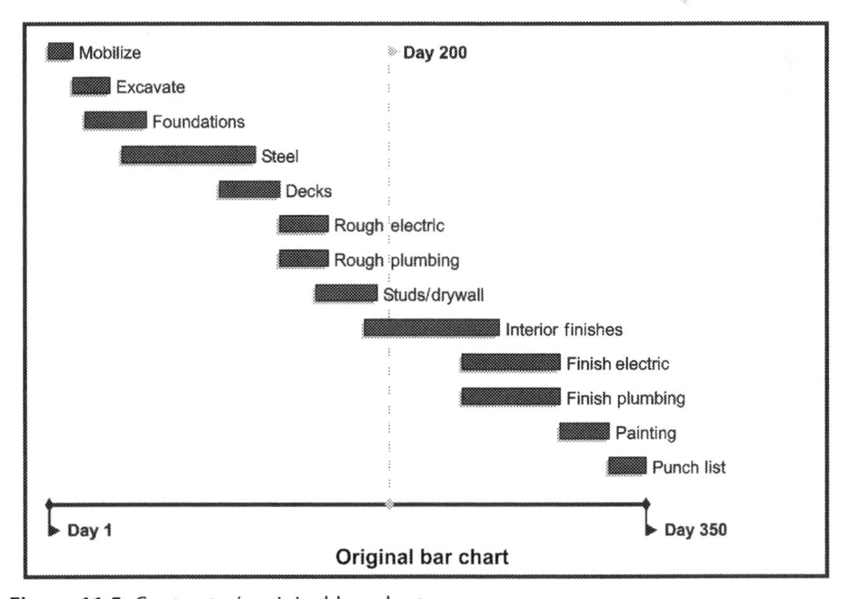

Figure 11.5 Contractor's original bar chart.

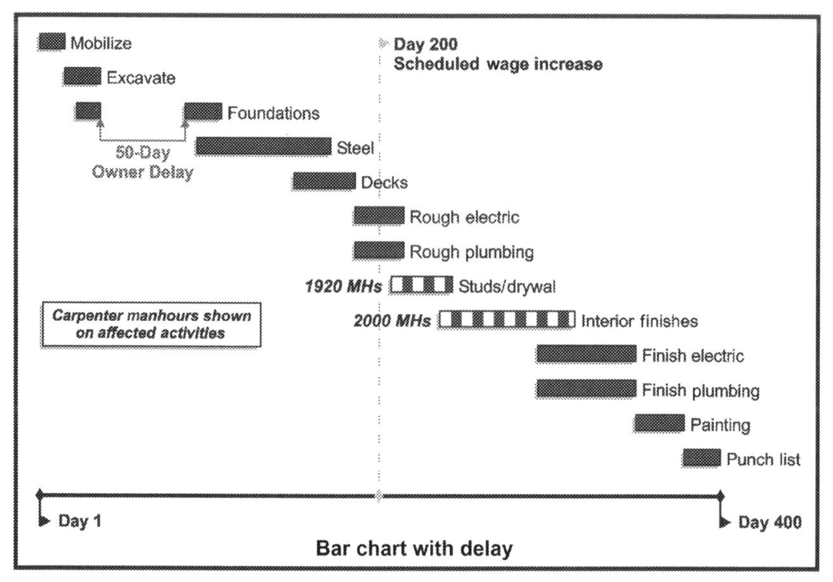

Figure 11.6 Actual bar chart.

activity. These labor hours were taken from the contemporaneously recorded project daily reports.

In order to calculate the cost escalation, the contractor uses its original bar chart (Fig. 11.5) and applies the actual carpenter labor hours to each respective task. This is shown in Fig. 11.7.

A comparison of Figs. 11.5 and 11.6 shows that there were 2740 carpenter labor hours shifted into the more expensive time frame. Therefore the contractor's cost escalation for this trade is $2740.

$$\$1 \text{per hour} \times (1920 \text{ hours} + 820 \text{ hours}) = \$2740$$

It should be noted that this conclusion assumes the delay was the only reason the work was shifted into the more costly time period and that no labor inefficiency resulted.

Equipment costs

Example 11.4. An owner is constructing a wastewater treatment plant addition. The project site is an inactive landfill. The excavation subcontractor is required to excavate approximately 200,000 cubic yards of material and dispose of it at a nearby active landfill. Prior to beginning

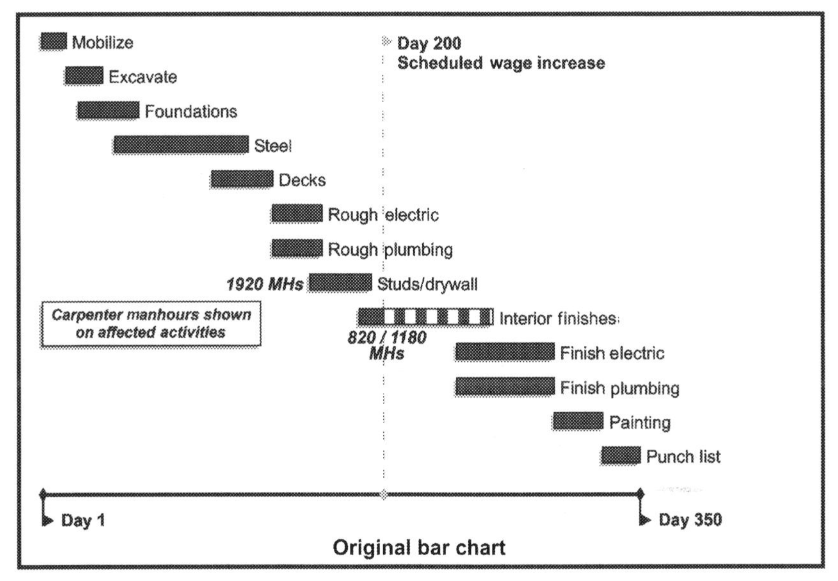

Figure 11.7 Actual carpenter hours applied to contractor's original bar chart.

excavation, the owner is advised that the site may contain hazardous waste materials. Consequently the owner stops the contractor's excavation work to perform toxicity testing. The testing takes 3 months to complete, at which time the excavation subcontractor is allowed to again proceed with the work as originally specified. Because of the delay, the subcontractor experiences changes to its equipment cost and availability.

Originally the subcontractor planned to excavate 150,000 cubic yards with scrapers, and the remaining 50,000 cubic yards with hydraulic excavators and 20-ton dump trucks. Due to the delay, the scrapers were no longer available, as they had been committed to another project. Consequently the subcontractor had to perform all of the excavation work without the use of scrapers. The subcontractor experienced two problems because of the delay. First, the excavators it intended to rent increased in cost. Second, the unit cost for the excavation work was higher using the dump trucks than if scrapers had been used. Based on these factors, the subcontractor calculated the increased costs as follows:

1. Escalation of rental equipment cost: Cat hydraulic excavator —quoted rate: $2800 per week (quote attached)—new rate: $2900 per week (invoice attached). Difference of $100 per week. Planned equipment time was 4 weeks (2 excavators for 2 weeks). Plus, 12 additional weeks

because excavators were used instead of scrapers. Extra cost: $100 per week × 16 weeks = $1600

2. Increased excavation cost: Excavation with scrapers: $4.50 per cubic yard (calculations attached). Excavation with dump truck: $7.20 per cubic yard (calculations attached). Difference of $2.70 per cubic yard. Extra cost: $2.70 per cubic yard × 150,000 cubic yard = $459,000
3. Total cost from delay: $406,600 = $405,000 + $1600

Of course, the contractor must always show that it complied with its obligation to mitigate costs. In this case, the subcontractor may need to demonstrate that the premium to rent scrapers over and above the cost of the planned scrapers was greater than the cost that resulted from the change in work methods.

The contractor must also ensure that it has not accounted for any of the cost components more than once, which is known as "double-dipping." In this example, that would require the contractor to demonstrate that the $7.20 per cubic yard excavation cost using the excavator and dump trucks was based on an excavator cost of $2800 per week, not the higher rate of $2900 per week. There is not enough information provided for this example to determine if there is a problem, but if the $7.20 rate is based on the escalated cost of the excavator, then the contractor's cost proposal would be double dipping by seeking both the escalated cost of the excavator and the higher cost of using an excavator and trucks where the excavator is priced at the higher rate of $2900 per week.

Material costs

Example 11.5. A project was originally scheduled to take 300 calendar days. The project experienced a 90-day, owner-caused delay. The delay occurred at the beginning of the project and forced the contractor to pay more for concrete. The contractor can demonstrate it had a purchase order for concrete at $80 per cubic yard from the beginning of the project until Day 100. After Day 100, by the terms of the purchase order, the cost of concrete increased to $85 per cubic yard. Due to the delay the contractor purchased more concrete material at $85 per cubic yard than it originally planned.

A review of the schedule shows that the contractor planned to have placed concrete on six of eight floors before the price increase took effect. A quantity takeoff establishes this as 2000 cubic yards of concrete. In the actual performance of the work, the contractor placed only 1200 cubic yards of concrete before the price increase took effect. The material cost increase is directly attributable to the excusable delay. Therefore the associated cost of the delay will be at least $4000 ($5 per cubic yard × 800 cubic yards) due to the escalated price of concrete.

2000 cubic yards planned − 1200 cubic yards actual = 800 cubic yards
Additional cost of $5 per yard × 800 yards = $4000 added cost due to the delay

Example 11.6. A project has a contract duration of 400 calendar days. The contractor estimates the construction schedule to last the full 400 calendar days, as shown in Fig. 11.8. The project is delayed 200 calendar days and finishes on Day 600 (See Fig. 11.9). Of the 200 days of delay, it was determined that the owner caused 150 days of delay and the contractor was responsible for the remaining 50 days. The owner has issued a two-part change order for its delays. Part I issued a time extension of 150 days. The owner is now in negotiation with the contractor over the compensation for the delay.

Fig. 11.10, a summary of delays to the schedule, shows that the owner delayed the start of the project by suspending the work for 50 days. The

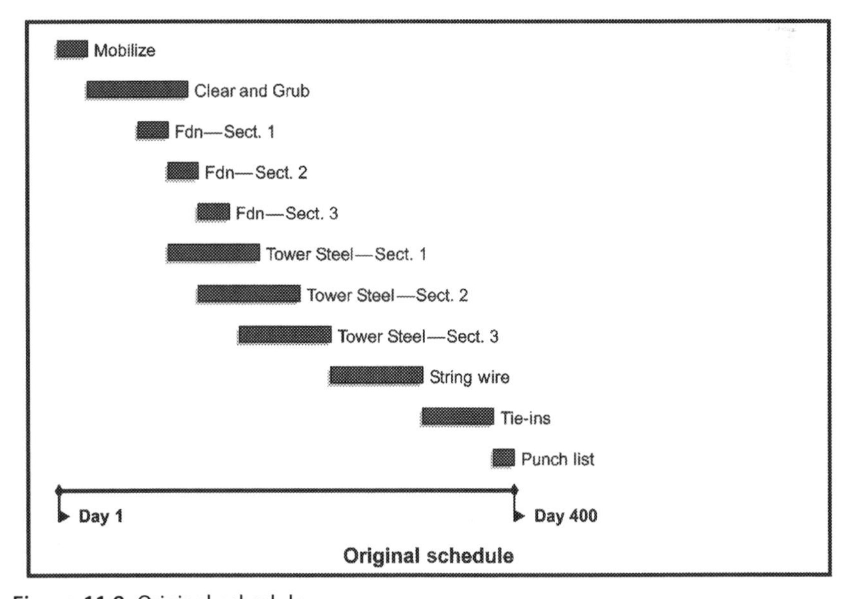

Figure 11.8 Original schedule.

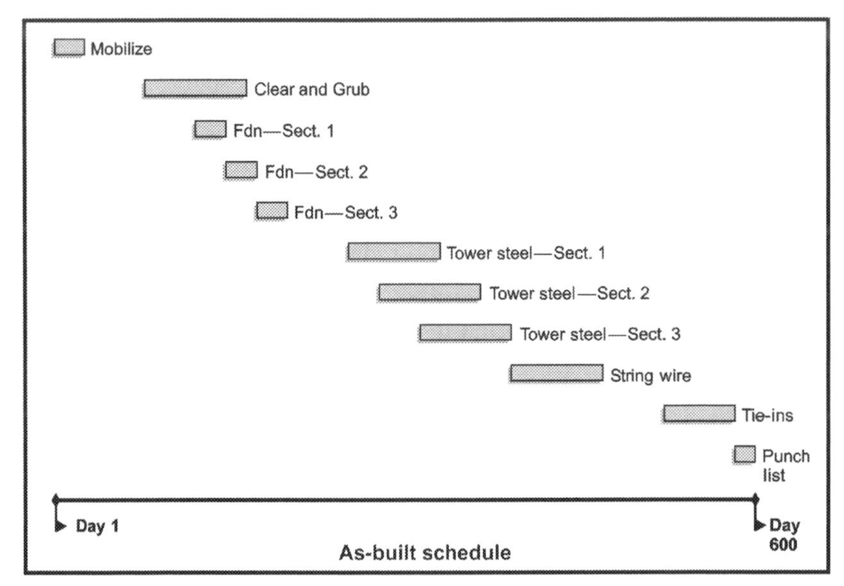

Figure 11.9 As-built schedule.

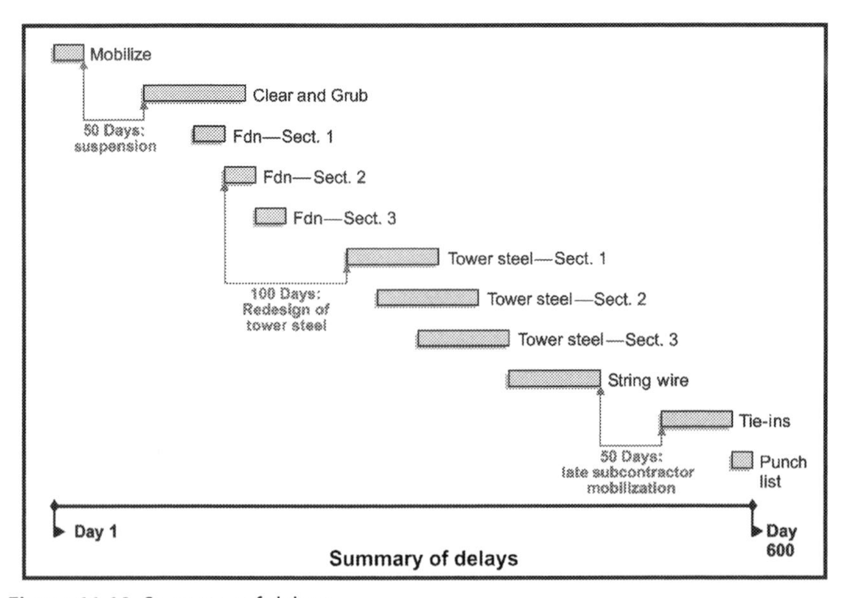

Figure 11.10 Summary of delays.

next delay was also caused by the owner, who, by requesting design revisions, delayed the project an additional 100 days. The final delay was 50 days, attributable to the contractor for failing to mobilize a subcontractor to perform work.

The presentation shown in Fig. 11.11 was made by the contractor to support the additional compensation it requested.

Request for equitable adjustment

Ace Construction Company hereby requests the following compensation for the 150 days of delay caused by factors beyond our control. In Part 1 of Change Order No. 7, the owner has granted a time extension of 150 calendar days. Because of this 150-day delay, Ace Construction accrued the following extra costs.

Extended field overhead:

Labor:				
Project manager:	150 days @ $600per day x 0.25	=	$22,500	Note: Certified payrolls attached. Project Manager is a direct job charge, but for only 25% of the time.
Superintendent:	150 days @ $520per day	=	$78,000	
Clerk:	150 days @ $200per day	=	$30,000	
Master mechanic:	150 days @ $480per day	=	$72,000	
Subtotal labor:	150 days @ $1350per day	=	$202,500	
Equipment:				Note: Invoices for equipment attached.
Superintendent's truck:	150 days @ $25per day	=	$3750	Superintendent's truck costed by normal company accounting practices.
Portable toilet:	5 mos. @ $150per month	=	$750	
Project trailer:	5 mos. @ $160per month	=	$800	
Tool trailer:	5 mos. @ $120per month	=	$600	
Subtotal equipment:			$5,900	
General conditions:				
Utilities:	Electricity: 5 mos. @ $150per month	=	$750	
			$500	
Dumpsters:	2 dumpsters: 5 mos. @ $150 ea/month	=	$1500	
Copy machine:	$150 ea/month	=	$600	
Office supplies:	5 mos. @ $120per month	=	$400	
Scheduling updates:	5 mos. @ $80per month	=	$5000	
Subtotal general conditions:	5 mos. @ $1000per month	=	$8750	

Figure 11.11 Contractor's additional compensation presentation.

Total Extended Field Construction:		=	**$217,150**	
Construction Equipment Costs:				
Extended Equipment:				
100-ton crane:	3 extra mos. @ $6,000/month	=	$18,000	Note: Applicable timesheets and certified payrolls attached.
Escalation of Equipment:				
D-7 dozer:	2 mos. @ increased rental rate of $400/month	=	$800	Union agreement and calculation of rates to include burden and overhead attached.
Idle Equipment:				
Scrapers:	327 hours @ $70/hour	=	$22,890	
D-7 dozer:	240 hours @ $75/hour	=	$18,000	
Total Construction Labor Costs:			**$59,690**	
Idle Labor:				
Carpenters:	725 hours @ $37/hour	=	$26,825	
Laborers:	816 hours @ $25.50/hour	=	$20,808	
Ironworkers:	420 hours @ $37/hour	=	$15,540	
Subtotal Idle Labor:			$63,173	
Escalation of Labor:				
Carpenters	Escalation: 700 hours @ $1.60/hour	=	$1,120	
Laborers:	Escalation: 250 hours @ $1.50/hour	=	$375	Note: Applicable timesheets and certified payrolls attached.
Ironworkers	(1st increase) escalation: 320 hours @ $1.20/hour	=	$384	
	(2nd increase) escalation: 650 hours @ $1.50/hour	=	$975	Union agreement and calculation of rates to include burden and overhead attached.
Subtotal Labor Escalation:			$2,854	
Total Labor Costs:			**$66,027**	
Material Costs:				
Escalation of Concrete:	1200 yards @ $3.00/yard	=	$3,600	
Storage of rebar:	Supplier Charge	=	$650	
Total Material Costs:			**$4,250**	
Grand Total:			**$347,117**	

Figure 11.11

CHAPTER TWELVE

Home Office Overhead

WHAT IS HOME OFFICE OVERHEAD?

In the event of an owner-caused delay, a contractor may seek to recover delay costs associated with home office overhead. For many reasons, despite being a commonly sought element of delay costs, home office overhead remains a contentious issue. The reasons given for opposition to payment of home office overhead as a cost of delay are many, but most can be categorized as follows:

- Some do not agree that there is a home office overhead cost incurred by the contractor as a result of a delay or that any cost incurred is the contractor's risk and not the owner's responsibility.
- Some believe that the home office costs associated with a delay are fully compensated by the markup for overhead provided on the cost of the change that caused the delay or on the other delay costs being sought.
- Some disagree with the formulas that are often used to calculate home office overhead costs.

Adding to the potential for conflict is the question of what constitutes home office overhead, a term that can take on different meanings given the financial structure of the contractor's organization and the accounting principles employed. Commonly, however, home office overhead consists of the contractor's fixed costs of operating its principal or home office. It is in the home office that executive and administrative functions are performed on behalf of the contractor's entire organization, including individual projects. Examples of such costs are shown in Fig. 12.1.

Specifically excluded from this definition of home office overhead are the direct costs of labor, equipment, and materials expended to construct and manage a specific project. The cost of providing a job site trailer, e.g., is not a home office cost as it is incurred specifically to support a particular project.

Construction Delays.
DOI: http://dx.doi.org/10.1016/B978-0-12-811244-1.00012-4

Rent
Utilities
Furnishings
Office equipment
Executive staff
Support and clerical staff not assigned to the field
Estimators and schedulers not assigned to the field
Equipment yard
Maintenance shop
Mortgage costs
Real estate taxes
Nonproject-related bond or insurance expenses
Depreciation of home office equipment and other assets
Security costs
Cleaning services
Office supplies (paper, staples, etc.)
Advertising
Marketing
Interest
Accounting and data processing
Professional fees and registrations

Figure 12.1 Example of home office costs.

The auditing or accounting standards employed by the owner can further restrict the definition of home office overhead. For example, under contracts governed by Federal Acquisition Regulation (FAR) cost principles, certain costs, such as those expended for marketing and entertainment may not be recoverable as home office overhead costs. As a result, on Federal projects, or on projects that rely on the FAR for the purposes of cost accounting, these marketing and entertainment costs end up being covered by whatever amounts the contractor is able to mark up its costs for profit, if at all.

Regardless of the exact nature of its home office overhead, a contractor must pay for these costs by drawing funds from the projects it performs. There are many ways for a contractor to do this. Many apportion home office costs to projects based on some appropriate measure. For example, home office costs might be allocated to a project based on the project's revenues or billings. Another contractor might apportion home office costs based on the contractor's project labor costs.

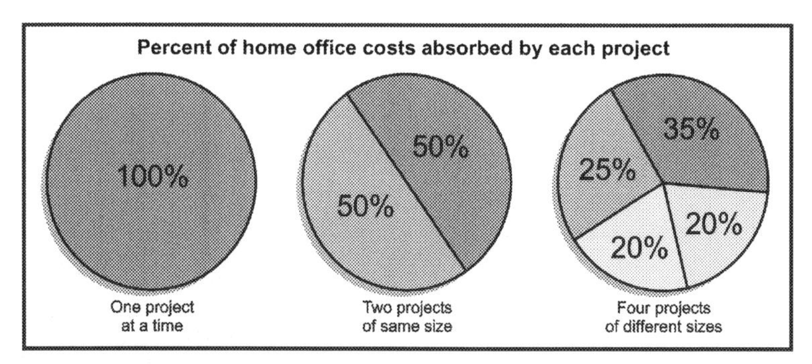

Figure 12.2 Percent of home office overhead costs absorbed by each project.

Normally the contractor includes home office overhead costs as a part of its pricing for each project. For projects that require the contractor to prepare a lump sum bid price, the contractor often calculates the final bid price by adding a markup percentage to the direct costs in its bid to provide for home office costs. The exact markup depends on the contractor's historical home office overhead costs, which the contractor typically evaluates on a quarterly or an annual basis. When reviewing a contractor's bid documents, this markup is often identified as "G&A" for "General and Administrative" costs. It might also be identified as "G&A&M," where the "M" stands for marketing expenses.

The number of projects the contractor has under construction at any one time may also affect the markup. For instance, if a contractor works on only one project at a time, 100% of the home office costs for the period of construction would have to be covered by the project. As the number of projects increases, the percentage allocated to each job would be reduced, as shown in Fig. 12.2.

EFFECTS OF DELAYS ON HOME OFFICE COSTS

As just mentioned, a contractor typically includes in its bid price for a particular project a percentage markup through which it will recover some portion of its home office overhead. If this project were to then experience a delay, project revenues may be earned over a longer period of time, disrupting the basis under which the contractor originally allocated its home office overhead costs. This effect can best be demonstrated through a series of examples.

One project at a time

For the contractor who performs one project at a time, the effect of a delay is relatively straightforward. As shown in Example 12-1, the contractor did not receive additional compensation for extra work on change orders to offset the increase in home office costs to be covered by this project. If, however, the 1-month delay was the result of extra work with direct costs of only $10,000, and the overhead markup allowed by the contract was 10%, the contractor would receive $1000 for overhead as compensation for the extra work. This overhead markup would have to cover both field and home office overhead. If actual home office overhead costs were $4000 per month, the contractor would be $3000 short in recovering home office costs and have nothing left to cover field overhead costs. Of course, if the value of the change was $100,000, the 10% markup would be greater than the overhead costs incurred. By their nature, markups are not intended to compensate the contractor for its actual overhead costs. Sometimes compensation provided by markups will be higher and sometimes lower than the actual cost incurred. The difference between delays caused by a suspension and those that result from extra work are addressed later in this chapter.

Example 12-1. A contractor has home office costs of $48,000 per year and has one, 1-year contract with a value of $1,000,000. The project experiences a 1-month delay during which time home office costs of $4000 accrue. (Many home office costs are roughly constant month to month. If the contractor's home office costs are $48,000 per year, then they are likely averaging approximately $4,000 per month.) If the delay were the result of a full suspension of project work and the delay was compensable, the contract amount of $1,000,000 would remain the same, but the contractor would now have to cover a total home office cost of $52,000 over the term of the contract, an increase of $4000.

Contract amount: $1,000,000

Yearly home office costs: $48,000

Unchanged contract amount: $1,000,000

New home office expense, including extending the project 1 month: $52,000

Increase in home office costs to be covered by this project: $52,000−48,000 = $4000

Based on this example, the contractor would seek $4000 to cover its home office costs for the delay.

Multiple projects

As shown in Example 12-2, the contractor has multiple ongoing projects. Clearly as the number of projects increases, the calculations become more difficult, as is apparent from the numerous books, legal briefs, articles, and web entries related to this subject. In characterizing home office overhead costs, some choose to distinguish between the terms *extended home office overhead* and *unabsorbed home office overhead*. The calculation of costs for each may differ, but in practice, courts and boards have not always maintained a clear distinction between these terms. This chapter will treat these costs as being essentially the same. The remainder of this chapter presents methods for calculating delay costs associated with home office overhead.

Example 12-2. This contractor has home office overhead costs of $1,000,000 per year. The contractor normally has four projects in progress at any one time, for a total yearly volume of $20,000,000. The contractor allocates home office costs by adding a 5% markup to each bid. Each of the four projects is valued at $5,000,000. What would happen if one project were to experience a delay?

Originally each project absorbed $250,000 of home office costs, or about $21,000 per month. If one project experiences a 1-month delay due to a full suspension of work on the project, then the contractor would receive no additional revenue to cover home office costs from that project, and would therefore suffer a net loss in recovered or "absorbed" home office overhead costs of $21,000. If the delay were attributable to added work rather than a suspension, then the contractor might receive compensation from the overhead markup allowed on the change order. However, in the absence of extra work, the project experiencing this delay may underabsorb its share of the home office costs, leaving the other projects to overabsorb these costs.

EICHLEAY FORMULA

The difficulty in precisely establishing the home office overhead costs due to delay has led to the development of formulaic approaches to determining the appropriate compensation to be provided to reimburse a contractor for its unabsorbed or underabsorbed home office overhead costs. One such method is the Eichleay Formula. The formula originated from a decision by the Armed Services Board of contract Appeals in 1960 (Eichleay Corporation, ASBCA 5183, 60-2 BCA 2688). In its appeal before the Board, the Eichleay Corporation proposed a formula for calculating its costs for home office overhead during a delay. The Board accepted this formula, which has since become known as the Eichleay Formula, as a reasonable method of calculating such costs.

The Eichleay Formula is a simple three-step formula. First, the total contract billings are divided by the total actual company billings for the contract period, as shown in Fig. 12.3. The quotient is then multiplied by the total home office overhead costs during the actual contract period to produce the allocable overhead. Next the allocable overhead is divided by the actual number of days of contract performance, including the delay. This produces the daily allocable overhead rate (Fig. 12.4). Finally, in Fig. 12.5, the daily allocable overhead rate is multiplied by the number of days of compensable delay to produce the home office overhead costs.

A short example illustrates the application of the formula. A company has total billings of $50,000,000 during the contract period. The contract billings total $5,000,000. Total home office overhead is $1,000,000 during the contract period. The actual project duration and the contract period total 200 days, and compensable delays total 30 days. The first calculation to produce the allocable overhead of $100,000 (Fig. 12.6) is as follows:

The next calculation produces the daily allocable overhead rate (Fig. 12.7). And, finally, in Fig. 12.8, the daily allocable overhead rate is

$$\frac{\text{Total contract billings}}{\text{Total company billings}} \times \text{Total home office overhead} = \text{Allocable overhead}$$

Figure 12.3 Eichleay Formula: Allocable overhead calculation.

$$\frac{\text{Allocable overhead}}{\text{Number of days of contract performance (including delay)}} = \text{Daily allocable overhead rate}$$

Figure 12.4 Eichleay Formula: Daily allocable overhead rate calculation.

$$\text{Daily allocable overhead rate} \times \text{Number of days of compensable delay} = \text{Home office overhead damages}$$

Figure 12.5 Eichleay Formula: home office overhead damage calculation.

$$\frac{\text{(Contract billings)}}{\text{Total Co. billings}} \times \text{(Home office OH)} = \text{(Allocable OH)}$$

$$\frac{\$5{,}000{,}000}{\$50{,}000{,}000} \times \$1{,}000{,}000 = \$100{,}000$$

Figure 12.6 Eichleay Example allocable overhead calculation.

$$\frac{\text{(Allocable overhead)}}{\text{No. of days of contract performance}} = \text{Daily allocable overhead rate}$$

$$\frac{\$100{,}000}{200 \text{ days}} = \$500/\text{day}$$

Figure 12.7 Eichleay Example daily allocable overhead rate calculation.

$$\text{Daily allocable overhead} \times \text{Number of days} = \text{Home office overhead damages}$$

$$\$500/\text{day} \times 30 \text{ days} = \$15{,}000$$

Figure 12.8 Eichleay Example home office overhead damage calculation.

multiplied by the number of days of compensable delay to yield home office overhead costs of $15,000.

Problems with the Eichleay Formula

While the Eichleay Formula is simple to apply, some have questioned its accuracy. The Eichleay Formula is an estimated allocation and may, therefore, yield results that are either too high or too low. Despite this limitation, the Federal courts and Boards have, for the most part, accepted the formula as a reasonable approximation of the damages sustained. With less case law to draw upon, state courts have been less receptive to outright acceptance of the formula.

Debate over the use of the formula has led to refinements regarding its application. For example, in Excavation-Construction, Inc., ENG BCA 3851 (1984), the Engineering Board of Contract Appeals (ENG BCA) recognized the use of the Eichleay Formula to determine the cost not only of a suspension of work but also of a delay caused by extra work. The Board's opinion is noted in the following quote. The Board did, however, subtract from the Eichleay calculation the amount of overhead that was already being paid by virtue of the markup on the change order.

The Board believes in general that an Eichleay-type approach is the preferred way to determine home office overhead in a suspension situation and that a markup of direct costs is the preferred way to determine home office overhead for a change. However, no automatic approach can be applied in avoidance of a careful scrutiny of the facts. In this appeal, the changed work on the retaining walls is only part of the reason for extension of the contract period. Much of the delay and disruption occurred because the change was not timely made. The parties have agreed that the net effect was to extend the period necessary for performance by 99 days. Measurement of the effect on home office overhead by the costs alone is likely in these circumstances to understate the amount to which E-C is entitled. Therefore, the Board considers this appeal to be a proper one for application of the Eichleay Formula.

The Armed Services Board of Contract Appeals' response in George E. Jensen Contractor, Inc., ASBCA No. 29772 (1984), is as follows:

Finally, the Government argues that the home office or extended overhead costs are fixed costs which would have been incurred even if there had been no delay. Its argument continues that to allow relief by utilizing the Eichleay Formula would permit recovery of overhead costs much greater than the direct costs incurred during the periods of delay. This argument misses the point. Home office expenses are indirect costs usually allocated to all of a contractor's contracts based upon each contractor's incurred direct costs. When a government initiated delay causes a contractor's direct costs to decline greatly, that contract does not receive its fair share of the fixed home office expenses. The Eichleay Formula is one method approved by boards and courts over a long period of time which corrects this distortion in the allocation of these indirect expenses.

The most common argument against the use of the Eichleay Formula is that the contractor receives compensation for home office overhead by virtue of the markup on a change. However, a criticism of this argument is that a contractor likewise receives this same markup whether the change causes a delay or not.

Unless the markup clearly contains an allocation for home office overhead, the argument that Eichleay should not be used may not be valid. The following quote shows how the ASBCA addressed the point in Shirley Contracting Corporation, ASBCA No. 29848 (1984):

We find no support for this position. The Corps' Area Engineer testified that he did not know the composition of the 15 percent allowed but nonetheless approved it as a matter of course. He also admitted that the 15 percent overhead was allowed even 'on modifications that involved no delay at all.

Arguments against the broad application of Eichleay are provided in P.J. Dick Inc., Appellant, v. Anthony J. Principi, Secretary of Veterans

Affairs, Appellee. Nos. 02–1290, 02–1401, April 07, 2003. The United States Court of Appeals decided in part, as follows:

In short, a court evaluating a contractor's claim for Eichleay damages should ask the following questions: (1) was there a government-caused delay that was not concurrent with another delay caused by some other source; (2) did the contractor demonstrate that it incurred additional overhead (i.e., was the original time frame for completion extended or did the contractor satisfy the Interstate three-part test); (3) did the government CO issue a suspension or other order expressly putting the contractor on standby; (4) if not, can the contractor prove there was a delay of indefinite duration during which it could not bill substantial amounts of work on the contract and at the end of which it was required to be able to return to work on the contract at full speed and immediately; (5) can the government satisfy its burden of production showing that it was not impractical for the contractor to take on replacement work (i.e., a new contract) and thereby mitigate its damages; and (6) if the government meets its burden of production, can the contractor satisfy its burden of persuasion that it was impractical for it to obtain sufficient replacement work. Only where the above exacting requirements can be satisfied will a contractor be entitled to Eichleay damages.

On the subject of suspension, the Court wrote:

Our case law clearly requires that the contractor must show a suspension, whether formal or functional, of all or most of the work on the contract.

Explaining this statement further, the Court added:

As to PJD's factual arguments, those too must fail. The Board found that "[t]he evidence before us shows conclusively that PJD was able to progress other parts of the work during the time periods it alleges it was suspended." Although PJD argues the evidence shows its direct billings were greatly diminished during the delays, substantial evidence supports the Board's finding. The evidence, at worst, shows that in one of these delay periods PJD billed 53% less than it had the month before. There is, however, no evidence whatsoever that PJD's direct billings were less than they would have been absent the suspensions-that is the controlling test. A comparison of pre- and intradelay billings or intra- and post-delay billings is not the test. Regardless, PJD's direct billings during the delay periods can hardly be characterized as "minor." See Altmayer, 79 F.3d at 1134. At worst, PJD's direct billings during one of the periods of suspension were 47% of what they were in prior months; thus, PJD continued to perform substantial amounts of work on the contract during the suspension periods. On these facts, which are supported by substantial evidence, we must affirm the Board's conclusion that PJD was not on standby.

Based on these examples, recovery of unabsorbed home office overhead costs using an Eichleay Formula is not automatic. Like all claims for delay costs, to recover home office overhead costs, the contractor must first establish that the delay was indeed compensable. This typically entails establishing that the delay was on the critical path, delayed the project completion date, was the owner's fault or responsibility, and could not have been reasonably anticipated or otherwise mitigated by the contractor. In addition, many courts that have accepted the Eichleay Formula have attached other prerequisites to its application. These may include the contractor establishing (1) an owner-imposed suspension of all or nearly all of the work on the project, (2) an owner requirement that the contractor be on standby during the associated delay, and (3) proof that while on standby, the contractor was unable to take on additional work.

When to apply the Eichleay Formula

Based on the examples presented, it may be evident that the Eichleay Formula is a calculation applied at the end of the project after all of the delays have occurred and the work is completed. If the parties attempt to resolve the question of home office overhead during the project, some form of a modified Eichleay Formula may be appropriate. One approach is to apply the Eichleay Formula from the beginning of the project up to the point of negotiations. In this situation, the total contract billings, the total company billings, the total home office overhead costs, and the total number of days from the start of the project up to the approximate date of the calculation may be used. However, there has been reluctance on the part of the Federal Boards and courts to accept the use of a "modified" Eichleay Formula. But, by the time such disputes get to an actual trial, the project will have long been completed and there is no longer a need to use a modified formula.

As noted earlier in this chapter, when using the Eichleay Formula in Federal government cases, because of the FAR, some costs (such as advertising, entertainment, interest, etc.) that might be included as home office overhead costs are not allowable. The government will normally disallow these costs during its audit procedures. In its contract, the Federal government has the right to audit a contractor's records. For delay claims in excess of $100,000, the government will often perform an audit.

Other jurisdictions, such as at the state and municipal level may or may not allow recovery for unabsorbed home office overhead or may or may not allow the use of the Eichleay formula. For example, New York State Courts have rejected the Eichleay formula and prefer, instead, a formula known as the Manshul formula. In addition, many jurisdictions throughout the country have no applicable case law addressing the contractor's entitlement to recovery of home office overhead or the use of the Eichleay formula. Because of this, for many owners and contractors, it may make sense to address the recovery of home office overhead by referencing or providing applicable formulas in the contract. At a minimum, at least for public owners, referencing the applicable FAR regulations governing the determination of allowable home office overhead costs may be advisable. The exclusion of marketing, entertainment, and other precluded costs is not without reason. Most public agencies have formal procurement systems that do not require or allow contractors to entertain or market government officials in order to get work. For this reason, it is not unreasonable for public owners to object to paying even a portion of such costs in the event of an owner-caused project delay.

Home office overhead costs in a delay situation can represent a significant percentage of the overall delay costs. Owners should carefully consider this fact in drafting their construction contracts. Some owners address this issue by defining in the contract the allowable costs for delays, including a computation for home office overhead, if any.

CANADIAN METHOD

An alternative method of calculating home office overhead costs for a delay is known as the Canadian Method. The Canadian Method uses the contractor's actual markup for overhead in its calculation. This markup is based on either the project bid documents or an audit of the contractor's records. An audit would reveal the historical percentage markup for home office overhead applied to each project. Fig. 12.9 shows that the percentage markup is multiplied by the original contract amount and then divided by the original number of days in the contract. This yields a daily overhead rate based on the amount the contractor bid. This rate is then applied to the number of days of compensable delay, as illustrated in Fig. 12.10.

If the delay is significant (a couple of years, for example), then the daily rate may be escalated to account for an inflationary increase in

$$\frac{\text{Percentage Markup X Original Contract Sum}}{\text{Original number of days in the contract}} = \text{Daily overhead rate}$$

Figure 12.9 Canadian Method: Daily overhead rate calculation.

$$\frac{\text{Daily}}{\text{overhead}} \times \text{No. of days of compensable delay} = \text{Compensation for home office overhead}$$

Daily overhead X No. of days of compensable delay = Compensation for home office overhead

Figure 12.10 Canadian Method: Home office overhead damage calculation.

Example of the Canadian Method used to estimate overhead cost associated with a 50-day delay		
Project bid		= $5,000,000
Contract duration		= 500 days
Overhead from bid papers		=10%
Compensable delay		= 50 days
Daily rate =	10% × $5,000,000 / 500 days	= $1000per day
Costs = $1000per day × 50 days		= $50,000

Figure 12.11 Canadian method calculation example.

overhead over time. For example, say a project is bid for $5,000,000, and the contracted duration is 500 days. Based on the original bid documents, the overhead markup is 10%. Multiply the contract amount, $5,000,000, by the percent overhead, and divide by the number of days to determine the daily overhead rate, $1000 per day, as shown in Fig. 12.11. The daily overhead rate of $1000 per day is then multiplied by 50 days to determine the overhead for the delay period, $50,000.

Because typically no consideration is given to unallowable costs, the Canadian Method is simpler to apply than the Eichleay Formula. However, despite this simplicity, it has not been widely used in the United States.

A variation of the Canadian Method, known as the Hudson Method, has been used. In this method, the percentage markup portion of the formula includes a profit allocation. As with any method, the contractor must demonstrate that the underlying markup and cost assumptions are reasonable before recovery under these alternate methods will be allowed.

ALLEGHENY FORMULA

The Eichleay Formula, Canadian Method, and similar formulas and methods have their most obvious application to a delayed contractor working on a construction project. However, these formulas can be problematic when applied to a manufacturer or fabricator. Though the issue is still underabsorption, the effect of the delay may be more difficult to evaluate.

To understand this problem, consider a steel fabricator. The fabrication facility represents a huge capital investment with substantial ongoing operating costs that must be recovered through the fabrication and delivery of structural steel components to many construction projects. Typically, fabricators "schedule" their fabrication facilities, with each project being assigned a "window" of time during which fabrication is planned to occur. If the approval of the steel shop drawings is delayed because of a design error or for some other reason that is the owner's fault or responsibility, the fabricator may not be able to fabricate the affected project's steel in the anticipated window. This may result in the fabricator rescheduling work in its fabrication facility. In the best case, this rescheduling is simple—the project that is not ready for fabrication is moved later in the year and another project is moved up to fill the "hole" left by the rescheduled project. In the end, the fabrication facility is fully utilized and the costs of owning and operating this facility are covered by the amounts bid for the work.

But it is also possible that bumping the delayed project's steel fabrication work to a later fabrication window will leave the fabrication facility underutilized during the project's planned fabrication window, particularly if there is a lot of steel to be fabricated. The Eichleay Formula is not designed to calculate the resulting underabsorption. It is more applicable to the evaluation of the underabsorbed home office overhead costs of a project. Also, the issue is not really so much one of delay as it is of underabsorption. The problem for the fabricator is not that the fabrication window slipped a month, the problem is that its fabrication facility and workers were idle or operating at reduced capacity during the originally scheduled window.

Note, also, that the issue is further complicated by the possibility that the fabricator may have over-scheduled its fabrication facility or taken on replacement work to keep the fabrication facility busy, mitigating the underabsorption.

For manufacturers and fabricators, the more appropriate formula to use for this set of facts is likely to be the Allegheny Formula. That is because the Allegheny Formula is based on a comparison of overhead rates, not on the duration of the delay. In fact, the duration of the delay is not a number used when calculating underabsorbed home office overhead using the Allegheny Formula.

To understand how to use the Allegheny Formula, consider the following example:

Example 12-3. A steel fabricator is awarded a project. In accordance with the schedule for the project and its agreement with the general contractor, the fabricator schedules the fabrication window for the structural steel for the project for the month of July. Because of delays that are the owner's fault and responsibility, steel cannot be fabricated during the July fabrication window and is pushed to the next available window, which is September of the same year. Though the fabricator is able to reschedule some of its fabrication work into the July window vacated by the delayed project, the actual volume of work performed during the month of July is substantially less than the fabricator's usual volume and also less than the volume of work that it could have completed had the delayed project not missed its July fabrication window.

During the July fabrication window, the planned period of performance, the overhead absorption rate was 20%. This means that the actual overhead costs divided by the actual cost of labor and materials was 20%. During an "actual" period of performance (which could be the period during which the steel was actually fabricated), the overhead rate was 15%. The overhead rate is lower because the overhead costs are being spread over the "normal" volume of work, not the reduced volume of work actually performed during the original fabrication window in July.

The underabsorbed overhead cost is calculated by taking the difference between the overhead rate calculated for the actual period of performance and the normal period of performance, which is 5%. This difference is than multiplied by the cost of the delayed contract steel work. This calculation captures the underabsorbed portion of the home office overhead costs. If the labor and material cost of the contract steel work was $200,000, then the resulting underabsorbed amount is $10,000.

CALCULATION USING ACTUAL RECORDS

Some owners are reluctant to include home office overhead costs in compensation for delays. This is particularly true when these costs are based on a formula (Eichleay, Canadian, Manshul, Allegheny, and many others) that provides only an approximation of costs. The contractor

Calculation based on actual records

A project is suspended for 50 calendar days. The home office staff consists of 8 people, including the CEO who does not maintain a time sheet. Staff time working on this project during the delay was:

Position	Project hours	Total hours
Secretary	25	288
Estimator	24	288
Accountant	96	288
Project Mgr.	288	288
Project Mgr.	0	288
General Super	40	288
Clerk	10	288
	543	2016

543 project hours/2016 total hours = 26.9%

Home office costs during suspension = $51,000
26.9% × $51,000 = $13,719

Figure 12.12 Home office overhead calculation based on actual records.

may be able to strengthen its argument by maintaining accurate records in the home office that would support its specific claim for costs.

For example, assume a contractor has a home office staff of 20 people. The staff includes project managers, estimators, schedulers, clerical workers, and accountants. The company requires that all employees maintain accurate time sheets by activity and by project. This documentation may be useful in supporting the contractor's request for home office overhead costs when a delay occurs. In this approach, the contractor can use the records to determine the staff's percentage effort expended, either throughout the project or during the specific delay period, if appropriate. The time sheets show the hours the home office staff expended on the delayed project. This percentage can be applied to the other fixed home office costs to apportion those to the delayed project. Alternatively, the computation in Fig. 12.12 could also be performed on a salary basis to generate the percentage of salary expended on the delayed project, as opposed to time.

SUMMARY

Regardless of the method used to calculate home office overhead costs resulting from delay, it is necessary to demonstrate two fundamental points. The first is that the contractor actually incurred a cost related to its home office that was not covered by the contract price due solely to an excusable and compensable delay. Generally that means that the contractor's period of performance was extended with no commensurate increase in revenues from which home office overhead costs for the extended period were recovered. In some venues, this may be considered to have occurred only in a pure suspension period. In other venues, the contractor may be able to demonstrate that a large delay during which only minimal change order or contract work occurred still resulted in a loss.

The second point is that the method chosen to calculate the underabsorption of home office overhead costs results in a reasonable estimate of the loss incurred. The most appropriate method will be a function of many factors, including the type of business the contractor's company is engaged in and the historical trends of the company's revenues and costs. As with any cost being claimed, the contractor has the burden of demonstrating that it actually incurred a loss of the degree being claimed.

CHAPTER THIRTEEN

Other Categories of Delay Costs

The calculation of delay costs is as much an art as it is a science. This is because, while an understanding of the theory that leads to the types of additional costs that may germinate from a delay is straightforward, the appropriate cost calculations for each project situation will vary considerably. As a result, no book can address every combination and permutation. This chapter addresses certain points that may not be obvious, but are important to defining the breadth of damages that may result from delays.

DAMAGES ASSOCIATED WITH NONCRITICAL DELAYS

Thus far, we have focused on delays to the critical path or delays that are associated with a delay to the overall project. However, activities that are not on the critical path can be delayed and such delays can result in additional costs without ultimately affecting the completion date. For example, a contractor has a contract for the construction of a hospital complex. The complex consists of three buildings. Two buildings (A & B) already exist and must be renovated. The third (C) is new construction.

The contractor schedules the project using the critical path method. The schedule shows that the critical path of the overall project is controlled by the construction of the new building, C. The other two buildings need only be completed within that overall duration and have 12 months of float. (Fig. 13.1)

Early in the project, the owner discovers asbestos in Buildings A and B. The contractor cannot proceed until the project Architects develop a method for the safe removal of the asbestos. The hold on the two buildings remains in effect for 10 months, at which time an acceptable method is devised and the contractor is released to continue work. In the interim,

Construction Delays.
DOI: http://dx.doi.org/10.1016/B978-0-12-811244-1.00013-6

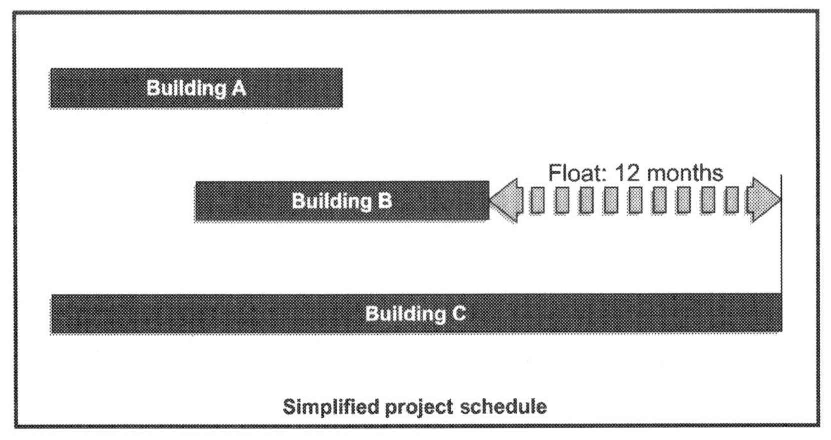

Figure 13.1 Schedule showing 2 months of float for building A and B.

the contractor works on the construction of Building C. The owner and contractor meet to negotiate a change order for the delay and the extra work associated with removing the asbestos in buildings A and B. The direct cost of the asbestos removal is a straightforward calculation. The owner and contractor can agree on the work involved and the cost of the work. The contractor, however, requests additional compensation for the delay to Buildings A and B. The owner argues that the overall project was not delayed and, therefore, the contractor is not due any delay damages or any additional compensation associated with the delay.

Did the contractor experience any damages from the delayed start of these two buildings? To answer this question, the contractor must establish the effect of the delay and the corresponding additional cost.

The contractor explains that the work sequence in the original schedule would have been more efficient and that the delay affects these three areas:

1. Escalation of Labor
2. Additional Supervision
3. Reduced Efficiency

Escalation of labor

The contractor had originally planned to work each trade through the three buildings in a consecutive sequence. For instance, the drywall crew was to work Building A first, then move to Building B, and then move

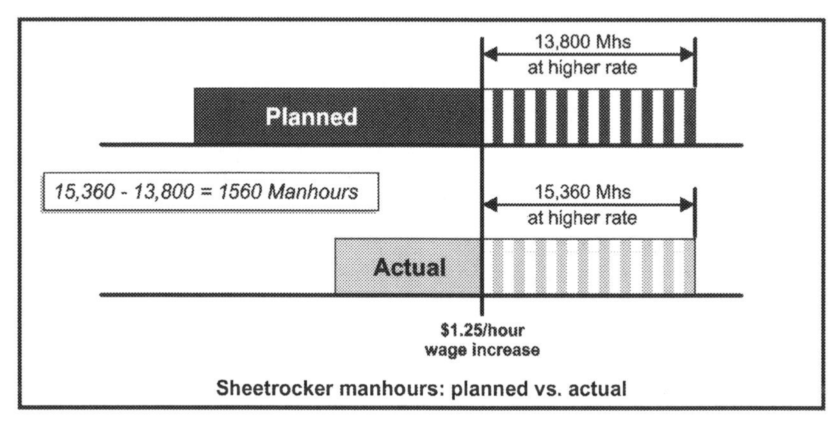

Figure 13.2 Sheetrocker manhours: planned vs. actual.

into Building C. This same sequence was planned for electrical, HVAC, plumbing, millwork, flooring, and painting.

Because of the delay, the contractor was required to work several crews in all three buildings at the same time to meet the project completion date, instead of moving the same crew from building to building. As a result, the distribution of labor shifts to a later time frame. The contractor can document a wage increase of $1.25 per hour for the sheetrockers. The shift in labor is plotted on a graph as shown in Fig. 13.2. The graph shows that 1560 man-hours have been shifted to a later time frame.

For the drywall crew, the contractor claims damages of $1950 (an increase of $1.25 per hour for labor for 1560 manhours). The contractor performs the same analysis for each trade affected by the change in sequence. These calculations would be similar to that shown for the sheetrockers in Fig. 13.2.

Additional supervision

To allow for the inclusion of supervision costs, the contractor explains that when crew sizes increase, additional nonworking foremen must be added. The contractor calculates the additional supervision for the affected trades and summarizes the claim as shown in Fig. 13.3.

Reduced efficiency

Finally, the contractor argues that if it had been able to use the same crew throughout all the buildings, it would have absorbed the initial

ADDITIONAL SUPERVISION

Trade: *Sheetrockers*

Crews planned: 2
Duration planned: 23 months
Thus, 1 non working foreman for 23 months

Crews actual: 4
Actual duration: 15 months
Thus, 2 non working foreman for 15 months each

Additional supervision:
30 man months - 23 man months = 7 man months

Damages: 7 man months X $2950/month = $20,650

Figure 13.3 Additional supervision calculation.

mobilization and learning curve to become most productive. Because each crew had to go through the mobilization/learning curve and each had different foreman and crews with varying capabilities, the result is a lower productivity level than that originally planned. The contractor argues that reduced productivity represents an increase in cost.

Unfortunately, productivity is a very difficult item to document. In order to raise the owner's level of confidence, the contractor has used the best crew for each trade as a benchmark to measure the learning curve/ startup effect and the difference in productivity caused by added crews. Fig. 13.4 is an example the contractor developed for the drywall crews.

Equipment

The preceding example addressed labor costs. If the added crews needed additional equipment to complete their work, the contractor would be forced to mobilize an additional piece of equipment. The contractor would be entitled to reimbursement for the mobilization and demobilization costs of this added equipment, only, since the cost of using that equipment (once mobilized) should be included in the contract sum (Unless the cost to rent the additional equipment was more than the cost of the original piece of equipment. In this case, the contractor may also be entitled to recover the difference in cost between the hourly rate of the second piece of equipment and the hourly rate of the original piece of equipment for the hours that the second piece of equipment was used.)

Loss of efficiency	
Demonstrated productivity by benchmark crew:	1200 sf per crew day (based on daily reports and labor tickets)
Productivity of additional crews:	1000 sf per crew day (based on daily reports and labor tickets)
Loss of efficiency:	1200 −1000 = 200 sf per crew day 200/1200 = 16.7%
Additional crews' gross costs:	$9600 per month X 28 months = $268,800
Damages:	16.7% X $268,800 = $44,890

Figure 13.4 Loss of efficiency calculation.

CONSULTING AND LEGAL COSTS

In general, the costs associated with legal counsel and consultants are not recoverable in a claim situation. This does not mean that they cannot be collected in a settlement or that they can never be awarded in litigation or arbitration. The parties involved in a dispute may include legal and consultant costs as valid elements of their claims, recognizing the limited chance of recovery.

Although these costs are typically disallowed in litigation, the contract may allow reasonable costs of experts expended to support a change order under the changes clause. In this case, the consulting costs would typically be connected to the administration, monitoring, or completion of extra work.

Therefore, while the recovery of legal fees is generally precluded, the contract or legal statutes may provide for recovery.

LOST PROFITS/OPPORTUNITY COSTS

Contractors and owners may each seek to recover damages associated with lost profits and lost opportunities.

In general, these costs are extremely difficult to recover due to their speculative nature. In order to achieve recovery, the contractor or owner must prove that the lost profits or lost opportunity costs are directly caused by the delay. A well-supported claim that can clearly establish an economic loss, such as a loss in bonding capacity, or prove that anticipated profits are not speculative, may support the recovery of these types of damages.

In the absence of a liquidated damages clause in the contract, an owner may assert that a contractor's delay caused it to lose the use of the project and the associated profit. In that scenario, the owner may argue that the facility could have produced a certain amount of revenue had it been completed on time. However, because lost profits are usually considered to be highly speculative and subject to many diverse factors, such damages may be difficult to prove. Accordingly, courts have demonstrated a reluctance to award lost profit damages to owners. It is, therefore, recommended that owners consider including a liquidated damages clause in the contract, rather than relying on recovery of these damages through litigation.

A contractor's additional costs on other projects may also be difficult to recover. For example, a contractor may intend to use its owned excavator on a follow-on project. However, when a delay is encountered, the contractor-owned excavator is needed on the delayed project for a longer period and the contractor must rent another excavator for the follow-on project. In its delay claim, the contractor includes the cost to rent the excavator for the other project. While this argument may seem viable, these types of costs are usually not recoverable for a variety of reasons including the fact that such damages are not reasonably contemplated by the owner as a consequence of a delay or at the time the project was executed.

Interest

A contractor may incur interest or financing costs as a result of borrowing funds to finance the construction costs, including the damages identified in the claim. Although these interest costs may be a real cost or damage resulting from delays, contractors often are not successful in their recovery, as interest claims may be barred by the contract or by statute. When allowable, interest may be claimed either on the overall value of the claim or as a component of the claim that represents the recovery of the cost of borrowed monies used to fund the work.

The interest claimed on the value of the claim may face several obstacles to recovery. Generally, interest on an unresolved claim represents prejudgment interest and is often excluded as an allowable damage either by the contract or by statute. Contract agreements often preclude such interest charges, but typically will allow interest to be paid once a claim amount has been successfully litigated or resolved. Generally, a negotiated settlement to a claim excludes interest. If a claim is litigated, interest may also be regulated by State and Federal laws, and interest rates may be set by statute.

When interest is included as a specific cost of the work, or the cost of funds borrowed to perform the work, the chances of recovery are increased. The contract must support such a claim by identifying that the increased borrowing and the interest cost is based on the actual financing cost incurred. If such interest is demonstrated to be an actual cost, recovery may be allowable, providing the contract has supplied adequate support.

When calculating interest, the period of interest may vary based on the nature of the claim. The start date may be the date a payment became due or the date a claim was filed. Ending dates may be the date of recording the judgment or the date of actual payment. Other factors to consider include the type of interest allowable (compound or simple interest).

Many interest claims are not recoverable because the contractor has not provided adequate support or has taken shortcuts in perfecting the claim. A successful interest claim must adhere to the contract agreement and be adequately supported.

CHAPTER FOURTEEN

Inefficiency Caused by Delay

In addition to increasing the cost of the project, there are other delay-related effects that may further increase the cost of the work. One of these is a decrease in the contractor's efficiency caused by delays. This effect is typically referred to as either "inefficiency" or "lost productivity." A delay may either directly cause inefficiency or be caused by inefficiency.

WHAT IS INEFFICIENCY?

Perhaps the best way to define "inefficiency" is to start with a definition of "productivity." Productivity is a measure of units of work performed per units of resources consumed to perform that work. Inefficiency, which may also be referred to as a loss of efficiency or lost productivity, is a relative measurement. An operation is inefficient when performance of a unit of work consumes more units of resources than should have been consumed.

This chapter is not intended to explain every type of inefficiency or to present techniques for measuring productivity. Rather, it will show how a delay can affect the productivity of the work and discuss how that reduction in productivity can be measured when accurate, contemporaneous records are maintained.

WAYS THAT DELAY CAN LEAD TO INEFFICIENCIES

There are a variety of ways that a delay can contribute to a loss of efficiency. To provide a basic understanding of the relationship between delay and productivity, some of the more common instances of delays contributing to inefficiency are presented in the following examples.

Construction Delays.
DOI: http://dx.doi.org/10.1016/B978-0-12-811244-1.00014-8

Shifts in the construction season

A delay to a project can shift work originally scheduled for one season into a different season. For example, work scheduled for late summer and early fall may be pushed into the winter months by a delay. The effect of the delay on the contractor's efficiency depends on the type of work being performed. Several examples follow.

Example 14.1. A contractor plans to complete all concrete operations before the winter season. A delay forces the contractor to continue concrete work through the winter months in a cold-weather environment. As a consequence, the concrete crews do not work as efficiently as they would under the more ideal conditions of the summer and the fall. The result is that the contractor experiences an increase in crew hours (both labor and equipment) for placing concrete. The contractor is also forced to change the concrete mix design to include accelerators, which further increases the unit cost of materials. Finally, the contractor must use winter concrete placing techniques, including extra winter protection and steam curing in some instances. All of these extra items were the direct result of the delay shifting concrete work from summer and fall into the winter.

Example 14.2. A highway contractor plans to complete all paving operations before the winter, which marks the seasonal shutdown of local asphalt plants. The project is initially delayed and, as a result, the contractor cannot finish paving before the winter begins. Because the asphalt plants shut down and the owner's specifications do not allow paving from November 1 to April 1, the initial delay is compounded by the winter shutdown period. The contractor must now finish the work during the next season. In this case, there may not be an inefficiency associated with the contractor's labor or equipment productivity, but the contractor may experience additional demobilization and remobilization costs. It is also possible that some loss of efficiency may result because new workers may have to be trained and, initially, may have a less productive period before reaching peak productivity levels.

As an alternative, the contractor may accelerate the work in Example 14.2 to complete paving before November 1, in which case there may be inefficiencies caused by the acceleration efforts. Acceleration is addressed in the next chapter of this book.

Example 14.3. A contractor performing a heavy-earth-moving operation is delayed such that work that was to be performed during a relatively dry season is forced into a wetter season. The earth-moving operation is adversely affected by muddy conditions in the cut and fill areas. In wet weather, the overall productivity of cubic yards per crew/equipment day is reduced.

Example 14.4. An HVAC contractor is scheduled to install the heating system in a building to be operational by March. Because of some initial delays, the work is resequenced and the HVAC contractor must accelerate the work to ready the system for operation by the end of November. Because the project is in a cold-weather climate and the subsequent crews will be working inside by December 1, the building will now be a heated structure in which to work, which would not have been the case according to the original schedule. The result should be an increase in productivity.

Numerous other scenarios could develop from a shift of work from one season to another. The important issue for the delay analyst is to assess whether a delay caused the operations to shift into another season and if that shift had any effect on productivity. When work is shifted into adverse seasonal conditions, the analyst should also evaluate the work that was shifted from that season into possibly more favorable conditions.

Availability of resources

At times, delays can affect the availability of resources, such as manpower, subcontractors, or equipment. The following examples illustrate the effects of unavailable resources.

Example 14.5. A contractor plans to complete a project in April. Because of a delay, the work extends into the summer. However, because the construction workload in that location is at its peak during the summer, there is less available labor from which to draw. Therefore, the contractor may not be able to obtain enough labor to finish the work by the revised schedule for completion. This is particularly true of weather-related work such as exterior painting, site work, and landscaping. The inability to complete the work according to the original schedule may be a compounding delay and not have a component of inefficiency. Conversely, the contractor may hire less-experienced crews or "travelers" and have either reduced efficiency for a portion of the work or a higher unit labor cost for the work.

Example 14.6. An earth-moving contractor plans to excavate several hundred thousand cubic yards of material using scrapers. The project is delayed at the beginning. By the time it gets under way, the scrapers are committed to another project and are no longer available. Consequently, the contractor must either rent scapers at a higher cost than the owned equipment it planned to use, or use different equipment such as loaders and dump trucks, which is less efficient in moving the material. Because the productivity resulting from the use of loaders and dump trucks is significantly lower than that originally planned based on the use of scrapers, the excavation operation may be more costly.

Example 14.7. A contractor constructing a bridge must schedule a portion of the work during a specific interval because of the availability of certain equipment—for instance, the use of a snooper crane. Because of a delay to the project, the work shifts into a period in which the equipment is no longer available. The contractor must now perform the work using a new method, thereby reducing its efficiency and increasing its project costs. When a delay occurs, the analyst must look closely at exactly what the effects are on resources, such as equipment and manpower, and how to quantify those effects.

Manpower levels and distribution

Certain types of delays affect the level of manpower and its distribution on a project. These changes may occur in the form of additional manpower, erratic staffing, or variations in crew size. Any of these situations may affect the efficiency of the work.

Additional manpower

Delays to specific activities may force the contractor to work on more activities than planned at one time or may increase the levels of manpower significantly for a specific trade. As a result of union work rules, additional manpower may also require more foremen or support crews, such as master mechanics.

Also, as the contractor increases the crew size, it is not uncommon for the added personnel to be less productive than the original crew.

Contractors often say that as they draw more personnel from the union hall, they see a decline in productivity.

Erratic staffing

In the face of a delay, a contractor may staff a project erratically in order to address specific needs as they arise. Theoretically, a contractor would like to staff a project in a bell curve fashion: starting with a small crew, building up to optimum size, and then tapering down toward the end of the project.

Constant fluctuations in the size of the crew on the site are not desirable. However, the contractor may, in some circumstances, be forced to man the project erratically to achieve required schedule dates. In such situations, there may be a measurable reduction in efficiency.

To demonstrate the negative effect of a forced change in labor distribution, a contractor can use the original schedule to graphically portray the planned distribution of labor and then plot the actual distribution of labor caused by the delay, and compare the two.

Preferred/optimum crew size

Another factor that should be considered is the preferred or optimum crew size. For example, a finish contractor has a standing force of 8 carpenters employed through the year. Because the crew works together throughout the year, they have established a smooth and efficient routine. If a delay now causes the contractor to accelerate its work by increasing its staff above its optimal crew, there can be some measured loss of efficiency as the original crew assimilates the new personnel and brings them "up to speed."

Sequencing of work

Delays to critical and noncritical activities can also force a contractor to resequence the work. The resequencing itself may not be a problem, but its effects may reduce the contractor's productivity in a number of ways. The contractor's crew may be hampered in their work by the presence of another trade, or the crew may be obstructed by material stockpiled in the work area. With such interferences, workers may experience some reduction in productivity.

QUANTIFYING INEFFICIENCY

There are many ways in which the efficiency of a contractor's work can be affected because of changes to the work schedule. The delays may cause these problems directly or indirectly. The delays may be to critical or noncritical items. The contractor must be able to measure and demonstrate how the delays adversely affected the workers' productivity if it is to be compensated for the additional costs. There are several methods for quantifying productivity loss. The delay analyst should be aware of each of these techniques. The following list ranks the different methods by their reliability and persuasiveness:

1. Compare the productivities of unimpacted work with impacted work.
2. Compare the productivities of similar work on other projects with the impacted work on the project in question.
3. Use statistically developed models.
4. Compare the productivity of unimpacted work with the contractor's bid productivity
5. Use expert testimony.
6. Refer to industry published studies.
7. Use the total cost method.

Compare the productivities of unimpacted with impacted work

The impacted versus unimpacted method, usually referred to as a measured mile, is the preferred method to measure losses in productivity. When utilizing this method, the contractor compares the productivity of the work that was impacted with the same type of work that was performed while the work was unimpacted or unaffected by the delay. For example, if a contractor's work is shifted into a cold-weather season, the contractor would compare the productivity of the work performed during the cold-weather season with the productivity attained during the more favorable weather. Of course, the comparison must be made on the same type of work, with no or limited variation in crew makeup.

Example 14.8. A contractor plans to set reinforcing steel during the summer. A delay pushes this activity into the winter months. The contractor's records show that during the favorable weather, the work crews were able to set 2 tons of steel per crew-day. During the less favorable weather, however, the same crews were able to set only 1.5 tons of steel per crew-day. Thus, the demonstrated inefficiency was 25%; the formula is the change in productivity divided by the unimpacted productivity. $(2.0 - 1.5 = 0.5,\ 0.5/2.0 \times 100\% = 25\%)$

To measure productivity in this manner, all information must be recorded in a form that can be converted into productivity units.

Total cost method

The total cost method is the least desirable method to quantify inefficiency and should only be used as a last resort, when the other methods cannot be performed. The total cost method is explored in the following example.

Example 14.8. In the total cost method, a contractor argues that it estimated a certain cost for its work. Because of the delay and the subsequent inefficiency of a shift in work seasons, the actual cost was higher. The contractor claims the difference as added costs. This method is carried out as follows:

Actual cost of paving operation: $1,975,000

Estimated cost of paving operation: $1,250,000

Damages claimed because of inefficiency: $725,000

This method assumes that the contractor's estimate was accurate. It also assumes that the contractor in no way contributed to the reduced efficiency and that all additional costs are solely attributable to the delays cited. All of these assumptions may be challenged.

This chapter is not intended to be a treatise on the subject of inefficiency or on the techniques for measuring productivity. Rather, the intent is to point out that a delay may adversely affect productivity on the project. Also, it must be recognized that detailed, accurate, and contemporaneous information must be maintained in order to measure inefficiencies associated with a delay.

QUANTIFYING THE COSTS OF INEFFICIENCY

The costs associated with inefficiency are direct costs. Because we are discussing delays as the catalyst for the inefficiency, the indirect costs related to the cause of the inefficiency are primarily addressed by the added costs of the delay itself. Therefore, the costs associated with the inefficiency are the direct costs for labor, equipment, and materials. If the analyst can reasonably measure the magnitude of the loss of efficiency, the cost calculations are straightforward. In essence, the inefficiency factor, similar to the inefficiency calculated earlier in this chapter, would be multiplied by the actual labor and equipment hours expended to perform the impacted work.

Contract provisions related to inefficiency

Though rare, some public agencies have developed contract provisions related to inefficiency. Here are a couple:

> *J. Inefficiency*
>
> *The Department will compensate the Contractor for inefficiency or loss of productivity resulting from 1402, —Contract Revisions. Use the Measured Mile analysis, or other reliable methods, comparing the productivity of work impacted by a change to the productivity of similar work performed under unimpacted (unchanged) conditions to quantify the inefficiency. The Department will pay for inefficiencies in accordance with this section (1904).*

Note that this example focuses attention on a measured mile comparison of the impacted and unimpacted work.

The challenge of performing a measured mile analysis comes when there is no "unimpacted work"—no measured mile. When this situation occurs, and it is not uncommon, the contractor should strive to find a productivity comparison that provides the same confidence as the measured mile approach. Often, this means a comparison to similar projects by the same contractor crews working under the same conditions, or a comparison to the contractor's bid productivity with additional supporting information supplied to validate the reasonableness of the contractor's bid.

CHAPTER FIFTEEN

Acceleration

WHAT IS ACCELERATION?

The Merriam-Webster, online dictionary defines the word "accelerate" as "the act or process of moving faster or happening more quickly." This definition applies directly to the progress of a construction project. A construction project is accelerated when the contractor works to complete its original scope of work, or some aspect of that scope of work, in less time than planned. In this case, the plan we are speaking of is the plan to complete the remaining work as reflected in the current project schedule. Usually, this plan is predicting that the project will complete later than was originally planned or later than required by contract; hence the need to accelerate. Acceleration may be performed by a contractor to recover its own delays or to recover delays caused by the owner. When the owner directs the contractor to accelerate to recover owner-caused delays, it is typically considered to be a change to the contract.

A construction project may also experience acceleration when the scheduled completion date is unchanged. For example:

1. Performing additional work on the critical path of the project within the same contract performance period.
2. Performing noncritical items of work in less time than planned.

The following example illustrates acceleration of noncritical work. The example also makes the important point that acceleration and increased costs resulting from acceleration are not limited to work on the critical path.

Example 15.1. The contractor's May 1, 2007 schedule shows a noncritical path of work through the construction of a bridge pier in a river that is scheduled to finish on September 25, 2007. For environmental reasons, the contract prohibits work within the river from October 1, 2007, to April 30, 2008. After starting pier work in May 2007, a differing site condition suspended work on temporary cofferdams for 10 calendar days.

Construction Delays.
DOI: http://dx.doi.org/10.1016/B978-0-12-811244-1.00015-X

Although the suspension was not a critical delay, the delay would push the scheduled completion of the pier until after September 30, 2007. The delay increases the risk that the cofferdams would have to be repaired due to damage over the winter. To eliminate the risk of damage, the owner may choose to direct the contractor to accelerate the pier work by 5 calendar days to finish the river pier before October 1, 2007. In doing so, additional costs may be incurred.

When a contractor requests a change for performing acceleration to recover owner-caused delays, the contractor must show that the acceleration on a project was a change in accordance with the contract clauses. After establishing that acceleration was a change, the contractor must also show the acceleration had some definable effect and resulted in additional costs. The following two examples illustrate the concept of change, effect, and damages.

Example 15.2. A contractor planned to perform two, 5-day activities sequentially working 8-hour days, expending 40 crew-hours on each activity. However, to recover a 2-day delay caused by the owner, the contractor actually performed the two activities sequentially working 12-hour days, thus accelerating the work and expending 48 crew-hours per activity. By performing the work in 8 days rather than10, the contractor mitigated the owner's 2-day delay.

The scope of work for the two activities was unchanged, and there was no delay to the project. However, the work was accelerated in that it was performed at a faster rate than planned at the direction of the owner to recover owner-caused delay. The effect of the acceleration was that each activity required more man-hours than planned (48 man-hours vs 40 man-hours), some of which were at premium rates. The additional costs presented to the owner would be the cost of the additional hours and the premium costs incurred for some of the labor.

Example 15.3. A contractor planned to perform two, 5-day activities working sequentially. To recover a 5-day delay caused by the owner, the contractor actually performed the two activities at the same time, thus accelerating the work. The contractor worked on both activities for 5 days, recovering the owner's 5-day delay.

The scope of work for the two activities was unchanged and there was no delay to the project. However, the work was accelerated in that it was performed in a different sequence than planned at the direction of the owner to recover the owner-caused delay. The contractor was able to complete each activity in the planned duration without additional labor and equipment costs. The acceleration did not affect the contractor's work, did not increase the contractor's resources, and the contractor did not incur any increased costs.

WHY IS A PROJECT ACCELERATED?

A project is accelerated when there is a need for the contractor to complete some portion of the work in less time. The most common reasons a project is accelerated relate to money. This includes saving money by avoiding delay costs or reducing overhead costs. It also may include making more money by allowing the project to earn income earlier or by freeing the contractor to begin other work. Sometimes, acceleration is required to meet some other need, such as the early completion of a facility to avoid, or take advantage of, a change in regulations or business climate. However, the majority of the time, projects is accelerated because they are forecast to finish later than the required project completion date. Simply put, a project is accelerated when it is necessary for the project, or a portion of the project, to complete more quickly than it would otherwise.

CONSTRUCTIVE ACCELERATION

While owners should grant time when it is due, sometimes an owner will not accept or resolve a legitimate time extension request. If the contractor is due an extension to the contract time, but is not provided one, and later accelerates its work in order to finish in the time provided, the contractor may have been constructively accelerated. Constructive acceleration, similar to a constructive change, is subtle and less readily recognized by an owner. Let us look at two similar situations to help distinguish between directed acceleration and constructive acceleration.

Example 15.4. The contractor experiences an excusable 20-day delay associated with revised sitework drawings. The contractor properly notifies the owner and requests a time extension, which the owner grants. Later, the owner directs the contractor to accelerate the project by 20 days. The contractor accelerates the work, makes up the 20 days of delay and finishes the project on time. This is directed acceleration. The owner is likely to recognize and assume liability for the costs associated with the directed acceleration.

Given the same excusable 20-day delay, let us now assume that the owner refuses to grant the requested time extension. In addition, the owner requires the contractor to finish within the original contract duration. The contractor accelerates the work despite protesting that it is due to the need for a time extension, makes up the 20 days of delay, and finishes the project on time. In this instance, the contractor may have been constructively accelerated. Often, owners do not recognize and assume liability for the costs associated with constructive acceleration.

Constructive acceleration is a legal theory of recovery. While the exact requirements may vary by jurisdiction, for constructive acceleration, the contractor will typically be required to show that it:

1. Experienced an excusable delay, which is a delay for which the contractor is entitled to a time extension.
2. Properly requested a time extension in accordance with the contract.
3. Had its time extension request denied or ignored.
4. Was directed by the owner to finish the project by the required contract completion date.
5. Provided notice that it was going to accelerate to mitigate the denied time extension request.
6. Accelerated the work on the project.
7. Incurred additional costs in accelerating the work.

If most or all of these requirements have been met, the contractor may be entitled to recover for the additional costs incurred. Note that it may not be necessary for the contractor to complete the project by the required date in order to recover costs. In the case where constructive acceleration is believed to have occurred, the parties should consult qualified legal counsel on how to proceed.

HOW IS A PROJECT ACCELERATED?

In accelerating a project, the critical work must be performed more quickly or the sequence of critical work must be changed to allow more critical work to occur at the same time. In terms of the project schedule, the longest path must be shortened. There are several ways to accomplish this, including changing the sequence of activities in the schedule, increasing manpower, adding equipment, changing the materials used, changing the method of construction, or improving productivity. These different acceleration techniques can be combined. For instance, you can increase both equipment and manpower to accelerate a project. While each of these techniques should, in theory, serve to shorten the critical path, none are guaranteed to do so. When accelerating a project, it is necessary to monitor the effectiveness of each technique that is employed.

A common acceleration method is to add manpower. This may be in the form of increased daily work hours, added workdays, added shifts, more workers, or any combination of these. The hope is that, if manpower is increased, the rate of work accomplished would increase as well. However, this may not be the case and, while the rate of work produced may increase with increased resources, the relationship may not necessarily be linear. In other words, if you double the resources, you may not double the output. This occurs because the productivity of the work can be negatively affected by the addition of manpower to a project. While production—the number of units produced in a given time period—might improve, the addition of manpower may decrease the productivity of each worker. Lower productivity may occur for a variety of reasons, including less available workspace, overlapping of trades, worker exhaustion, learning curves for new employees, lack of supervision, or introduction of new supervision. If productivity is reduced by the addition of more manpower, it can limit the effectiveness of the acceleration effort.

It is also very important that the additional manpower increase production of the critical path work. If the materials are not being distributed to the floor on which they are needed because the material lift is at maximum capacity, then increasing the workforce will have little chance of accelerating that work. When attempting to accelerate through increased manpower, it is important to monitor productivity and make sure that the critical work will benefit from increased manpower.

The addition of more equipment is another way to accelerate. More equipment, assuming the crews are available, should result in increased production. Time savings from equipment will be limited to the work activities affected by the equipment, and more equipment will often necessitate additional crews. Thus, having an extra backhoe helps to accelerate if the work affected by the backhoe is critical, and there are enough workers available to support it. When using equipment to accelerate, recognize that only certain areas of work may be affected, and that when additional equipment is used, additional manpower may also be required.

Materials can be changed to accelerate the work. One common example is the use of accelerators in concrete or high-early-strength concrete. Other examples might include using expansion anchors in place of epoxy anchors, ordering stock marble instead of custom, using laminate floors instead of marble. The use of different materials accelerates the work either by being easier to incorporate into the project or by being more readily available. Much like the previous example, the use of different materials will only affect certain work activities on the project and may require specialized training or different skills than currently available on the project.

The method of construction can be changed to accelerate the work. Some examples may include changing from welded to mechanical connections, the addition of more falsework to limit or remove an interim phase, changing an application method from rolled to sprayed, and using precast concrete members instead of cast in place. Changing the method of construction to accelerate is effective only to the extent that the new method of construction affects the critical path of work and takes less time than what was originally planned.

A project can also be accelerated through improved productivity. While practical limits always exist, productivity can be improved through better equipment and tools, optimizing crew sizes, incentives, better supervisors, training, proper work sequencing, a specialized workforce, clear direction, and good planning. As with every method used to accelerate, this will only be successful to the extent it affects the critical work.

Though not necessarily a way to accelerate in and of itself, planning and managing acceleration is an essential part of the process. While it is easy to change the sequence of work activities from sequential to concurrent, such changes must be properly planned and managed to ensure that sufficient labor, equipment, and materials are available and can be properly supervised to occur in the new sequence.

CRITICAL PATH SHIFTS DUE TO ACCELERATION

Throughout this chapter, we have stated that a time savings from acceleration will not occur unless the accelerated work is on the critical path. Perhaps the most important element of acceleration to understand is how acceleration affects the critical path. As we have said, in order to accelerate a project, in effect, the longest path of the project must be shortened. But, what happens when you shorten the longest path such that it is shorter than another work path? If we have done our job well explaining the concept of CPM scheduling throughout this book, your answer will be that the shortened path is no longer the longest path. In fact, it ceased being the sole longest path when its length was shortened to that of another work path. At that point, two paths were critical and further shortening of that same path would no longer shorten the critical path. Instead, to accelerate beyond the point where the critical path has been shortened to the length of another path, both paths must now be shortened. And as those paths are further shortened to the length of another path, that new path must also be shortened. As a result, it may be necessary to accelerate multiple paths of work as the acceleration effort is needed to save more time. A successful acceleration effort will require the identification of the critical path of work and a focus on selectively applying resources toward shortening the critical path's duration. By modeling the acceleration in the project schedule, critical path shifts can then be identified to ensure that the work being accelerated has the potential to improve the completion date of the project.

QUANTIFYING THE TIME SAVINGS ASSOCIATED WITH ACCELERATION

In quantifying the time savings gained through acceleration, it may be tempting to compare the project's original completion date to the actual completion date to determine the effect of any acceleration efforts. Unfortunately, if you just measure the difference in the end dates, not only do you have to wait until the end of the project, but you are likely to end up with the wrong answer. If a project is accelerated to make up for existing delays or if delays or improvements are experienced after the acceleration effort is put into effect, then the difference in the end dates would be the aggregation of the acceleration effort, project delays, and perhaps even some project improvements. This would not be an accurate measure of the time savings associated with acceleration.

To isolate the time savings associated with acceleration, it is necessary to determine the effect the acceleration had on the length of the project's critical path. Because the schedule is the tool used to identify the critical path, it is the schedule that is used to quantify the time savings associated with accelerating. In practice, the schedule is updated, and then revised based on the acceleration plan. The difference in length of the critical path being accelerated prior to and after the schedule revisions is the planned time savings associated with acceleration. At the end of each update period, the actual progress of the work is compared to the plan in the same manner described earlier in Chapter 7, Delay Analysis Using Critical Path Method Schedules. Evaluating progress on a periodic basis will demonstrate the true time savings being realized and will allow for adjustments in the acceleration plan to accommodate the actual progress of all of the work. For example, a delay on another path of work may render the current acceleration plan worthless and a new plan may become necessary. When updating the schedule for this purpose, while accelerating the project, it may be prudent to update the schedule weekly.

QUANTIFYING THE COSTS OF ACCELERATION

The cost of acceleration is the difference between what it would have cost to do the work as originally planned versus what it will or did cost to do the work in the accelerated time frame. If acceleration is being planned, the projected costs can be put together based on detailed estimates. After acceleration has occurred, costs can be evaluated using actual data. When quantifying acceleration costs, there are several cost categories that need to be evaluated, including additional material costs, labor premiums, inefficiency, additional equipment costs, and other miscellaneous expenses.

Additional material costs are simply the difference in the cost of the materials that would have been required to execute the work prior to the acceleration plan compared to the cost of the materials needed to perform the work after the acceleration plan is put in place. If a concrete additive is used to accelerate the concrete cure time, then the additive is an additional material cost. If the plan is changed to include temporary falsework, the falsework is an additional material cost. Any cost for more or better material, that is a direct result of the acceleration plan, would be considered an additional material cost.

Labor premiums are additional costs associated with the manpower needed to accelerate. Overtime and holiday pay are examples of labor premiums. Others include having to use higher paid employees, or higher cost subcontractors. On some projects, a large-scale acceleration effort can affect the prevailing rate paid to local labor as the demand for workers exceeds the supply. When the average cost of an hour of labor is increased as a result of an acceleration effort, the difference in the old and new hourly rate is part of the labor premium.

One thing owners should be cautious of when accepting an overtime premium is the flat 50% markup on the accepted or average burdened labor rate provided to them previously on the project. While it is true that an employee will typically receive time-and-a-half for overtime, the increase is applied to base salary, and may not affect the overhead and benefits package that are included in the burdened rate originally provided.

Let us start with an example of labor premium costs that will be expanded to include other acceleration costs through this section. A framing subcontractor originally planned to frame the second floor with two, four-man crews working 8 hours per day for 10 workdays. Instead, the subcontractor will increase to three, four-man crews working 12 hours per day, and complete the work in 5 days. Prior to acceleration, the average burdened labor rate was $15 per hour and after acceleration the average burdened labor rate was expected to be $20 per hour. In addition, a third supervisor was required for the acceleration effort. Thus, the cost of each hour of labor was expected to increase $5 per hour due to the acceleration, and a third supervisor will be needed for a week.

In many cases, as you increase your manpower, the productivity of the workers suffers. In our example, the contractor had originally planned to frame the second floor with two, four-man crews working 8 hours per day for 10 workdays (640 man hours), but will instead increase to three, four-man crews working 12 hours per day to complete the work in 5 days (720 man hours). As a result, the same work is expected to take 80 man hours longer to perform. This 80 hours is inefficient time associated with the acceleration.

Another common cost of acceleration is additional equipment. In our example, to support the additional crew, the framing contractor had to purchase two new nail guns and rent a third compressor. The additional rental cost of a compressor and any associated delivery charge may be a legitimate acceleration cost. The nail guns are questionable, since the contractor may keep them and gain the full benefits of their use over time. On the other hand, this contractor may have no need for new nail guns and no desire to run three crews on future projects. In that case, the nail guns may also be a legitimate cost of the acceleration.

Miscellaneous costs may include the costs of using express mail, the housing of additional staff and labor, the administrative costs of planning the acceleration and revising the schedule, additional cleaning costs, running additional temporary power, per diem costs, and other miscellaneous expenses, along with markups for profit and overhead, that would not have been incurred if the work had not been accelerated. Depending upon how they are paid and the tasks they perform, additional supervisors may be captured as either a labor expense or identified as an overhead item.

One aspect of costs that is sometimes overlooked is the savings associated with acceleration. The acceleration effort will decrease the amount of time required to complete the work and, as a result, the time-related costs associated with the decreased time are avoided. Continuing with our example, the contractor may be able to return the two compressors it originally rented a week (5 workdays) earlier. Likewise, the supervisors, who are salaried employees, should also be finished with this job a week earlier than originally planned.

The acceleration costs for our example of the framing contractor accelerating its work from 10 workdays to 5 workdays are estimated in Fig. 15.1.

Item	Cost
Labor premium	
$5/hour for 640 hours	$3200
Additional supervisor for 1 week	$1465
Inefficiency	
$20/hour for 80 hours	$1600
Additional equipment	
One week air compressor rental	$116
Two nail guns	$246
Miscellaneous	
Evening meals, 5 days @ $50/day	$250
Savings	
Supervisor A 1 week	$(946)
Supervisor B 1 week	$(916)
note that the original supervisors were less costly than the late addition	
Air compressor rental (2 @ 1-week each)	$(232)
Subtotal	$4783
Profit and overhead @ 15%	$717
Total	$5500

Figure 15.1 Acceleration cost example.

Managing acceleration

Before accelerating a project, it is essential to carefully plan what tasks will be accelerated and how the acceleration will be implemented. All too often, when a decision is made to accelerate a project, the response is to go to overtime or increased work weeks for every facet of the project. This is the wrong approach and is often an unnecessary or excessive expenditure of effort and money. The focus on any acceleration is to shorten the remaining duration of the critical path at the lowest cost. One of the preferred approaches to determine the most cost-effective way to accelerate work is to use a cost slope calculation. The basic equation is:

The Crash Time is the absolute fastest time that the activity can be performed. The Crash Cost is the cost of performing the activity within the Crash Time. The Original Costs and Original Time are those that existed prior to any consideration for acceleration. Using our previous example:

The Cost Slope identifies the incremental costs per day for any specific activity to shorten that activity by 1 day. The Cost Slope of the individual work activities and the project schedule can be used to plan the project's acceleration. This is demonstrated in the next example.

Example 15.5. *Using the schedule to accelerate intelligently*

The project in Fig. 15.2 has 9 remaining work activities. The critical path is highlighted in *red* (dark gray in print versions). The project is currently expected to finish in 32 days. The owner has directed the contractor to accelerate the work so that the project finishes in 25 days. Therefore, the contractor must shorten the overall duration by 7 days.

Fig. 15.3 shows the daily incremental cost (cost slope) and the maximum number of days that each of the work activities can be reasonably accelerated.

In looking at the chart, the most affordable work activity to accelerate along the critical path would be Activity G. We can accelerate this by 2 days, before the critical path shifts to include another path of activities. After accelerating Activity G by 2 days, our schedule would look like Fig. 15.4.

Thus, after the second day of acceleration, it will be necessary to accelerate both the path that begins with Activity A, and the path that

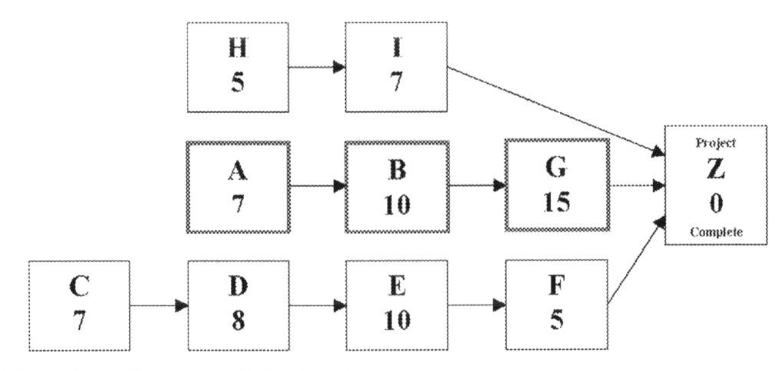

Figure 15.2 Schedule network logic diagram.

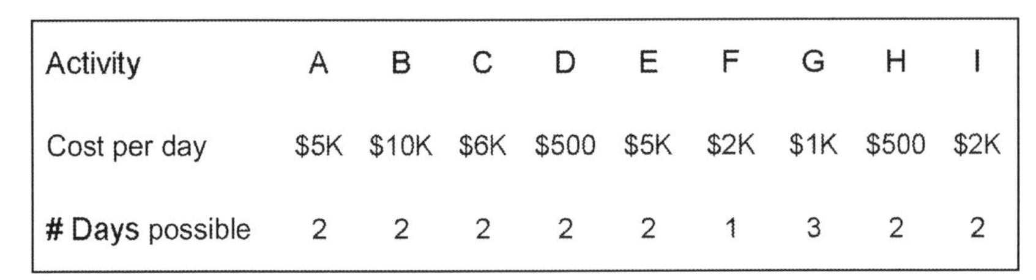

Activity	A	B	C	D	E	F	G	H	I
Cost per day	$5K	$10K	$6K	$500	$5K	$2K	$1K	$500	$2K
# Days possible	2	2	2	2	2	1	3	2	2

Figure 15.3 Work activity cost per day.

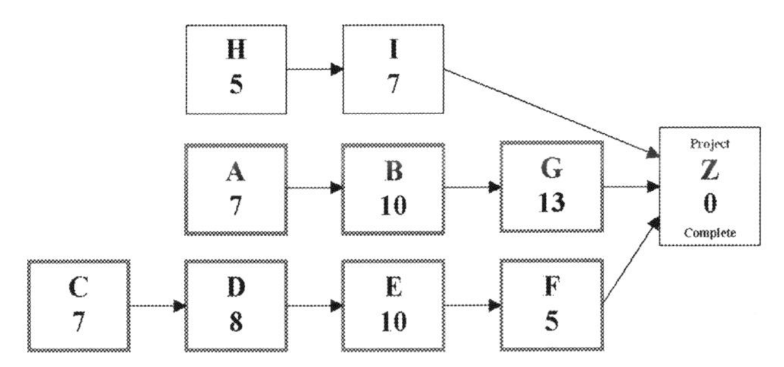

Figure 15.4 Critical paths with Activity G accelerated.

begins with Activity C in order to improve the completion date. The activities accelerated and the cost to do so are summarized in Fig. 15.5. The total cost to accelerate the project by 7 days is $46,000.

This example demonstrates a phenomenon that occurs on most projects, that being that each day of acceleration may become more expensive than the preceding day. Accelerating some areas of work will be a waste of money, such as Activities H & I in the previous example. Also, sometimes the work that is currently ongoing should not be

From Day	To Day	Activity(ies) accelerated	Cost
32	31	G	$1K
31	30	G	$1K
30	29	G & D	$1.5K
29	28	A & D	$5.5K
28	27	A & F	$7K
27	26	B & E	$15K
26	25	B & E	$15K
		Total	**$46K**

Figure 15.5 Estimated cost to accelerate project 7 days.

accelerated in favor of accelerating future, less costly work, such as Activity C in the previous example. By applying a reasoned plan to accelerate, the project will avoid unnecessary expenses and wasted effort.

CHAPTER SIXTEEN

Determining Responsibility for Delay

In addition to determining the activities that were delayed, the magnitude of the delays, and the general nature of the delays (extended duration, late start, etc.), the analyst must also determine the party responsible for the delay. Often, responsibility will be determined after the analyst has identified and quantified the delay. This is because specific delays are often identified only after the project schedules are analyzed. But even if the delay has not yet been quantified, the basic concepts associated with determining responsibility are the same.

Note, also, that "responsibility" may have to be determined on several levels. First, there is the basic question as to which party is responsible for the delay, or if responsibility is shared. Once responsibility is determined, the contract may then provide guidance as to how the consequences of the delay will be addressed. For example, is the delay excusable or compensable? The types of delay—excusable or nonexcusable, compensable or noncompensable—were discussed in detail earlier in this book.

CONTRACT REQUIREMENTS

When determining who was responsible for a delay, it is essential to first refer to the contract. The contract will be determinative in identifying which party was responsible for the performance of the work that led to the delay. In some cases, such as unusually severe weather, neither party will be "responsible."

The General Conditions or General Provisions of the contract will generally be the most informative and helpful regarding the contractor's entitlement to a time extension. Keep in mind that contracts include references to contract provisions that are not physically included in the contract, but are part of the contract by reference. With respect to delays, it is important to focus on the changes clause or clauses (differing site

Construction Delays.
DOI: http://dx.doi.org/10.1016/B978-0-12-811244-1.00016-1

condition and suspension clauses, e.g., are "changes clauses") and on the scheduling and time extension provisions of the contract. These are most commonly the clauses that address delays and time extensions. Also, pay attention to exculpatory clauses like no-damage-for-delay clauses that may seem unfair but may be enforceable depending upon the applicable jurisdiction and case law.

Delays that are the contractor's responsibility will generally not entitle the contractor to any time or monetary relief and will typically expose the contractor to liquidated or actual damages for completing the project late.

Delays that are the responsibility of the owner are generally caused by some form of change, either directed or constructive. These may include a change in the design, an error or omission in the contract documents, a differing site condition, failure to provide timely direction or approvals, failure to timely answer questions needed to progress the work, inspecting the work to a higher standard than specified, or a stop work order.

Some contracts include time extension provisions that provide a detailed list of the types of delays for which a contractor would be entitled to a time extension and reimbursement of delay costs. Some contracts, however, do not even have a time extension provision. When the contract is silent in this regard, it is generally prudent to assume that, based on industry norms, the contractor will be entitled to a time extension for project delays that are not the contractor's fault or responsibility and that could not reasonably have been anticipated or mitigated by the contractor. Similarly, if the contract is silent with regard to compensation, absent contract language, or case law to the contrary, it is generally prudent to assume that delays are compensable when they are excusable and they are the owner's fault or responsibility.

For example, critical project delays caused by industry-wide strikes are generally considered to be excusable, but not compensable. That is because such strikes are not seen as being the fault or responsibility of the contractor and could not have been foreseen or easily mitigated. Strikes are usually not compensable, because they are also not the owner's fault or responsibility.

In contrast, a critical project delay caused by the owner's unwarranted suspension of the contractor's work would typically be seen as being compensable. This is because the delay was the owner's fault and responsibility and not the contractor's.

Most construction contracts have some kind of notice provision related to project delay. Though there is a great variation in how these provisions are written, they generally require the contractor to provide timely written notice to the owner that the contractor is experiencing a delay for which the contractor believes it is entitled to a time extension. There are at least three reasons that owners require notice. The first is to give the owner an opportunity to investigate the problem. The second is to give the owner and the contractor time to work together to develop a strategy to mitigate the delay, and, third, to give the owner an opportunity to document the delay contemporaneously. Even if the contract does not have a notice provision, the contractor would be well advised to prepare contemporaneous written documentation regarding the delay and to timely submit this information to the owner.

The contract may specify a set number of days within which time extension requests must be submitted to the owner. The contract may further state that if the required written documentation is not submitted within the time specified, then the contractor forfeits the right to recover any additional time or money. In addition to specifying time limits, the contract may also specify exactly the information that the contractor must submit when requesting a time extension. Such information would include a clear statement of how much time is being sought and an explanation as to why the delay was the owner's responsibility; references to the specific contract clauses that apply; a clear definition of the specific effects associated with delay (i.e., extra work, overtime, delay, etc.); and a detailed breakdown of the added costs associated with the delay, if any.

EVALUATING RESPONSIBILITY

In addition to what is referenced in the clauses of the contract, the process of assigning responsibility for a delay requires a thorough review of the project documentation to find out what factors influenced the performance of the delayed work. The facts required to assign responsibility will often come from bid package and prebid meetings and inspections; contract plans and specifications; preliminary schedules, baseline and schedule updates, and recovery schedules; change orders, modifications, or supplemental agreements; daily reports or diaries; general correspondence and emails; requests for information (RFIs); submittals; meeting minutes; photos and videos; and cost reporting data. For example, the project daily reports may show that the contractor experienced an equipment breakdown on two separate occasions, which caused the delays to a particular activity. Similarly, a comparison of the schedules and the submittal log may show that the late start of an activity was caused by the engineer's slow return of shop drawings.

For a specific delay, there may be no documentation that reasonably explains why an activity had an extended duration. Without documentation, it may be reasonable to conclude that the extended duration was the contractor's fault. For example, either the scheduled duration was too optimistic or inadequate resources were applied to the task to accomplish the work in the time scheduled. Unfortunately, it is common to find insufficient documentation to answer every question related to the cause of a delay.

It is also possible that another work activity was responsible for the activity delay being investigated. For example, an excavation activity may not have started because the contractor planned to use the area for crane access to erect steel and a steel design delay delayed the fabrication of that steel. In this example, the steel activity would likely not have been linked to the unrelated excavation. The project documents may not readily reveal the reason for the excavation delay. However, a more thorough analysis would reveal that the steel delay was the cause of the late start of excavation.

To arrive at the correct answer, it is necessary to collect all of the available data related to the work activity that was the source of the delay. Once the pertinent project documents have been identified, they should be organized chronologically so that they can be evaluated in the time period and sequence in which they occurred. It is sometime necessary to supplement this information with fact witness interviews as some pertinent information may not have been documented in the project documents.

WEATHER DELAYS

It is common for time to be "of the essence" in a construction contract. It is also common for "contract time" or the contract completion date to be terms defined by the contract. In addition, given the contract's role in defining the sharing of risk, especially the considerable risk associated with weather, it is common for the contract to specifically address how weather-related time extensions will be determined and administered.

It is important to know how your contract addresses time extensions for weather. Common contract language addressing time extensions for weather can vary from allowing recovery of 1 work day for every work day lost to allowing no time at all. Typically, weather-related time extension provisions only provide for time extensions in the event of unusually severe weather. When that is the case, the contractor is required to consider and account for anticipated weather in its bid and in its baseline schedule and schedule updates.

In requesting a time extension for weather delays, the ability to identify the number of days that the actual precipitation exceeded the expected frequency or levels or the number of days that the temperature fell well below that which should have been anticipated is only one part of the equation. The contractor must also demonstrate that the adverse weather experienced at the project site limited its ability to perform critical path work on those weather-affected days and, thus, delayed the project. To do so, the contractor will need to rely on the project schedule—in the best case, a Critical Path Method—to determine if the weather-affected work was, in fact, on the critical path and that the adverse weather resulted in a delay to the project.

To evaluate a weather-related time extension, as with any time extension request, we need to review the contract. The most common approach to addressing weather-related delays is for the contract to allow excusable, noncompensable time extensions for weather-related delays. Normally, the contract wording will refer to "unusually severe weather." A well-written contract should further define what "unusually severe weather" means. It may clarify this as weather beyond that which is ordinarily anticipated in that location based on the National Oceanic and Atmospheric Administration historical records. If this type of wording exists, all parties to the contract should refer to this standard in the resolution of weather-related time extension requests. Some public agencies go as far as checking the weather experienced at the end of each month, and if a number of days lost to weather exceeds the norm (as defined by the contract), the agency will unilaterally grant weather-related time extensions each month. The more normal practice is for the contractor to request weather-related time extensions.

Though the project may experience weather of a nature that would qualify for a time extension, the mechanism for providing a time extension must be spelled out in the contract and must be followed. For example, as noted previously, normal practice requires the contractor to request a time extension for weather-related delays. If the contractor fails to make this request, it is possible that no time will be granted. Therefore, all parties need to understand the process for weather-related time extensions and must follow it as prescribed.

Merely because a project experiences inclement weather that would qualify for a time extension does not necessarily mean that one is due or should be granted. For example, if the project is a midrise building and the building is completely enclosed and "water-tight," the occurrence of a significant rainfall may have no effect on the contractor's ability to perform critical finish work inside the structure. Consequently, no time extension may be due. Conversely, if the project is a heavy civil job involving earthwork, and significant rainfall occurs, it is possible that 3 days of severe rain may actually delay the project for more than just 3 days. Additional time may be required to clean up the site and dry the soil so that work may continue in a productive fashion. Therefore, it is necessary to verify that the specific work that is adversely affected by the weather is on the critical path of the project and will delay the project completion date. If a project experiences unusually severe weather that would qualify for a time extension, but the work affected had ample float with respect to the critical path, it is likely that a time extension is not due.

To properly request or evaluate a time extension request for weather-related delays, the analysis must assess the basic requirements of the contract, the severity of the weather occurrence with respect to the contract requirements, the specific activities that are affected, and the effect of the affected activities on the project's critical path.

CHAPTER SEVENTEEN

Delay—Risk Management

The construction process is one that includes considerable risks. One of the greatest areas of risk is time and the associated costs of delay. A successful project requires that the risks associated with time are well managed. By recognizing the risks that exist and planning for these by properly modeling them in the project schedule, these potentially costly risks can be minimized and controlled.

The companion of good planning and scheduling is good project documentation. The analytical processes presented in this book, which will enable project managers to identify risks early and adjust for them efficiently, cannot be performed if good documentation does not exist. At the top of the list are daily reports, which are enhanced by photographs and videos taken periodically over the course of the project. Detailed documentation of costs must also be maintained in a way that, as much as practicable, allows for segregation to discrete issues.

OWNER'S CONSIDERATIONS

Owners should begin to evaluate the potential risks associated with the duration of the project during the feasibility studies and initial planning stages of the project. One of the initial considerations of the owner is the external constraints concerning time. For example, does the facility need to be completed by a certain date to meet a critical production need, or for political reasons? These factors influence the way in which the owner pursues the project. Certain time requirements and needs may indicate that a fast-track approach to the project is required. The owner must consider the realities of finishing the project within the required time frame. Merely because external considerations require that a project be completed by a certain date does not mean that the project can, in fact, be completed by that date. The owner should consult with knowledgeable advisors to determine a reasonable project duration to specify in the contract documents. Bidders should themselves thoroughly investigate

Construction Delays.
DOI: http://dx.doi.org/10.1016/B978-0-12-811244-1.00017-3

the time constraints, but it is always best for the owner to point out any special considerations up front. For example, if the required duration can only be achieved by an accelerated effort, such as multiple shifts and 7-day work weeks, the potential need for these elements should be stated in the contract or at least discussed during the prebid meeting. The owner is far better off alerting bidders to the anticipated urgency at which the project will need to be constructed.

Scheduling clauses

To control and manage the project's duration, the contractor must have a reliable schedule that is shared with the owner. To ensure that the desired schedule is used, the contract should specify the requirements for both a reliable schedule and periodic updates. Depending on the owner's degree of participation, different scheduling requirements may be dictated by the contract.

Scheduling clauses should include the requirement to provide a baseline schedule before work will commence. The baseline schedule should be due within a prescribed period of time and there should be a specific time period for owner's review. These requirements often appear in contracts but enforcement varies greatly. As with any provisions, unenforced scheduling provisions accomplish nothing. An owner could consider allowing the contractor to submit a basic schedule to get the project started, but require more detail within a prescribed time frame. The same should be true of updates; there should be specific requirements about when to submit them and the submission should be tied to progress payments.

The scheduling clause may or may not require manpower, equipment, and cost loading. While this type of information may be useful to both the contractor and the owner, the owner should carefully consider if the benefits are worth the effort. Clearly, having resource and cost loading in a schedule allows for more precise tracking of progress and a means for a more objective evaluation of payment requests, but these enhancements complicate the schedule and can lead to confusion and other problems. It may be just as beneficial for the owner to specify a basic Critical Path Method (CPM) schedule with no resource loading but with the ability to require the contractor to submit such information for specific activities at the request of the owner.

A sample scheduling specification is provided in the last chapter of this book.

Liquidated damages clauses

When the contract is being prepared, the owner should also decide whether to include a liquidated damages clause. In doing so, the owner should carefully consider the potential damages if the project is delayed. When drafting liquidated damage provisions, the owner should determine not only the amount of potential damages that it may incur if the project were to be completed late, but also whether there are anticipated damages associated with any interim milestones that may be missed and if those should be liquidated in the contract, as well.

Disputes clauses

The owner should also establish time limits in the contract for the filing of claims by the contractor. The contract should clearly specify that claims for additional compensation must be submitted to the owner within a set number of days after the commencement of the event that gave rise to the claim. A period of between 30 and 90 days is a common time frame for the submission of claims. The clause should further state that if the required claim information is not submitted within the time specified, then the contractor forfeits the right to recover any additional compensation. While this is a useful provision, the enforceability of such a clause may depend on the case law in your specific jurisdiction.

Not only should the contract specify a time limit for the filing of claims, but it also should specify exactly what information must be included in the contractor's submission. Such information should include:

- A clear narrative of what the claim is with references to attached documents.
- An explanation of why the claim item differs from the work already required by the contract.
- References to the specific contract clauses that apply.
- An explanation of the cause (or liability) for the claim.
- A clear definition of the specific impacts associated with the claim (i.e., extra work, overtime, delay, etc.).
- A detailed breakdown of the damages or extra costs with supporting information that relate to each claim issue.

A sample clause addressing claims is shown in Fig. 17.1.

ADMINISTRATIVE PROCESS

RESIDENT ENGINEER. When disputes and disagreements arising out of or relating to the contract of any work performed pursuant to the Contract, including additional work required in a Change Order or written or oral order or direction, instruction, interpretation, or determination by the Resident Engineer occur, the Contractor shall immediately give a signed written notice of intent to file a construction claim to the Resident Engineer. If such notification is not given and the Resident Engineer is not afforded the opportunity by the contractor to examine the site of work or is kept from keeping a strict account of actual costs incurred to perform the disputed work or is not afforded the opportunity to review the Contractor's project records, then the Contractor shall waive all his rights to pursue the claim under the contract.

Unrelated claim issues will be processed as separate claims and therefore must be submitted as separate claims.

The Contractor shall supplement the written notice of claim within 15 calendar days of filing the notice of intent to file a construction claim with a written statement providing the following:

1. The date of the claim
2. The nature and circumstance which caused the claim.
3. The contract provisions that support the claim
4. The estimated dollar cost, if any, of the claim and how that estimate was determined.
5. An analysis of the schedule showing any schedule change, disruption, and any adjustment of contract time.

If the claim is continuing, the Contractor shall supplement the information required above in a timely manner.

The contractor shall provide to the Resident Engineer full and final documentation to support the claim no later than 60 calendar days following the date the claim has been fully matured. A claim fully matures when all the direct damages (money and/or time) resulting from the claim issue can be reasonably quantified. The possibility of impact damages should not delay the submittal of full and final documentation of claims for direct damages.

The full documentation of the claim, as presented in the administrative process shall, at a minimum, contain the following elements.

1. A detailed factual narration of events that details the nature and circumstances which cause the claim. This detailed narration of events shall include, but is not limited to, providing all necessary dates, locations, and items of work affected by the claim.
2. The specific provisions of the contract or laws which support the claim and a statement of the reasons why such provisions support the claim.
3. The identification and copies of all documents and the substance of any oral communications that support the claim. Manuals that are standard to the industry may be included by reference.
4. If an adjustment for the performance time of the contract is sought:
   a. The specific days and dates for which it is sought.
   b. The specific reasons the Contractor believes a time adjustment should be granted.
   c. The specific provisions of the Contract under which additional time is sought.
   d. The contractor's detailed schedule analysis to demonstrate the justification for a time adjustment.
5. If additional monetary compensation is sought, the exact amount sought and a breakdown of that amount into the following categories:
   a. Labor. Listing of individuals, classification, hours worked, etc.
   b. Materials. Invoices, purchase orders, etc.
   c. Equipment Listing detailed description (make, model, and serial number), hours of use and dates of use. Equipment rates shall be at the applicable Blue Book Rate, which was in effect when the work was performed, as defined in Subsection 109.03.
   d. Job Site Overhead.
   e. Home Office Overhead (General and Administrative).
   f. Other categories

Figure 17.1 Sample claims clause.

6. The above data shall be accompanied by a notarized statement form the Contractor containing the following certification:
   Under penalty of law for perjury or falsification, the undersigned,

   ____
   (Name)

   ____
   (Title)

   ____
   (Company)
   hereby certifies that the claim is made in good faith; that the supporting data are accurate and complete to the best of my knowledge and belief; that the amount requested accurately reflects the contract adjustment for which the Contractor believes the Department of Transportation is liable; and that I am duly authorized to certify the claim on behalf of the Contractor.

   ____
   (Dated)
   Subscribed and sworn before me this__________ day of
   ____________________,20__________.
   Notary Seal
   My commission expires:____________________
   All pertinent information, references, arguments, and data to support the claim shall be included.

Figure 17.1 (Continued).

Change order clauses

During the course of the project, the owner must carefully monitor and manage change orders. Every change order has two parts, time and money, and every change order should state whether or not additional time is warranted. This task is far easier if an up-to-date CPM schedule is maintained throughout the project that can be used to make a reliable determination of delay for every change order.

Delay damages clauses

Another risk for the owner is the liability for delay damages. The owner can insert a no-damage-for-delay clause in the contract, thus attempting to shift the burden of the risk for delays to the contractor. However, the use of this type of exculpatory language may increase the amounts bid by the contractors bidding on the work and there is still no guarantee that a dispute over delay damages will be prevented. The owner should research the use of a no-damage-for-delay clause with qualified counsel before including it in the contract as they may also be unenforceable in certain jurisdictions.

An alternative approach is to specify limits to what types of damages are allowable in the event of a delay. Some government agencies use this approach.

CONSTRUCTION MANAGER'S CONSIDERATIONS

Some projects include a construction manager, typically abbreviated by the letters "CM." The CM can be hired by the owner as its construction representative, in which case the CM is responsible for representing the owner and protecting the best interests of the owner. This arrangement is typically called "agency CM."

In other cases, the CM may have a financial interest in the project and may be performing the project at a preestablished maximum cost, often called a "guaranteed maximum price," or GMP. The CM may be working in a GMP arrangement with some sharing of the savings below the GMP. These types of arrangements are often called "CM at risk" or "CMAR."

Other variations in the relationships between the owner and the CM or the CM and the contractors exist, but the discussion of these relationships is beyond the scope of this book. More details about these arrangements can be found at the website of the Construction Management Association of America at cmaanet.org or the Design Build Institute of America at DBIA.org.

Construction manager and the project timetable

The CM's considerations regarding time begin at the inception of the project during the planning phase. The CM must ensure that the overall project schedule includes adequate time for all parties to perform their work, including time for the exchange of project performance information between the owner and designer during the design phase; for the careful preparation of contract documents, including the clauses that address schedule and time; for developing contractor or subcontractor interest in the project; for the preparation of responsive bids; for the owner to evaluate bids; and most importantly, for the project to be constructed under normal conditions.

The CM must manage all the project parties to make sure that the project stays on schedule. Just as the contractor is typically responsible for means and methods of construction, the CM is responsible for the means and methods to monitor and manage the performance of all the parties including the evaluation of delays.

Some project parties may not be used to strict time management by the CM. The CM must ensure that the contract language for all the parties includes time management provisions and procedures.

Even with the best planning, delays might occur. The CM must be able to foresee delays and take proactive measures to mitigate or recover from delays.

Often, project delays will be caused by the performance or lack of performance of one or more of the project participants. The CM must keep detailed and accurate records of the performance of all the parties so that it can evaluate liability for delays. Once again, the CM is expected to be the expert during the project who has the responsibility to sort out project delays for the benefit of all parties including the owner.

Construction manager responsibility to contractors and subcontractors

If the CM is in an "at-risk" arrangement, it is now responsible to the owner for managing the construction and to the contractors and subcontractors for administering the contracts. In this case, again, the CM is seen as the construction management expert that should be able to manage the performance of the project parties. Detailed and accurate performance records, clear contract provisions, and dispute resolution procedures must be developed and maintained by the CM. However, a CM "at risk" functions much like a general contractor. Consequently, the owner must ensure that its contract with the CM is appropriately structured so that the owner's risks are managed.

Construction manager responsibility for managing changes

Along with managing time and schedule, the CM must manage changes and the change order process. Change orders must address the additional cost of the changed work and the time required to perform the changed work. By properly maintaining a current and up-to-date CPM schedule and detailed performance records, and by seeking adequate information from the contractors and subcontractors, the CM should be able to evaluate time extensions and additional costs. All too often, only additional costs resulting from changes are addressed and additional time or time extension requests are left out of the change order. The "postponing" or "delaying" of the management and assessment of time is not recommended and usually results in unnecessary disputes. The CM should be proactive in preventing this common problem.

Construction manager responsibility for delay analysis

Construction projects involve many variables in terms of project needs, the participants' motivations, and the need to build a project under sometimes unpredictable circumstances.

As a result, it is not uncommon for project delays to be caused by a combination of actions or lack of actions by more than one project party, changes, or unforeseen conditions. Therefore, even with the best records, it may be difficult to identify and evaluate responsibility for delays. Some CMs may be more experienced than others with delay analysis and evaluation. It may be necessary to engage the services of a scheduling and delay analysis specialist to augment the CM's services and to assist the owner to evaluate the delays on a project. If that is the case, it is recommended that a schedule/delay consultant be retained as soon as a problem is perceived.

Construction manager responsibility for quality, safety, and environment

The CM may also be retained as the owner's representative for ensuring quality project work, project safety, and/or compliance with environmental regulations. Again, because the CM is expected to be the expert in construction and representative of the owner, the choice of CM should only be made after careful consideration of previous experience in these areas. Consideration should be given to the experience of the CM firm and the credentials of the individuals that the CM commits to assign to the project.

GENERAL CONTRACTOR'S CONSIDERATIONS

Like the owner, the general contractor also must assess the risks of delays to the contract completion. These considerations parallel those of the owner, but from a different perspective.

Assess the time allowed in the contract

The contractor should assess the time allowed in the contract to perform the work to determine if enough time is provided to perform the work without the use of extraordinary resources. If necessary or if required to do so as directed in its contract, the contractor must include in its bid the cost for and additional or extraordinary effort, such as overtime, added supervision, and additional or special equipment required to meet the contract completion date.

Assess exculpatory language

If the contract contains exculpatory language, especially in the area of no-damages-for-delay, the contractor should carefully consider accepting the risks involved. Some projects are not worth the risk of bidding. The contractor should consult with qualified counsel before entering into a highly restrictive contract with exculpatory language. Again, if the risk is too great, the contractor may consider not bidding on the project or pricing in the risk accordingly.

Not only must contractors assess the risk of exculpatory language in a contract, but they must also read, understand, and comply with the contract provisions, particularly with respect to changes and claims. For instance, the contract may specify a time limit for issuing a notice of a change or for filing a claim. The contractor must comply with these requirements. Also, the contractor should make sure to submit all information and documentation required by the contract.

Critical path method schedules

CPM schedules required by owners on projects can be effective tools for managing a project. Many contractors resisted using this time management tool for many years, but most professional builders now realize the usefulness of these schedules. Aside from managing the work, the CPM schedule, properly updated, has become the most respected and reliable document for recording a project's as-built history. Trying to recreate the progress on a project, after it is completed, is far more laborious than contemporaneously updating the project schedule.

CPM schedules consider not only time but also resources. It is the efficient use of resources that will allow the contractor to maximize its profits. If delays arise on the project, the CPM schedule is one of the most effective tools that the contractor has to demonstrate the delays that occurred to both critical and noncritical activities.

Risk to subcontractors

The general contractor may pass some of its project risk on to its subcontractors. The amount of risk and responsibility is dictated to some extent by the terms of the contract. Passing risk to the subcontractors is not always as easy as including a general pass-through clause. This type of clause incorporates by reference all the conditions of the general contract into the subcontract agreement. For example, a subcontract may include the following general language: "All of the conditions of the contract between the owner and the general contractor are incorporated herein by reference and are binding upon the subcontractor."

With a general pass-through clause, if a subcontractor delays a project, the damages assessed against the subcontractor may be limited to the liquidated damages amount specified in the general contract. Yet the general contractor is liable for that same amount of liquidated damages to the owner, plus its own additional costs.

For example, if the general contract has a Liquidated Damages amount of $200 per calendar day, the incorporation by reference of the general contract may effectively limit any subcontractor's liability for delays to the $200 per day. Yet, the damages to the general contractor may exceed that amount. Therefore, the general contractor may not simply want to pass through all of the prime contract provisions. Rather, it may want to include language stating that if the subcontractor causes a delay, it can be held liable for delay costs from the general contractor, including the liquidated damages to the owner and delay damages from other subcontractors.

Consider early finish

General contractors should try to complete projects earlier than the time allowed in the contract. By reducing time on the site, the contractor reduces its general conditions costs and thereby realizes greater profit. At the planning stage, the general contractor should approach every contract with the intent of early completion. If the contractor plans to finish the project early, the project schedule should so state. There is no sense in having two schedules on the job_one for the early completion (the actual schedule), and one that is shown to the owner reflecting the full contract duration.

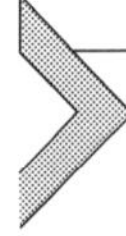

SUBCONTRACTOR'S AND SUPPLIER'S CONSIDERATIONS

Subcontractor considerations

Normally, the general contractor dictates the subcontractor's schedule. At times, the schedule requirements incorporated into subcontract agreements may be undefined or unreasonable. For example, subcontract agreements commonly state that the subcontractor will perform its work in accordance with the general contractor's schedule and will adjust accordingly so as not to delay the project work. Since this statement leaves the work period undefined, the subcontractor may have to accelerate its work for the entire duration of the job.

Subcontractors can insist on a lot of things regarding their contracts, but few subcontractors can expect much success with this. What the subcontractor does need to make clear is that it bid the project for the contract start and completion dates in the bid for its work. If asked upfront to bid an accelerated schedule, this request needs to be documented in the subcontract. The subcontractor will have its own idea of how long it needs to perform the work, and if the general contractor requires a shorter duration or a work sequence that the subcontractor did not contemplate in its bid, the subcontractor should carefully examine its ability to perform and request a contract increase before agreeing to a change in the schedule.

Specific schedule

To reduce risks associated with acceleration costs, the subcontractor should clearly communicate to the general contractor the time and schedule to which the bid applies. The subcontractor should insist that schedules be included in the terms of the subcontract agreement.

Contract language

Subcontractors should seriously consider whether to bid on contracts with extensive exculpatory language that shifts the general contractor's risk to the subcontractor. As noted earlier, some projects are not worth bidding. Many general contractors have their own specific contracts written to suit their interests and strongly resist any changes to the language. A subcontractor should review the general contractor's agreement before bidding and find out if the general contractor is amenable to contract changes. If not, then the subcontractor needs to make a decision to live with the language and account for it in the bid, or not bid the project. The subcontractor is better off knowing what it is getting into in advance before spending the time and money to estimate the project, be the low subcontractor bidder, and be faced with an onerous contract.

Subcontractors must become thoroughly familiar with all clauses in their contracts. For instance, it is common to see clauses that state that for subcontractor claims, the general contractor will "pass through the claim to the owner." The subcontractor will accept whatever damages the general contractor is able to collect and will also share in any costs for litigation or arbitration. Clauses of this nature may not sound "fair," but they are common.

It is also common to see clauses that state that if the subcontractor is delayed by another subcontractor, it must seek compensation directly from that subcontractor. Obviously, the clause is not desirable and creates legal problems for the subcontractor who suffers the loss, vis-à-vis privity of contract.

Clearly, the general contractor may not always be motivated to act in the subcontractor's best interests. Therefore, the subcontractor should attempt to include the following in its agreement with the general contractor:

- An equitable breakdown of awards or settlements for claims involving more than just the respective subcontractor.
- The right to pursue damages only against the general, and not the other subcontractors.
- Proportional legal and administrative costs in claims actions.

DESIGN CONSULTANT'S CONSIDERATIONS

Designer considerations

Normally, the designer provides input on the duration of the project. Such observations should not be formed or communicated casually. It is a subject that should be analyzed in a careful, detailed manner to determine the time required to perform the work considering the type of construction, the local site subsurface conditions, and local geographical considerations, such as temperature variations and weather.

Designer as owner's representative

Designers who act as the owner's representative during construction should require the contractor to submit a detailed schedule for construction. The designer then reviews the schedule submitted by the contractor and monitors progress during construction. It is advisable to establish procedures ahead of time with the contractor for schedule monitoring.

Changes

If the designer, as the owner's representative, makes a change to the project because of owner's decisions or errors and omissions in the plans and specifications, it should assess the time effect, if any, that results from that change. Designers should resist the tendency to deny any time extension simply because they fear it might negatively reflect upon them. If a change necessitates a valid extension to the duration of the project, the problem can be resolved much sooner and usually at far less cost if assessed fairly and impartially as soon as it arises.

Being the project designer and the owner's representative is a difficult dual role because there is a perceived, if not an actual, conflict of interest. Virtually all projects have design problems. This is one of the reasons construction contracts are unique in having a changes provision. If a contractor submits a change request because of a design problem, the designer must take all necessary measures to ensure that its decisions are fair and impartial.

REAL-TIME CLAIMS MANAGEMENT

On larger, time-sensitive construction programs, many forward looking public agencies, private owners, and contractors have instituted coordinated measures aimed at preempting and mitigating claims and disputes. These claims-focused risk management programs have proven to be extremely cost effective.

For owners, program goals are established and monitored. These goals can include initiatives to: (1) make each project "claim-resistant," (2) mitigate potential cross-contract problems during construction, (3) remove contract administration inconsistencies for multiple project programs, and (4) facilitate the timely resolution of any claim issues.

For contractors, goals can include initiatives to: (1) implement best practices for scheduling and change administration, (2) implement effective documentation practices, and (3) satisfy contract requirements for changes, claims, and time extensions.

In both cases, the effective handling of changes and claims calls for advanced preparation and a coordinated strategy in four areas:

- Claim Avoidance
- Claim Mitigation
- Claim Evaluation
- Claim Resolution

The basic concept of Real-Time Claims Management is to establish, before the project starts, a program that is aimed at identifying problems as early as possible and resolving them as quickly as possible. It requires that the project participants "war game" the project before it starts and identify as many potential problem areas as they can. Once the potential problem areas are identified, the participants then develop a game plan and specific approaches to prevent these problems from occurring and to manage them expeditiously if they do occur. The overall attitude is that the project participants will resolve their problems rather than seek a resolution from a judge or jury.

In the context of delays, the Real-Time Claims Management approach focuses on time-related areas of possible problems. As an example, the program should include the following:

- Type of schedules to be required by contract
- Independent review of all schedules and schedule updates
- Precise requirements for identifying time extensions associated with changes timely
- Procedure for independent review of time extension requests
- Scheduled meetings dedicated to review of project progress and project schedules
- Independent review of initial analysis of changes and associated time extension requests

The practice of Real-Time Claims Management has shown that having independent input, review, and analysis before and during the project results in fewer problems and more expeditious resolution of problems. Most often, independent assistance is required when a delay dispute goes to arbitration or trial. But with that timing, the expertise applied becomes a determination of liability and a measure of the delays. Applying the same level of expertise much earlier becomes a tool for prevention and resolution. This is a far more cost-effective approach to the problem of construction delays.

CHAPTER EIGHTEEN

Delays and the Contract

This chapter addresses the Triple Crown of contract provisions that relate to construction delays, these being the provisions that govern the development of the project schedule, the evaluation of delays, and the pricing of delay-related costs. For the purposes of this chapter, we refer to these as the project schedule provision, the time extension provision, and the delay cost provision.

Each of these provisions is addressed separately and should be kept separate in the contract. Often, the project schedule and time extension provisions are comingled. This comingling should be avoided as each of these provisions has a different purpose. The project schedule provision provides a detailed description of the contract's scheduling requirements. The time extension provision addresses the sharing of the risk of delay. Though these subjects are related, the purposes of each are sufficiently different to warrant separation in the contract.

PROJECT SCHEDULE PROVISIONS

This book is primarily focused on identifying and measuring critical project delays using the project's Critical Path Method (CPM) schedule. This focus is based on the presumption that the project has a schedule that can be used as the basis for the analysis of project delays. Though many contractors will produce a good schedule without being asked, some will not. Consequently, it is prudent to include a project scheduling provision in the contract. This helps ensure that even when the contractor is reluctant to prepare a good schedule, there will still be a good schedule on the project. Owners should not expect to be provided with a good schedule if one is not specified.

Construction Delays.
DOI: http://dx.doi.org/10.1016/B978-0-12-811244-1.00018-5

There are several important attributes of a good project schedule provision. From the perspective of identifying and measuring project delays, two of the more important requirements are as follows:

- The requirement to develop and submit for approval a detailed baseline CPM schedule. This schedule should reflect the project team's plan for executing the project at the project's inception, before any work has been performed.
- The requirement to prepare and submit periodic schedule updates that record the actual performance of the work and that also capture the project's ever-evolving plan for completion.

The following is a project schedule provision developed for a large, public owner. This provision embodies many of the characteristics we advocate for a good project schedule provision. Note that the specification is based on the use of Oracle's Primavera scheduling software products. (Alternatives and commentary are provided in parentheses in the provision.)

Please also note that contract language is specific to a contract. Do not use contract language provided in this book without first consulting with competent and experienced legal counsel or consultants.

PROJECT SCHEDULE

PART 1 GENERAL

1.01 DESCRIPTION

This work consists of preparing, maintaining, and submitting a Critical Path Method (CPM) Schedule, referred to as the Project Schedule, that depicts the sequence and timing of all Project Work.

1.02 PURPOSE

A. The purpose of the Project Schedule and this specification is to:

1. Ensure that the Contractor has a detailed plan to complete the Project in accordance with the Contract Documents
2. Ensure that the Project Schedule is regularly updated and revised to accurately depict the Contractor's plan
3. Provide a means of monitoring the Work
4. Aid in communication and coordination of activities among all parties

1.03 DEFINITIONS

Activity—A discrete, identifiable task or event that has a definable start date and finish date, a planned duration, and that can be used to plan, schedule, and monitor a project task.

Activity, Controlling—The first incomplete activity on the Critical Path.

Activity, Critical—An activity on the Critical Path.

Actual Start Date—The date that meaningful work on an activity actually starts.

Actual Finish Date—The date that meaningful work on an activity actually finishes.

Bid Date—The date that bids are submitted.

Business Days—Any calendar day that is not a Saturday, Sunday, or a recognized holiday.

Completion Date, Contract—The date specified in the Contract for completion of the Project or a revised date resulting from a properly executed Time Extension.

Start Date, Contract—The day following approval of the Contractor's Performance Bond. (This date might also be triggered by the Notice to Proceed or other appropriate official indication of the start of the Contract depending on the owner's administrative procedures.)

Completion Date, Scheduled—The date forecasted by the Project Schedule for the completion of the Project. (Note the important distinction made between the Contract and Scheduled Completion Dates for the project.)

Constraint—A restriction imposed on an activity that may modify or override an activity's relationships.

Contemporaneous Period Analysis Method—Also known as the Contemporaneous Schedule Analysis, it is an observational schedule analysis technique used to identify and quantify critical delay. The method is identified and described in AACEI's Recommended Practice for Forensic Schedule Analysis, Method Implementation Protocol (MIP) 3.4.

Critical Path—The longest continuous path of activities that determines the Scheduled Completion Date or other Contract Milestones, also called the Longest Path.

Critical Path Method (CPM)—A method of planning and scheduling that relies on activities, activity durations, activity relationships, and network calculations to forecast when activities will be performed. This method also allows for the identification of the Critical Path of the Project.

Data Date (DD)—The date entered in the Project Details, in the Dates tab, which is used as the starting point to calculate the schedule. For the As-Planned Schedule, the Data Date shall be the Contract Award Date; for Progress Schedule Update submissions, the Data Date shall be the date up to which the Contractor is reporting progress (generally the day after the last working day for the corresponding contract payment period).

Duration, Original (OD)—The original estimated number of working days during which the Work associated with an activity is expected to be performed.

Duration, Remaining (RD)—The estimated time, expressed in working days needed to complete an activity that has started, but has not finished.

Early Dates—The earliest dates an activity can start or finish based on activity relationships, durations, and the activity's placement within the network. These dates are calculated by the software during the forward pass.

Early Completion Schedule—A Project Schedule that forecasts a Scheduled Completion Date(s) that is earlier than the Contract Completion Date(s).

Float—The amount of time an activity or work path can be delayed and not delay the Project.

Float, Free (FF)—The time in workdays that an activity can be delayed without delaying its successor activity(s).

Float, Total (TF)—The calculated difference in workdays between either the activity's early and late start dates or early and late finish dates.

Float, Sequestered—Float hidden in activity durations or consumed by unnecessary or overly restrictive logic.

Fragnet—A "fragmentary network" that consists of an activity or collection of activities that represents work added to the Contract. A fragnet representing added work may be inserted into the Project Schedule to estimate or predict a delay, if any, in a Time Impact Analysis.

Lag—An amount of time, measured in workdays, between when an activity starts or finishes and when its successor activity can start or finish. It is often used to stagger the start or finish dates of activities that are linked to another with Start-to-Start or Finish-to-Finish logic relationships.

Late Dates—The latest dates an activity can start or finish based on activity relationships, durations, and the placement of the activity in the network. These dates are calculated by the software during the backward pass.

Longest Path—The sequence of activities in the Project Schedule network that calculates the Scheduled Completion Date.

Milestone—An activity with zero duration that represents a significant event. For example, the beginning and end dates of the Project specified by the Contract or a revised date resulting from a properly executed Change Order.

Narrative Report—A descriptive report that accompanies each Project Schedule submission. The required contents of this report are set forth in this specification.

Open End—The condition that exists when an activity has either no predecessor or no successor, or when an activity's only predecessor relationship is a finish-to-finish relationship or only successor relationship is a start-to-start relationship.

Predecessor—An activity that is defined by the schedule logic to precede another activity.

Project Scheduler—The person designated by the Contractor and approved by the owner who is responsible for developing and maintaining the Project Schedule.

Project's Planned Start Date—The date entered in the Project Details, in the Dates tab, that reflects the Contractor's planned start of work based on contract requirements, which is the Notice of Award. Note that this date is not considered during the calculation of the Project Schedule. See the Contract Start Date for the start date of the project's duration. (This date might also be triggered by the Notice to Proceed or other appropriate official indication of the start of the Contract, depending on the owner's administrative procedures.)

Project's Must-Finish-By-Date—A date constraint entered in the Project Details, in the Dates tab, that reflects the Contract Completion Date specified by the Contract or revised date resulting from a properly executed Change Order.

Relationships—The interdependence among activities in the network. Relationships link an activity to its predecessor(s) and successor(s). Relationships are defined as:

- **Finish-to-Start (FS)**—The successor activity can start only when the predecessor activity finishes.
- **Finish-to-Finish (FF)**—The successor activity can finish only when the predecessor activity finishes.
- **Start-to-Start (SS)**—The successor activity can start only when the predecessor activity starts.
- **Start-to-Finish (SF)**—The successor activity can finish only when the predecessor activity starts.

Resource, Contract Pay Item—Identify Contract Pay Item Resources as a Material Resource Type. When required, Contract Pay Item resources are developed for each Pay Item in the Contract, with the Resource ID matching the Contract Pay Item and the Resource Name matching the description of the Contract Pay Item.

Resource, Equipment—Identify Equipment Resources as a Non-labor Resource Type. A unique identifier shall be used in the Resource Name or Resource Notes to distinguish the piece of equipment from a similar make and model of equipment used on the Project.

Resource, Labor—Identify Labor Resources as a Labor Resource type. Labor Resources shall identify resources at the Crew level.

Schedule, As-Built (AB#)—This schedule is typically the final Schedule Update for the project and records the completion of all Contract work. This schedule shall be submitted for final payment in accordance with specification section 01 3300—Submittal Procedures.

Schedule, Final As-Planned (FAP)—This schedule fully details the plan to complete the Project in accordance with the Contract Documents. Once the Final As-Planned Schedule is accepted, it shall be archived and a copy of it shall be used as the basis to create Schedule Update No. 1. (Note that other names for this schedule might be used. It is sometimes known as the "baseline schedule," or the "initial schedule.")

Schedule, 90-day As-Planned (90D)—This schedule fully details all of the work to be completed in the first 90 days after the Performance Bond is approved. The Work after the first 90 calendar days is depicted with summary activities. The 90-day As-Planned Schedule shall be used as the basis to develop the Final As-Planned Schedule and will be used to track and manage the Work until the Final As-Planned Schedule is accepted. (Note that other names for this schedule might be used. It is sometimes known as the "preliminary schedule," or the "initial schedule." When the term "initial schedule" is used, it should be clearly distinguished from the baseline or Final As-Planned Schedule.)

Schedule, Look-Ahead—These are short excerpts from the Project Schedule that are presented in construction meetings and used for coordination purposes.

Schedule, Project—This term will be used when referring generally to the Project's CPM Schedules. This term only refers to the 90-day As-Planned Schedule, Final As-Planned Schedule, Look-Ahead or, Schedule Update, Recovery Schedule, or any other schedule submitted to the owner and used by the contractor.

Schedule, Recovery (RS#)—A schedule that is developed to recover or mitigate the forecasted project delay depicted in an unaccepted Schedule Update.

Schedule Update (SU#)—A version of the Project Schedule that reflects the status of activities that have started or have finished prior to the Data Date. This schedule depicts the activities' actual start dates, actual finish dates, and remaining durations as of the day before the Data Date.

Scheduling/Leveling Report—The report generated by the software application when a user "schedules" the Project Schedule. It documents the settings used when scheduling the project, along with project statistics, error/warning, scheduling/leveling results, exceptions, etc.

Substantial Completion Date—The date that the facility and the components of the facility are sufficiently complete to initiate the 60-day test and for its full use. (This date might have several different definitions, depending on the needs of the owner. For example, transportation departments might define this date as the "open-to-traffic" date. Generally speaking, this date is the date that the owner can obtain "beneficial use" or "beneficial occupancy" of the Project and use it for the purpose it was intended.)

Successor—An activity that is defined by schedule logic to succeed another activity.

Work Breakdown Structure (WBS)—A deliverable oriented grouping of project elements that organizes and defines the total scope of the project. Each descending level represents an increasingly detailed definition of project components or work packages.

Working Day—A Working Day is a calendar day designated as a day that work can occur in the work calendars of the Project Schedule.

1.04 RELATED SECTIONS

A. XX XXXX—Time Extensions

B. XX XXXX—Delay Costs

C. XX XXXX—Web-based Project Management System

D. XX XXXX—Submittal Procedures

1.05 REFERENCES

A. Oracle Corporation:

1. Primavera P6 Project Management Software

B. The Association for the Advancement of Cost Engineering International (AACEI):

1. AACE International Recommended Practice No. 37R-06, SCHEDULE LEVELS OF DETAIL—AS APPLIED IN ENGINEERING, PROCUREMENT AND CONSTRUCTION.

2. AACE International Recommended Practice No. 52R-06, TIME IMPACT ANALYSIS—AS APPLIED IN CONSTRUCTION.
3. AACE International Recommended Practice No. 29R-03, FORENSIC SCHEDULE ANALYSIS.

PART 2—PRODUCTS

2.01 SCHEDULING SOFTWARE

A. Contractor will obtain and use Oracle's Primavera P6 Project Management software, or a newer release, for its own use to develop, maintain, and submit the Project Schedules required by this Contract. The Contractor will establish the naming convention as further outlined in this specification for all Project Schedule submissions. (While other software products are available, this specification is written to Oracle's Primavera P6 Project Management software. As an alternate, the Contractor could be given the option of using other software products, but obligated to provide an appropriate number of licenses to the owner for the use of these alternative products.)

PART 3—EXECUTION

3.01 FILE NAME CONVENTION

A. Use the naming convention in Table No. 1, Schedule Filename Convention, as the basis for naming all Project Schedules.

Table No. 1
Schedule Filename Convention

Project Schedules	Initial Submission	Resubmission
90-Day As-Planned Schedule	XX-XXX-XX 90D-R0	XX-XXX-XX 90D-R1
Final As-Planned Schedule	XX-XXX-XX FAP-R0	XX-XXX-XX FAP-R1
Schedule Update Month No. 1	XX-XXX-XX SU1-R0	XX-XXX-XX SU1-R1
Schedule Update Month No. 2	XX-XXX-XX SU2-R0	XX-XXX-XX SU2-R1
1st Recovery Schedule	XX-XXX-XX RS1-R0	XX-XXX-XX RS1-R1
2nd Recovery Schedule	XX-XXX-XX RS2-R0	XX-XXX-XX RS2-R1
As-Built Schedule	XX-XXX-XX AB1-R0	XX-XXX-XX AB1-R1

* XX-XXX-XX is the Contract Number.

3.02 PROJECT SCHEDULE SUBMITTALS

A. Project Scheduler:

1. Designate an individual, with the title "Project Scheduler," who will develop and maintain the Project Schedule. Ensure that the Project Scheduler is present at the Preconstruction Schedule Meeting and attends all meetings, or is knowledgeable of the meeting minutes that outline schedule-related issues during those meetings, that may affect the CPM schedule, including but not limited to those between the Contractor and their Subcontractors and between the Contractor and owner. The Project Scheduler must be knowledgeable of the status of all aspects of the Work throughout the duration of the Contract, including but not limited to, original Contract Work, alterations or additions, suspensions, and all unanticipated circumstances.

2. Provide a Project Scheduler with the following minimum qualifications:
    a. The Project Scheduler shall have at least one (1) year of experience using Oracle's Primavera P6 Project Management software in an enterprise environment. A Project Scheduler with less experience may be acceptable if they can document the completion of at least three (3) days of training in Oracle's Primavera P6 Project Management from a certified instructor and has one year of experience in the use of Oracle's Primavera Project Planner (P3) software.
    b. The Project Scheduler's duties should not be shared by more than one person at any time. It may be a full or part-time position or may be filled by a Consultant.
    c. The Contractor may fill the Project Scheduler position using a person who is not on the Project, except for meetings and other times when the Project Scheduler's presence is required on the Project to satisfactorily fulfill the Project Schedule requirements of the Contract.
    d. The Contractor may not name the Project Manager as the Project Scheduler, but may identify a Project Scheduler with other part-time responsibilities.
    e. The Contractor's submittal proposing the Project Scheduler shall contain a resume and other documentation sufficient to establish the Project Scheduler's compliance with the requirements of this specification. The owner will review the submittal and indicate its approval, ask for additional information regarding the proposed Project Scheduler's qualifications or other responsibilities, or reject the Contractor's proposed candidate. The owner will not accept the Contractor's Project Schedule submission before the Contractor has submitted, and the owner has accepted, the Contractor's proposed Project Scheduler.

B. Preconstruction Schedule Meeting:
1. The purpose of this meeting is to discuss all essential matters pertaining to the satisfactory scheduling of the Project.
2. The Contractor is required to submit the 90-day Project As-Planned Schedule that demonstrates how the Project Scheduler's entire proposed alphanumeric coding structure and the activity identification system for labeling work activities in the Project Schedule will conform to the detailed requirements of this specification.
3. At the Preconstruction Schedule Meeting, ensure that the Project Scheduler is prepared to discuss the following:
    a. Construction plan describing the proposed sequence of Work and means and method of construction.
    b. How the Contractor plans to depict its planned sequence of Work in the Project Schedule.
    c. The proposed hierarchal Work Breakdown Structure (WBS) for the Project Schedules. The Project Scheduler shall provide a paper copy of the proposed WBS at the meeting.

d. The proposed project calendars.
e. The proposed project activity codes and code values for each activity code. The Project Scheduler shall provide a paper copy of this information at the meeting.
f. The Project Scheduler shall provide an outline for the content of the Narrative Report for future Project Schedule submissions.

4. Schedule meetings as necessary to discuss schedule development and resolve schedule issues until the Final As-Planned Project Schedule is accepted.

C. Project Schedule Submissions:
1. Submit the Project Schedule for review and acceptance. Ensure that the filename conforms to the requirements of Table No. 1. Ensure that all submissions meet the requirement of specification section XX XXXX—Submittal Procedures.
2. Schedule submittals will be immediately rejected without review for the following reasons:
a. All submittal items required by subsection 3.02 D are not included in the submittal package.
b. The data date is not correct for the type of schedule being submitted. For schedule updates, the data date should be the first calendar day following the period for which progress was recorded.

D. Project Schedule Submittal Requirements:
1. 90-day As-Planned Schedule and Final As-Planned Schedule Submittal Requirements.
a. The .XER export file of the Project Schedule (archived copy).
b. A Critical Path plot of the Project Schedule with the critical activities sorted by Finish Date in ascending order, grouped by WBS, and with the "Longest Path" filter applied. This plot shall provide a clear longest path from the Data Date to the last activity in the schedule. This plot should consist of the following columns in the following order from left to right: Activity ID, Activity Name, Calendar ID, Original Duration, Remaining Duration, Start, Finish, and Total Float. Then, to the right of these columns the plot should include the Gantt Chart that shows the entire critical path from the data date until the last activity in the schedule.
c. A Scheduling/Leveling Report.
d. A Narrative Report.
e. A Look-Ahead Schedule.
2. Schedule Update and Recovery Schedule Submittal Requirements
a. The .XER export file of the Project Schedule (archived copy).
b. A Critical Path plot of the Project Schedule with the critical activities sorted by Finish Date in ascending order, grouped by WBS, and with the "Longest Path" filter applied. This plot shall provide a clear longest path

from the Data Date to the last activity in the schedule. This plot should consist of the following columns in the following order from left to right: Activity ID, Activity Name, Calendar ID, Original Duration, Remaining Duration, Start, Finish, and Total Float. Then, to the right of these columns the plot should include the Gantt Chart that shows the entire critical path from the data date until the last activity in the schedule.

c. A Scheduling/Leveling Report.

d. A Narrative Report.

e. A Look-Ahead or Schedule.

3.03 PROJECT SCHEDULE DEVELOPMENT

A. General

Project Schedules are developed based on the Contractor's knowledge and understanding of the project, the Contractor's means and methods, the Contract, and the needs and obligations of the other parties involved with the Project.

In addition to the requirements of this General subsection, prepare CPM schedules that comply with good scheduling practice as described in AGC's Construction Planning & Scheduling Manual. The requirements of the Contract take precedence over the recommended practices set forth by this Manual.

Develop and maintain a computer-generated Project Schedule utilizing Oracle's Primavera P6 Project Management software.

Use the Project Schedule to manage the work, including but not limited to the activities of subcontractors; fabricators; the owner; other involved local, state, and Federal governments, agencies and authorities; other entities such as utilities and municipalities; and all other relevant parties involved with the project.

No work, other than installation of the owner's Field Office, mobilization, procurement and administrative activities, installation of construction signs, installation of erosion control and pollution protection, clearing and grubbing, field measurements, and survey and stakeout, will be permitted to start until the 90-day As-Planned Schedule has been submitted and the owner determines there are no deficiencies consistent with those identified in Subsection 3.02.C.

At the regularly scheduled Project Progress Meetings, explain the nature of, and the reasons for, all changes to the Project Schedule, either before or immediately after making them. Only changes that are made to better model the Contractor's plan may be made to the Project Schedule. All such changes must be identified and explained in the schedule update narrative.

Save all schedule "Layouts," "Filters," and "Report" formats that the Contractor develops for the various Project Schedules submissions and submit this information in a format that will allow importation of these into the owner's version of the software.

In scheduling and executing the work, the Contractor shall:

1. Sequence the work commensurate with the Contractor's abilities, resources, and the Contract Documents. The scheduling of activities is the responsibility of the Contractor.
2. Ensure that the Project Schedule contains all work constraints and Milestones defined in the Contract. Schedule the work using such procedures and staging or phasing as required by the Contract.
3. Ensure that the Project Schedules prepared by the Project Scheduler for submission comply with the Contract. This includes the requirement that the Project Schedule submissions and accompanying Narratives are timely, complete, and accurate.
4. Communicate to the Project Scheduler in a timely manner all Contract changes and all decisions made or actions taken by the Contractor, its subcontractors, fabricators, etc., that affect the Project Schedule to allow appropriate development, maintenance, and updating of the Project Schedule by the Project Scheduler.
5. Include and satisfactorily complete all Work contained in the Contract.
6. Ensure that the Project Schedule includes all work directed in writing by the owner in the next Schedule Update submission following such direction.
7. Ensure that the Schedule Updates reflect the actual dates that work activities started and completed in the field.
8. In Schedule Updates and Recovery Schedules, break a schedule activity into multiple activities to reflect a discontinuity in the work if a work activity is suspended in the field and restarted at a later date, whenever the interruption in the work is significant compared to the original activity duration.

The Contractor is responsible for the means and methods necessary to complete the Work required by the Contract and as depicted in the Project Schedule. Failure by the Contractor to include any element of work required by the Contract in the accepted Project Schedule does not relieve the Contractor from its responsibility to perform such work.

Errors or omissions on schedules shall not relieve the Contractor from finishing all work within the time limit specified for completion of the Contract.

B. Detailed Schedule Requirements
   1. Defining Project details and defaults—Within the Dates tab, the "Planned Start" shall be 30 calendar days after the contract signing and the "Must-Finish-By" date shall be the Contract Completion Date. Within the Settings tab, define the Critical Activities as the "Longest Path." (Alternatively, these same requirements could be accomplished using constraints. It may also be unnecessary to fix these dates.)
   2. Detail the Project Schedule to Level 4 in the Schedule Levels Requirement of AACEI's Recommended Practice No. 37R-06, SCHEDULE LEVELS OF DETAIL—AS APPLIED IN ENGINEERING, PROCUREMENT AND CONSTRUCTION. The appropriate number of activities will be largely dependent upon the nature, size, and complexity of the project. In addition to all site construction activities, the Project Schedule shall include activities necessary to depict the

procurement/submittal process, including shop drawings and sample submittals; the fabrication and delivery of key and long-lead procurement elements; testing of materials and equipment; settlement or surcharge period activities; sampling and testing period activities; cure periods; activities related to temporary structures or systems; activities assigned to subcontractors, fabricators, or suppliers; erection and removal of falsework and shoring; inspections; activities to perform punch list work; and activities assigned to the owner and other local, state, or Federal governments, agencies, and authorities, adjacent contractors, utilities, and other parties. The Project Schedule shall indicate intended submittal dates, and depict the review and approval periods in accordance with specification section 01 3300—Submittal Procedures.

3. Work Breakdown Structure (WBS). A multilevel-hierarchal WBS shall be incorporated. The levels (nodes) shall include, but not be limited to:
   a. Level 1, which is the project level.
   b. Level 2, which shall have three nodes: PRECONSTRUCTION, CONSTRUCTION, and POSTCONSTRUCTION.
   c. Level 3, which shall consist of the following at a minimum:
      - The PRECONSTRUCTION node shall have at least two subnodes: SHOP DRAWINGS and PROCUREMENT/FABRICATION
      - The CONSTRUCTION node shall be broken into subnodes for the PHASES of the Work
      - The POSTCONSTRUCTION node requires no subnodes.
   d. Level 4—The nodes for Areas of Construction activities should include subnodes for the various categories of work
4. Activity ID—Provide a unique identification number for each activity. Activity ID numbers shall not be changed or reassigned.
5. Activity Name—Clearly and uniquely name each activity with a description of the work that is readily identifiable to inspection staff. Each Activity shall have a narrative description consisting at a minimum of a verb or work function (i.e., form, pour, excavate, etc.), an object (i.e., slab, footing, wall, roof, etc.), and a location. The work related to each Activity shall be limited to one Area of the contract, one Stage of the contract, and one Responsible Party of the contract.
6. Milestone Activities—Include activities for all Milestones that define Time-Related Contract Provisions and other significant contractual events such as Notice of Award, Contract Signing/Bond Approval, Contractor Start Work Date, Substantial Completion, Final Completion, and coordination points with outside entities such as utilities, local and state governments or agencies, or Authorities.

   All milestone activities in the schedule shall be assigned the standard calendar named "Standard Milestone/365 Day/8 hour." This calendar should also be assigned to any activities for concrete curing.
   a. The Notice of Award milestone shall have a primary constraint of "Start On."
   b. The Final Completion milestone shall have a primary constraint of "finish on or before" the Contract Completion Date.

7. Activity Durations—Except for submittal and procurement activities, ensure that durations do not exceed 20 workdays and that durations for owner submittal reviews meet the requirements set forth in the Contract. If requested by the owner, the Contractor shall justify the reasonableness of planned activity time durations. The planning unit is workdays.
8. Activity Relationships—Clearly assign predecessor and successor relationships to each activity, and assign appropriate relationships between activities (Finish-to-Start, Start-to-Start, and Finish-to-Finish).
9. Ensure that there are no open-ended activities, with the exception of the first activity and last activity in the schedule.
10. Do not include inappropriate logic ties with Milestone activities (For example, for a finish milestone activity, a predecessor shall only be assigned a Finish-to-Finish relationship and a successor shall only be assigned a Finish-to-Start or Finish-to-Finish relationship. For a start milestone, a predecessor shall only be assigned a Finish-to-Start or Start-to-Start relationship and a successor shall only be assigned a Start-to-Start relationship).
11. Start-to-Finish relationships are prohibited.
12. Lag durations greater than 10 days are prohibited unless the Contractor requests and the owner authorizes the use of longer durations.
13. The Contractor shall not use negative Lag durations (leads).
14. Assign the "Contract Award Date" activity as a predecessor to all submittal preparation activities.
15. Activity Constraint Dates—The Contractor shall not constrain activity dates without providing a detailed explanation in the submittal narrative as to why such constraints are necessary to properly model the construction plan.
16. Activity Dates—With the exception of Milestone dates, "Actual Start" and "Actual Finish" dates and "Planned Start" and "Planned Finish" dates, ensure that activity dates are calculated by software. No Actual Start or Actual Finish dates shall be entered in the 90-day As-Planned Schedule or Final As-Planned Schedule.
17. Project Calendars—The contractor shall create the following Project Calendars at a minimum:
    a. 7-Day Work Week w/ no Holidays
    b. Standard Business Day, 5-Day Work Week w/Holidays
    c. If the Contractor needs additional calendars to model expected seasonal weather conditions (such as winter shutdown periods), inclement weather, environmental permit requirements, or other similar restrictions for the planning and scheduling of activities, the Contractor should create Project Calendars for the specific project. Do not incorporate an activity with a description of "Winter Shutdown" that requires constraints. Provide for the number of working days per week, holidays, the number of shifts per day, and the number of hours per shift by using the Calendar modifier in the P6 software.

d. All Calendars should be based on an 8-hour shift.
e. If the Contractor needs to perform work outside of the typical 8-hour workday, then the Contractor should request approval from the owner to work more than 8 hours per day. Upon receipt of owner's approval, create a Project Calendar with a title that describes the nonstandard working hours.
f. Ensure that calendars created for specific resources (i.e., a specific person or piece of equipment) are Resource Calendars, and that the Calendar name clearly identifies the resource.
g. Ensure that all calendars developed by a Contractor use the following naming convention: Contract No. XX-XXX-XX and describing the function (i.e., XX-XXX-XX—Concrete Calendar, XX-XXX-XX—Landscape Calendar, XX-XXX-XX—Painting Calendar, XX-XXX-XX—Contractor's 5 Day/8 Hour Workweek).
h. Assign activities for shop drawing reviews and other approvals by owner's personnel to the "Standard Business Day, 5-Day Work Week w/ Holidays," Calendar.
i. Neither the 90-Day As-Planned nor the Final As-Planned Schedule can include a calendar that reflects workers working more than 8 hours in any one calendar day or more than 5 days in any one week, without the written approval of the owner. Following Contract Award, the Contractor may submit a request for an overtime dispensation and, if approved, can add additional calendars in the next Schedule Update submission.

18. Include all activities for the owner, utility companies, adjacent contractors, and other entities that might affect progress and influence Contract-required dates in the Project Schedule. This includes dates related to all Permits or Agreements. Ensure that the Project Schedule includes special consideration to sensitive areas and indicates any time frames when work is restricted in these sensitive areas as outlined in the permits issued by the regulatory agencies.
19. Activity Resources—The owner may require manpower loading of activities by DOL or Union job classification.
20. Activity Codes—Include a well-defined activity coding structure that allows project activities to be sorted and filtered. Activity Codes shall include, but not be limited to, Responsible Party; Stage; Area of Work; CSI Code; Nonstandard Work Hours; and any additional codes, as required by the owner, to meet the needs of the Work to facilitate the use and analysis of the schedule.
    Establish the following Project Activity Codes and use these to the maximum extent practicable.
    a. RESPONSIBLE PARTY
    b. STAGE
    c. AREA

d. CSI CODE
e. PAY ITEM
f. NONSTANDARD CALENDAR
g. CHANGED (ADDED/DELETED) WORK
h. TIME-RELATED CLAUSES
i. DELAY

16. Activity Code Values—Ensure that each Activity Code contains individual Activity Code Values that are then assigned to activities.

C. Allowance for Inclement Weather

Make allowance in the schedule for the effect of inclement weather on the project work. This may be done in a variety of ways, depending upon the work type and the Contractor's typical practices for managing weather risk. Acceptable methods for managing the risk of inclement weather include Activity Durations that include allowances for anticipated inclement weather and Project Calendars that block out work days in anticipation of inclement weather. The Contractor shall identify the method(s) it has utilized to allow for the effect of inclement weather on the project work in its submittal narratives.

D. Float Ownership

Float, Total Float, Free Float, and Sequestered Float are all owned by the Project and are not for the exclusive benefit of the Contractor or the owner. Float of all types is available for use by both the Contractor and the owner until it expires.

3.04 90-DAY AS-PLANNED SCHEDULE

A. 90-day As-Planned Schedule Requirements

1. Submit the 90-day Project As-Planned Schedule at any time following the Notice of Award, but at least seven business days before the Preconstruction Schedule Meeting. The 90-day As-Planned Schedule shall depict the Work that will occur in the first 90 calendar days of the project to Level 4 as described in AACEI's Recommended Practice No. 37R-06, SCHEDULE LEVELS OF DETAIL—AS APPLIED IN ENGINEERING, PROCUREMENT AND CONSTRUCTION, and the remaining Work shall be represented by summary work activities.
2. Ensure that the 90-day As-Planned Schedule depicts the plan to complete the Project based on the Contract at the time of Award.
3. Ensure that the sequence of the Work meets the requirements of the Contract.
4. The parties will use the 90-day As-Planned Schedule to track and manage the work until the Final As-Planned Schedule is accepted.
5. Schedule meetings as necessary with the owner to discuss schedule development and resolve schedule issues until the Final As-Planned Project Schedule is accepted by the owner.
6. Ensure that the Data Date of the 90-day As-Planned Schedule is the Notice-of-Award date.

7. 90-Day As-Planned Narrative

   Include a Narrative that describes:

   a. The Contractor's general approach to construct the work depicted in the 90-day As-Planned Schedule. Address the reasons for the sequencing of work and describe any resource limitations, potential conflicts, and other salient items that may affect the 90-day As-Planned Schedule and how they may be resolved.
   b. The justification(s) for every activity with a duration that exceeds 20 working days in the first 90 calendar days after Award.
   c. The reason for all lag durations used in the first 90 calendar days after Award.
   d. The justification(s) for Contractor imposed activity constraints used in the 90-day As-Planned Schedule.
   e. A list of all calendars that are being used in the 90-day As-Planned Schedule, along with a general reason for their use.
   f. The Critical Path and challenges that may arise associated with the Critical Path.
   g. Anticipated coordination issues related to work by other entities that require additional information or action by the owner in the first 90 calendar days.
   h. Appendix A to the Narrative shall be the Schedule Log report created when the project was scheduled.

3.05 FINAL AS-PLANNED SCHEDULE

A. Final As-Planned Schedule Requirements

1. Within 60 calendar days of Award, submit the Final As-Planned Schedule. Ensure that the Final As-Planned Schedule depicts all of the Work to Level 4 as described in AACEI's Recommended Practice No. 37R-06, SCHEDULE LEVELS OF DETAIL—AS APPLIED IN ENGINEERING, PROCUREMENT AND CONSTRUCTION.
2. Ensure that this schedule depicts the Contractor's plan to complete the project based on the Contract as of Award.
3. Ensure that the sequence of the Work meets the requirements of the Contract.
4. Use the 90-day As-Planned Schedule as the basis to develop the Final As-Planned Schedule. The detailed activities depicted in the first 90 calendar days of the Final As-Planned Schedule should match the same activities in the 90-day As-Planned Schedule. If the activities in the first 90 calendar days of the Final As-Planned Schedule and the 90-day As-Planned Schedule differ, then the Contractor should provide an explanation(s) for the differences in the Narrative.
5. Once the Project Schedule has been accepted, all deviations from the schedule and changes to the logic of the schedule must be described in the narrative accompanying the next schedule.
6. Ensure that the Data Date of the Final As-Planned Schedule is the Bond Approval date, which is the same as the 90-day As-Planned Schedule. (Other dates may be used. For example, the Notice to Proceed date is a common data for the first complete project schedule.)

7. Final As-Planned Schedule Narrative

   Include a Narrative that describes:

   1. The Contractor's general approach to construct the work depicted in the Final As-Planned Schedule. Address the reasons for the sequencing of work and describe any resource limitations, potential conflicts, and other salient items that may affect the reasonableness or execution of the Final As-Planned Schedule and how they may be resolved.
   2. The justification(s) for each activity with a duration exceeding 20 working days.
   3. The reason for all lag durations used.
   4. The justification(s) for Contractor imposed activity constraints used in the Final As-Planned Schedule.
   5. A list of calendars that are being used in the Project Schedule, along with the general reason for their use.
   6. The Critical Path and challenges that may arise associated with the Critical Path.
   7. Anticipated coordination issues related to work activities by other entities that require additional information from or action by the owner.
   8. Appendix A to the Narrative shall be the "Schedule Log" report created when the project was scheduled.
   9. Include a written representation to the owner that the Contractor has determined and verified all data on the Schedule and assumes full responsibility for it, and that the Contractor, subcontractors, and suppliers have reviewed and coordinated the activities and sequences in the work schedule with the requirements of the Contract Documents.

D. List of Submittals—Submit a list of all submittals (i.e., Shop Drawings, required permits, Erection/Demolition plans, Health and Safety Plan, etc.) generated from the Final As-Planned Schedule for review and approval by the owner. Ensure that the Submittal List only includes prepare and submit activities and review and approve activities. Transmit the report to the owner with the Final As-Planned Schedule.

3.06 SCHEDULE UPDATE

A. Schedule Update Requirements

1. General—Maintain the schedule in a current state and prepare an update of the Project Schedule on at least a monthly basis. Submit a complete Schedule Update that includes all progress achieved and schedule revisions from the Data Date of previous Project Schedule submission through the last working day of the current Contract payment period. Submission of the schedule update is a condition precedent for payment unless waived by the owner.
2. Submit the monthly CPM Schedule Update prior to processing that month's application for payment. Payment applications are due on the 10th day of the following month. Ensure that the data date of the monthly Schedule Update is the first of the next month and that the Schedule Update depicts all progress achieved through the end of the current month. Schedule

Updates will be checked for adequacy based on the items listed in Section 3.02.C.2. After the 10th day of the following month, or as soon as time allows, and after the requirements listed in Section 3.02.C.2 have been met, the owner will review the schedule and determine if any deficiencies exist. If deficiencies exist, the Contractor shall correct those deficiencies prior to the following month's submission of the payment application.

3. The Contractor shall submit a monthly CPM Schedule Update every month regardless of whether the previous month's CPM Schedule Update was accepted by the owner.
4. Ensure that the Schedule Updates shall reflect the status of activities that have commenced or have been completed, including the following items:
   a. Actual dates in Activity Actual Start and Actual Finish columns as appropriate.
   b. Remaining Duration for activities that have commenced and not completed.
   c. If applicable, Suspension or Resume dates for activities that have commenced and not completed.
5. Project Schedule revisions include modifications made to activities and the schedule network as compared to the previous Project Schedule Submission for any of the following items:
   a. Activity Original Duration
   b. Changes in logic relationships between activities
   c. Changes in Constraints
   d. Changes to Activity Names
   e. Added or deleted activities
   f. Changes in Activity Code assignments.
   g. Changes in Activity Resource assignments.
   h. Changes in Calendar assignments.
   i. "Out-of-Sequence" activities noted in the SCHEDULE LOG shall be corrected to reflect the current construction operations.

   The Contractor shall minimize the number of revisions. Describe the reasons for each revision to the Schedule Update in the Narrative. Only revisions that are made to better model the Contractor's plan may be made to the Project Schedule. All Project Schedule revisions must be identified and explained in the schedule update narrative.
6. Additional Schedule Requirements—In addition to the schedule requirements detailed for the submission of the Final As-Planned Project Schedule, also provide:
   a. Activity Status:
      i. Durations: Do not change the Original Duration of an activity. The Contractor shall edit the Remaining Durations to reflect progress made on work activities, and shall not input, but shall allow the schedule to calculate, percent complete. If a proposed change to the Original Duration is due to additional or changed work to the Contract, the Contractor shall instead add an activity to reflect this additional work, and assign the appropriate Activity Code.

ii. Actual Start and Actual Finish Dates: For each activity where work was started during the update period, enter the date the work started. For each activity where work was completed during the month, enter the date the work finished.
iii. Calendars: To change a project calendar for activities scheduled in the future, copy the calendar and use a revised name that includes a reference to which Project Schedule Update the change was incorporated (i.e.,XX-XXX-XXX—Concrete Calendar should be revised to XX-XXX-XXX—Concrete Calendar 2). Document the reason for the change in the calendar in the Narrative.
iv. Notebook: For any activities on the Critical Path that are delayed, enter the dates the activity was delayed and the reason for such delay in the Notebook tab of that activity.

B. Incorporating Changes to the Contract

When the Contractor is performing additional work that has not yet been formally added to the Contract by an executed Change Order, the Contractor should obtain the owner's written acceptance before inserting a fragnet representing the additional work into the Project Schedule. When adding fragnet activities into the Project Schedule, they should assign the "At Risk Work" Activity Code Value in the CHANGED (ADDED/DELETED) Activity Code. Also, describe these changes in the Schedule Update Narrative. When the work represented by the new activities is formally added to the Contract by an executed Change Order, change the "At Risk Work" Activity Code Value in the CHANGED (ADDED/DELETED) Activity Code to "Change Order No. XX."

The owner's acceptance of added activities representing additional work to the Project Schedule is not sufficient to establish entitlement to a Time Extension.

C. Schedule Update Narrative

1. For each Project Schedule Update submission, the Contractor shall submit a Narrative that includes, but is not limited to:
   a. The Contract Number, project name, project location, and name of the Contractor.
   b. The Contract Award date, the current Contract Completion Date, and the Scheduled Completion Date.
   c. Any contact Interim Milestone dates (I/D, B-Clock, LD, etc.), and scheduled start and finish dates for those Milestone activities.
   d. In the event of the schedule shows the Project finishing after the Contract Completion Date, identify the activity that is the current primary delay and list all activities on the Critical Path (include Activity ID's and Activity Descriptions) where work is currently being delayed, and for each such activity provide detailed information including:
      i. The events that caused the delay.
      ii. The party(s) responsible for the delay event(s).
      iii. The number of days the activity has been delayed.

iv. The activities in the construction schedule affected by the events.
v. The reasonable steps needed to minimize the impact of the delay, and which party needs to take the action(s).

e. List any other problems experienced during this Project Schedule Update submission period, the party responsible for the problems, and the Contractor's approach to resolving the problems.
f. List all activities for procurement of long lead time materials that are behind schedule and the reason(s) why.
g. For major work items, describe the differences between the actual work performed and the work planned for the period as represented in the preceding Schedule Update submission, including explanations for deviations.
h. For all suspended work activities that could otherwise logically be progressed, identify the responsible party prohibiting the progression of the work, as well as the detailed reasons why.
i. Description of any changes to the Critical Path since the last Schedule Update submission and the impacts of such changes.
j. List of all added or deleted activities included in this Schedule Update submission, and the reason(s) for and the impact(s) of such changes.
k. List all changes in activity Original Durations, the justification for such change(s), and the impact(s) of such changes.
l. List all changes in relationships between activities included in this Schedule Update submission, and the reason(s) for and the impact(s) of such changes.
m. List the addition or deletion of activity or project constraints, and the reason(s) for and the impact(s) of such changes.
n. List all changes to the project calendars, and the reason(s) for and the impact(s) of such changes.
o. The major work elements, as defined in the WBS, to be accomplished during the next monthly update period.
p. Any potential problems that are anticipated for the next monthly work period and the proposed solutions to such problems. Identify potential problems or risks that either the owner or Contractor may be potentially responsible for. Explain what action the responsible party (i.e., owner or Contractor) needs to take and the date by which time the action needs to taken to avoid the problem.
q. Any planned acceleration of activities that the Contractor anticipates to undertake within the next monthly work period that are either owner directed, or that the Contractor believes are necessary.

3.06 RECOVERY SCHEDULE

A. Recovery Schedule Requirements

1. The owner may require the Contractor to submit a Recovery Schedule and written description of the plan to recover all lost time and maintain the Contract Completion Date or specified Interim Milestone Date(s) if the Scheduled Completion Date forecasts that the project will finish more than

14 calendar days later than the date required by the Contract, as adjusted, if appropriate.

2. Refusal, failure, or neglect by the Contractor to take appropriate recovery action or submit a recovery statement when required as specified herein shall constitute reasonable evidence that the Contractor is not prosecuting the work with all due diligence, and shall represent sufficient basis for the owner to increase retention monies by an amount equal to the amount of potential liquidated damages.
3. The Contractor shall not be entitled to any compensation or damages from the owner on account of any action undertaken by the Contractor to prevent or mitigate an avoidable delay or by the owner's determination to increase retention monies.
4. Ensure that the Data Date of the Recovery Schedule is the Data Date of the Schedule Update that forecasts the late completion.

B. Recovery Schedule Narrative
1. Describe the actions that the Contractor plans to implement to mitigate the forecasted delay. This includes describing:
    a. Any additional labor or equipment resources that it plans to use.
    b. Any re-sequencing of the work that it plans to follow.
2. Or if the Contractor believes that the forecasted delay is not its responsibility, then ensure that a Time Extension Request was submitted in accordance with the Contract.

3.07 REVIEW AND ACCEPTANCE OF THE PROJECT SCHEDULE

A. Immediate Rejection of Progress Schedule Submissions

If the Contractor's Project Schedule submission does not meet the requirement specified in Subsection 3.02.C., then the owner will immediately reject the submission, without further review, analysis, or comment.

B. Project Progress Meetings

One topic of the regular progress meetings held by the owner and attended by the Contractor will be a review of the Look-Ahead Schedule or generated from the Project Schedule. Ensure that the Contractor is represented by the Field Superintendent and Project Scheduler.

1. The review of the Status Report serves as the forum to discuss project progress and delays, suggested remedies, necessary revisions to the Project Schedule, coordination requirements, change orders, potential Contractor time extension requests, and other relevant issues. If Contract work is falling behind the Project Schedule, the responsible party shall be ready to discuss what measures it will take in the next thirty (30) days to put the work back on schedule so as to meet the Contract Completion Date specified in the Contract.
2. Items of discussion will include, but are not limited to project progress; schedule progress; near term and long-term schedule issues, including RFIs, Shop Drawing submittals, permit work, utility relocations, and mitigation work; project issues and risks; proposed solutions; and any relevant technical issues that are schedule related.

3. The Contractor shall keep minutes of this meeting, and shall compile an action item list that describes who is responsible for existing or pending issues and the date by which the issue needs to be resolved to avoid delays. The Contractor shall forward a copy of the meeting minutes and action item list to the owner within 2 business days following the meeting.

C. Review and Acceptance of Project Schedules.

The owner will review the Project Schedule submissions and will prepare a written response (Progress Schedule Review Report) to the Contractor's submission within five (5) Business Days following receipt of the Contractor's complete schedule submission. The owner will either "accept" the schedule, "accept as noted," or "reject" the schedule for re-submittal by the Contractor.

If the Project Schedule submission is not in compliance with the Contract, the owner may reject the submittal and forward any comments and requests for schedule revisions to the Project Scheduler with a copy to the Contractor. The Project Scheduler shall address all comments in writing or make the requested revisions and resubmit the revised schedule within three (3) Business Days of the owner's reply. If the owner determines the revised submission still does not meet the contract requirements, any further revisions required thereafter shall also be submitted for acceptance within three (3) Business Days of the request for revisions by the owner.

For schedules that are "accepted as noted" the owner shall forward any comments or requests for revisions, to the Contractor. The Project Scheduler shall address all comments in writing or make the requested revisions as part of the next Project Schedule submission.

The Project Scheduler shall make adjustments to the Project Schedule in accordance with the owner's comments and resubmit copies for review consistent with the requirements of this section.

By accepting the Project Schedule, the owner does not warrant that the Project Schedule is reasonable or that by following the Project Schedule the Contractor can complete the work in a timely manner. If, after a Project Schedule has been accepted by the owner, either the Contractor or the owner discover that any aspect of the Project Schedule is in error, the Contractor shall correct the Project Schedule in the next Project Schedule submission and describe this revision in the Narrative report.

Acceptance of Project Schedules by the owner shall not be construed to imply approval of any particular construction methods or sequence of construction or to relieve the Contractor from its responsibility to provide sufficient labor, equipment, and materials to complete the Contract in accordance with the contract documents.

Acceptance of the Project Schedule by the owner does not attest to the validity of assumptions, activities, relationships, sequences, resource allocations, or any other aspect of the Project Schedule. The Contractor is solely responsible for the planning and execution of the work.

Acceptance of the Progress Schedule by the owner shall not be construed to modify or amend the Contract or the date of completion. Completion dates can only be modified or amended by standard contractual means. (Usually a change order, modification, or supplemental agreement.)
Acceptance of the Progress Schedule by owner shall not be construed to mean that the owner accepts or agrees with the accuracy or validity of any historical information placed in the progress schedule relating to the Contractor's position on any claims.
If any resources are included in the Project Schedule, then the owner's acceptance of schedule does not represent an acceptance of the Contractor's planned resources. Resources included with the accepted Project Schedule shall not be misconstrued as a cost benchmark for the performance of planned or actual work.
Upon receipt from the Contractor of the Revised Project Schedule submission, a new review period by the owner of five (5) Business Days will begin.
The Contractor shall submit a monthly CPM Schedule Update every month regardless of whether the previous month's CPM Schedule Update was accepted by the owner.

TIME EXTENSION PROVISIONS

At the very least, every time extension provision should provide guidance on three topics. They are:

1. Defining excusable delays
2. Identifying instances when the contractor is due a time extension
3. Providing guidance regarding how the contractor is to establish the existence of and quantify the delay.

Defining excusable delays

Just as the circumstances under which every project is built are different, there is not one list of excusable delays that are applicable to every project. When an owner drafts its time extension provision or when an owner and contractor negotiate the terms of the construction contract, both parties should remember that the time extension provision is a risk-apportionment provision. At a minimum, the time extension provision should specify the types of delays that are considered excusable delays, which are delays for which the contractor is entitled to a time extension.

Earlier in this book, we described the typical classification of excusable, noncompensable; excusable, compensable; and unexcusable delays. This classification of delay is based on the party responsible for the delay.

Whether drafting or negotiating a construction contract, both parties should understand which delays are its responsibility and which are not, and the consequences of assuming responsibility for the different types of delays.

For example, in most contracts, when the project is delayed by unusually severe weather, the contractor would typically be entitled to a time extension, but not entitled to recover the costs of delay. Delays due to unusually severe weather are normally excusable, but not compensable. On some projects, however, owners may elect to shift the risk and responsibility for unusually severe weather to the contractor. Owners should recognize that one of the consequences of shifting additional risk to the contractor in this way may be an increase in the contractor's bid price. The reason for this increase is to cover the contractor's added risk.

Similarly, if the contract includes a no-damages-for-delay clause or other exculpatory clause, the resulting added risk may also result in an increase in the contractor's bid price. Again, the reason for the increase is to cover the contractor's added risk of performance. For some owners, the trade-off of risk for a higher bid price may be worth the added costs.

Identifying when the contractor is due a time extension

The next essential feature of a time extension provision is that it should explain the specific circumstances during which the contractor would be due a time extension. Most construction contracts do not provide clear guidance when a contractor is due a time extension. For example, many contracts state that simply that the contractor should submit a request for an extension of time when the contractor believes the project was delayed by circumstances that were beyond its control.

Unfortunately, this direction does not provide an objective basis to determine when the project is actually delayed or define the circumstance under which a time extension would be awarded. The time extension provision should clearly define and explain when the contractor is entitled to a time extension.

For example, the time extension provision should state that the "contractor is only entitled to a time extension when an excusable delay delays the project's critical path and the project's forecast completion date extends out to a date later than the current contract completion date."

If your contract does not provide clear guidance as to the specific circumstances under which the contractor would be entitled to a time extension, then an unscrupulous contractor might submit time extension request for all types of delays, whether they are excusable or critical or not.

Establishing the existence of and quantifying delay

As discussed in earlier chapters, there are a variety of ways to use the project schedule to demonstrate delays. Chapter 5, Measuring Delays—The Basics, introduced two basic types of schedule delay analyses, Prospective and Retrospective schedule delay analyses.

A "Prospective" schedule delay analysis allows the analyst to estimate or forecast the project delay resulting from added or changed work before that work is performed. As also acknowledged earlier, there is almost universal agreement that the Prospective Time Impact Analysis approach is the best analysis method to measure the delay caused by added or changed work before that work is performed.

A "Retrospective" schedule delay analysis approach identifies and measures project delay after the delay occurs or even after the project is complete. Despite a lack of universal agreement on the most appropriate schedule delay analysis method to use to identify and measure project delay retrospectively, there does appear to be a trend toward reliance on the use of the Contemporaneous Schedule Analysis, also called the Contemporaneous Period Analysis, approach, which is identified as MIP 3.4 Observational/Dynamic/Contemporaneous Split in the AACE International's Recommended Practice 29R-03 for Forensic Schedule Analysis.

Below is a Time Extension provision that we have written for a public owner that incorporates both the use of the Prospective Time Impact Analysis and Contemporaneous Schedule Analysis. (Options are noted in parentheses.)

1806 DETERMINATION AND EXTENSION OF CONTRACT TIME

1806.1 GENERAL

The Proposal Package will specify the Contract Time. The Contractor shall prosecute the Work continuously and effectively, with the least possible delay, to the end that all Work is completed within the Contract Time.

If the Department issues a Notice to Proceed, the Contract Start Date established in the Notice to Proceed takes precedence over the Contract Start Date specified in the Proposal Package.

The Department will not consider a plea by the Contractor that the Contract Time was not sufficient as a valid reason for an extension of the Contract Time.

If the Department grants an extension of the Contract Time, the extended time for completion will be in full force and effect as though it was originally specified.

The Department will only extend the Contract Time if an excusable delay, as specified in 1806.2.A, "Excusable, Non-Compensable Delays," or 1806.2.B, "Excusable, Compensable Delays," delays Work on the Critical Path as described in items 1, 2, 3, and 4, below.

Mitigation of delay, whether caused by the Department, Contractor, a third-party, or an event, is a shared contract and legal requirement. Mitigation efforts include, but are not limited to, re-sequencing work activities, acceleration, and continuation of work through an otherwise planned shutdown period. The Contractor and Engineer will explore and discuss potential mitigation efforts promptly and agree upon costs or cost-sharing responsibilities prior to the implementation of mitigation efforts.

The Department will not evaluate a request for extension of the Contract Time unless the Contractor notifies the Engineer as specified in 1403, "Notification for Contract Revisions," and provides the required analysis as follows.

The Contractor shall evaluate delays and calculate the appropriate time extension due based on the following:

(1) The Contractor shall base all evaluations of delay and all calculations of the appropriate time extensions due on the schedules submitted to and accepted by the Department. The Contractor shall not use schedules that did not exist on the project or create schedules after the delay has occurred to demonstrate entitlement to a time extension.

(2) The Contractor shall base evaluations and calculations related to the determination of extensions of time on the Critical Path as established by the schedules submitted to and accepted by the Department. The Contractor is not entitled to a time extension for delays that do not delay the Critical Path.

(3) The evaluations and calculations required to establish entitlement to a time extension will vary depending on the nature and timing of the delay and whether the Contract Time is measured in working days, calendar days, or based on a fixed completion date.

(4) The schedules relevant to the evaluation and calculation of time extensions are the most current schedules submitted to and accepted by the Department. For example, if the Department determines that Extra Work is required and the Supplemental Agreement adding this work

will be dated June 2, then the determination of the time extension due the Contractor will be based on the last schedule submitted and accepted by the Department prior to June 2 of the same year.

(5) The Contractor's evaluations and calculations shall comply with the following Recommended Practices published by the Association for the Advancement of Cost Engineering, International:

(5.1) Recommended Practice No. 52R-06, Time Impact Analysis As Applied in Construction. The Contractor shall use this Recommended Practice for delays that are in the future (prospective). The Contractor shall not use this recommended practice to evaluate delays that have already occurred (retrospective).

(5.2) Recommended Practice No. 29R-03, Forensic Schedule Analysis, MIP 3.4 Observational/Dynamic/Contemporaneous Split approach. The Contractor shall use MIP 3.4 when evaluating delays that have already occurred.

The Engineer will review the Contractor's evaluations and calculations and determine the time extension due, if any. The Engineer will measure extensions to the Contract Time in working days for Working Day Contracts and in calendar days for Completion Date and Calendar Day Contracts.

The Department will relieve the Contractor from associated liquidated damages, as specified in 1807, "Failure to Complete the Work on Time," if the Department extends the Contract Time under this section (1806).

1806.2 TYPES OF DELAYS

A. Excusable, Non-Compensable Delays

Excusable, non-compensable delays are delays that are not the Contractor's or the Department's fault or responsibility, and that could not have been foreseen by the Contractor. The Department will not compensate the Contractor for excusable, non-compensable delays.

Excusable, non-compensable delays include, but are not limited to:

(1) Delays due to fires, floods, tornadoes, lightning strikes, earthquakes, epidemics, or other cataclysmic phenomena of nature.
(2) Delays due to weather if the Contractor is entitled to a time extension for weather as specified in 1806.3, "Determination of Charges on Working Day Contracts," and 1806.4, "Extension of Contract Time Due to Weather on Calendar Day and Completion Date Contracts."
(3) Extraordinary delays in material deliveries the Contractor or its suppliers cannot foresee or avoid resulting from freight embargoes, government acts, or regional material shortages.
(4) Delays due to civil disturbances.
(5) Delays due to acts of the public enemy.
(6) Delays due to labor strikes that are beyond the Contractor's, subcontractor's, or supplier's power to settle and are not caused by improper acts or omissions of the Contractor, subcontractor, or supplier.
(7) Delays due to acts of the government or a political subdivision other than the Department.
(8) All other delays not the Contractor's or Department's fault or responsibility and which could not have been foreseen by the Contractor.

B. Excusable, Compensable Delays

Excusable, compensable delays are delays that are not the Contractor's fault or responsibility, and are the Department's fault or responsibility, or are determined by judicial proceeding to be the Department's sole responsibility.

Excusable, compensable, delays include, but are not limited to:

(1) Delays due to revised Work as specified in 1402.2, "Differing Site Conditions," 1402.3, "Significant Changes to the Character of Work," and 1402.5, "Extra Work."
(2) Delays due to utility or railroad interference on the Project Site that were not anticipated.
(3) Delays due to an Engineer-ordered suspension as specified in 1402.4, "Suspensions of Work Ordered by the Engineer."
(4) Delays due to the neglect of the Department or its failure to act in a timely manner.

C. Non-Excusable Delays

Non-excusable delays are delays that are the Contractor's fault or responsibility. All non-excusable delays are non-compensable.

Non-excusable delays include, but are not limited to:

(1) Delays due to the Contractor's, subcontractor's, or supplier's insolvency or mismanagement.
(2) Delays due to slow delivery of materials from the supplier or fabricator when the material was available in warehouse stock, or when delivery was delayed for reasons of priority, late ordering, financial considerations, or other causes.
(3) Delays due to the Contractor's failure to provide sufficient forces and equipment to maintain satisfactory progress.
(4) Delays caused by plant and equipment failure or delays due to the Contractor's failure to provide and maintain the equipment in good mechanical condition or to provide for immediate emergency repairs.
(5) Delays caused by conditions on the project, including traffic conditions that could be foreseen or anticipated before the date of bid opening. Weather delays are addressed in 1806.3, "Determination of Charges on Working Day Contracts," and 1806.4, "Extension of Contract Time Due to Weather on Calendar Day and Completion Date Contracts."

D. Concurrent Delays

Concurrent delays are independent critical delays that occur at the same time. When a non-excusable delay is concurrent with an excusable delay, the Contractor is not entitled to an extension of Contract Time for the period the non-excusable delay is concurrent with the excusable delay. When a non-compensable delay is concurrent with a compensable delay, the Contractor is entitled to an extension of Contract Time, but not entitled to compensation for the period the non-compensable delay is concurrent with the compensable delay.

DELAY COST PROVISIONS

Many contracts do a poor job of describing the costs that the contractor is entitled to be paid in the event of an excusable and compensable delay. Delay costs themselves are addressed in more detail in an earlier chapter of this book.

A delay cost provision has one basic goal, which is to describe in detail the costs the contractor is entitled to recover in the event of an excusable and compensable project delay. For large owners that do a lot of construction, a secondary goal is to provide a delay cost provision that can be administered by the owner's personnel in the field. A delay cost provision with these attributes follows:

1904.6 COMPENSATION FOR DELAY

A. General

For compensable delays as identified in 1806, Determination and Extension of Contract Time, the Department will pay for the costs specified in 1904.6.B, Allowable Delay Costs. The Department will not pay for non-allowable charges specified in 1904.5, Non-Allowable Charges, or duplicate payment made under 1904.2 through 1904.4.

The Department will not pay for delay costs before the Contractor submits an itemized statement of those costs. The Contractor shall include the following content for the applicable items in the statement.

B. Allowable Delay Costs

1. Extended Field Overhead

The Department will pay the Contractor for extended field overhead costs that include costs for general field supervision, field office facilities and supplies, and for maintenance of field operations. General field supervision labor costs include, but are not limited to, field supervisors, assistants, watchman, clerical, and other field support staff. The Contractor shall calculate these labor costs as specified in 1904.4.A, Labor. For salaried personnel, the Contractor shall calculate the daily wage rate actually paid by dividing the weekly salary by 5 days per week. Field office facility and supply costs include, but are not limited to, field office trailers, tool trailers, office equipment rental, temporary toilets, and other incidental facilities and supplies. The Contractor shall calculate these costs to provide these services on a calendar-day basis using actual costs incurred due to the delay. Maintenance of field operations costs include, but are not limited to, telephone, electric, water, and other similar expenses. The Contractor shall calculate these costs to maintain these services on a calendar-day basis using actual costs incurred due to the delay.

2. Idle Labor
   The Contractor shall calculate labor costs during delays as specified in 1904.4.A, Labor, for all non-salaried personnel remaining on the Project as required under collective bargaining agreements or for other Engineer-approved reasons.
3. Escalated Labor
   To receive payment for escalated labor, the Contractor shall demonstrate that the Department-caused delay forced the work to be performed during a period when labor costs were higher than planned at the time of bid. The Contractor shall provide adequate support documentation for labor costs, allowances, and benefits.
4. Idle Equipment or Equipment Mobilization and Demobilization
   The Department will pay the Contractor for equipment, other than small tools, that must remain on the Project during Department-caused delays at the idle Equipment rate calculated in 1904.4.C, Equipment. The Department will pay the Contractor's transportation costs to remove and return Equipment not required on the Project during Department-caused delays.
5. Materials Escalation or Material Storage
   The Department will pay the Contractor for increased Material costs or Material storage costs due to the Department-caused delay. The Contractor shall obtain the Engineer's approval before storing Material due to a delay.
6. Extended or Unabsorbed Home Office Overhead
   The Department will pay the Contractor for unabsorbed or extended home office overhead costs in accordance with the Federal Acquisition Regulations, specifically 48 C.F.R. § 31. The Department will audit all extended or unabsorbed home office overhead claims in accordance with 1721, ——Audits. The Department will compensate the Contractor using the standard Eichleay formula. To recover home office overhead, the Contractor's claim shall prove:
   (1) that the delay was caused by the Owner suspending the entire project, in accordance with 1402.4, Suspensions of the Work Ordered by the Engineer.
   (2) that the Owner required the Contractor to standby during the suspension period.
   (3) that it was impractical for the Contractor to obtain replacement work during the suspension period.
   (4) that the suspension caused the contractor to be unable to complete the contract within the original contract performance period, as extended by any modifications.
   (5) that the Contractor suffered actual damages as a result of the delay caused by the suspension.

Item 6 of this specification might be a challenge to administer in the field. To address this problem, an owner may consider the following delay cost provision that simplifies the calculation of the payment for home office overhead:

---

f. Home Office Overhead

The Department will pay the Contractor for home office overhead, unabsorbed home office overhead, extended home office overhead, and all other overhead costs for which payment is not provided for in 109.05.D.2.e, including overhead costs that would otherwise be calculated using the Eichleay formula or some other apportionment formula, in the following manner provided all of the following criteria are met:

(1) The Contractor has incurred an excusable, compensable delay that delays the Work at least ten Calendar Days beyond the original Completion Date. These days are cumulative throughout the project. (2) The delay for which payment of home office overhead is sought is only due to delays defined in 108.06.D.2, 108.06.D.3 and 108.06.D.5.

Payment will be made for every eligible day beyond the original contract completion date at the rate determined by 109.05.D.2.f.i.

(i) Home Office Overhead Daily Rate

Calculate the home office overhead daily rate using the following formula:

$$\text{Daily HOOP} = (A \times C)/B$$

Where: A = original contract amount

B = contract duration in Calendar Days

C = value determine as follows:

Original Contract Amount up to \$5,000,000, C = 0.08

For Original Contract Amounts from \$5,000,001 to \$25,000,000, C = 0.06

For Original Contract Amounts greater than \$25,000,000, C = 0.05

INDEX

Note: Page numbers followed by "*f*," "*t*," and "*b*" refer to figures, tables, and boxes, respectively.

D

E

M

N

O

S

T

Printed in the United States
By Bookmasters